BACTERIOLOGY RESEARCH DEVELOPMENTS

LISTERIA MONOCYTOGENES

FOOD SOURCES, PREVALENCE AND MANAGEMENT STRATEGIES

BACTERIOLOGY RESEARCH DEVELOPMENTS

Additional books in this series can be found on Nova's website
under the Series tab.

Additional e-books in this series can be found on Nova's website
under the e-book tab.

BACTERIOLOGY RESEARCH DEVELOPMENTS

LISTERIA MONOCYTOGENES

FOOD SOURCES, PREVALENCE AND MANAGEMENT STRATEGIES

EDMUND C. HAMBRICK

EDITOR

New York

For permission to use material from this book please contact us:
Telephone 631-231-7269; Fax 631-231-8175
Web Site: http://www.novapublishers.com

Library of Congress Cataloging-in-Publication Data

ISBN: 978-1-63117-054-6

Library of Congress Control Number: 2013958103

Published by Nova Science Publishers, Inc. † New York

CONTENTS

PREFACE

Listeria monocytogenes has emerged as one of the major foodborne pathogens, characterized by high hospitalization and case fatality rates in humans. In this book the authors present current research in the study of this foodborne pathogen. Topics discussed include the natural approaches for controlling *L. monocytogenes*; *L. monocytogenes* in ready-to-eat foods and intervention strategies; carbohydrate utilization by *L. monocytogenes* and its influence on virulence gene expression; ozone and atmospheric cold plasma for control of *L. monocytogenes*; the potential of visible light as a means of controlling *L. monocytogenes* in the food chain; spatial distribution of *L. monocytogenes* and pseudomonas fluorescens in mixed biofilms; the main hygienic aspects of the processing and marketing of *L. monocytogenes* and *Salmosalar*; detection of *L. monocytogenes* and *Listeria spp.* in food and feed products; prevalence and control of *L. monocytogenes* in food processing environments; antimicrobial resistance, and growth kinetics of *L. monocytogenes* in ready-to-eat foods; novel technologies for controlling *L. monocytogenes* in ready-to-eat foods; the effect of soil abiotic and biotic factors on the preservation and reproduction of *L. monocytogenes*; the structure and function of the pathogen; and the prevalence of *L. monocytogenes* and occurrence of Listeriosis from ready-to-eat fresh fruits and vegetables.

Chapter 1 – *Listeria monocytogenes* has emerged as one of the major foodborne pathogens, characterized by high hospitalization and case fatality rates in humans. The ubiquitous distribution of *L. monocytogenes* in the environment along with its ability to form biofilms results in its frequent persistence in food processing and packaging facilities, thereby contaminating a variety of foods, especially ready-to-eat meat products. The drug of choice for the treatment of listeriosis in humans has been antibiotic, however, there have been reports of development of antibiotic resistance in *L. monocytogenes*, with isolation of multidrug resistant strains from cattle carcasses and meat. This increasing antibiotic resistance in *L. monocytogenes* and growing concerns over the use of synthetic chemicals in the food industry has ignited an interest in exploring the potential of various natural approaches as an alternative strategy to prevent food contamination and control listeriosis in humans.

This chapter discusses the feasibility, effectiveness and advantages of alternative approaches using plant derived antimicrobials and probiotic bacteria in controlling *L. monocytogenes* in food processing environments, high-risk foods, and humans.

Chapter 2 – *Listeria monocytogenes* has emerged as one of the major foodborne pathogens, characterized by high hospitalization and case fatality rates in humans. The ubiquitous distribution of *L. monocytogenes* in the environment along with its ability to form biofilms results in its frequent persistence in food processing and packaging facilities, thereby contaminating a variety of foods, especially ready-to-eat meat products. The drug of choice for the treatment of listeriosis in humans has been antibiotic, however, there have been reports of development of antibiotic resistance in *L. monocytogenes*, with isolation of multidrug resistant strains from cattle carcasses and meat. This increasing antibiotic resistance in *L. monocytogenes* and growing concerns over the use of synthetic chemicals in the food industry has ignited an interest in exploring the potential of various natural approaches as an alternative strategy to prevent food contamination and control listeriosis in humans.

This chapter discusses the feasibility, effectiveness and advantages of alternative approaches using plant derived antimicrobials and probiotic bacteria in controlling *L. monocytogenes* in food processing environments, high-risk foods, and humans.

Chapter 3 – The Gram-positive bacterium *Listeria monocytogenes* possesses a large number of transport systems for sugars and sugar derivatives. The uptake of the triol glycerol occurs in an energy-independent process via facilitated diffusion. In contrast, ion-driven transporters, ATP binding cassette (ABC) transporters and the phosphoenolpyruvate carbohydrate phosphotransferase system (PTS) require energy for the carbohydrate uptake reaction. Rhamnose and several sugar-phosphates are taken up via ion-driven permeases. *L. monocytogenes* contains several ABC transporters one of which was found to transport maltose. However, most carbohydrates used by *L. monocytogenes* are transported by the PTS. Among others they include glucose, fructose, mannose, cellobiose, gentiobiose, trehalose, D-arabitol and several β-glucoside type heterosides (salicin, arbutin, esculin, amygdalin). For many PTS the substrate has not yet been determined. Expression of several genes encoding PTS of unknown specificity, such as *lmo1974-1968*, was found to be upregulated during intracellular growth, suggesting that *L. monocytogenes* uses certain carbon sources transported by the PTS for intracellular proliferation. Certain carbohydrates taken up via the PTS strongly repress the expression of virulence genes. Virulence gene expression depends on the transcription activator PrfA (positive regulatory factor A). Its activity was found to be strongly inhibited by the presence of several PTS substrates. The underlying repression mechanism is not yet understood. It differs from common carbon catabolite repression, which in firmicutes is mediated by the PTS protein seryl-phosphorylated HPr bound to the catabolite control protein A. PrfA activity is probably regulated by a PTS component or a metabolic intermediate derived from an efficiently utilized carbohydrate.

Chapter 4 – *Listeria monocytogenes* is the causative agent of listeriosis which can survive under a wide range of adverse conditions. This gram positive organism is of major concern to the food industry due to the regulatory environment and concerns for consumer food safety. Due to its ubiquitous presence in the environment and psychrotolerant nature, its presence is anticipated even under refrigerated conditions. Adaptation to environmental stresses such as heat, acid, salt, cold provides protection to *Listeria* cells against wide range of other stresses. Biofilm formation by this organism presents a major risk specifically in the food industry, further introducing the possibility of post-process contamination. Thus, this pathogen is of great concern to manufacturers due to its resistance to a number of food preservation practices. In this chapter the authors describe the survival strategies employed by *L. monocytogenes* and how the novel non-thermal approaches of ozone and atmospheric cold plasma (ACP) may prove useful for control and inactivation of this pathogen. Studies presenting the effects of system and process control parameters associated with *Listeria* inactivation efficacy and the resulting control measures that may be considered for effective processing applications are highlighted. Further exploration of the potential to optimise both ozone and ACP for control of *Listeria* spp is given in the context of the wider research in the area.

Chapter 5 – The ability of *Listeria monocytogenes* to withstand processes employed by the food industry to inhibit or limit bacterial growth has become a major health and economic problem worldwide. The bacterium's survival is due, in part, to the action of the alternative sigma factor σ^B. This alternative transcription factor is responsible for regulating defence mechanisms in response to stresses such as extremes of temperature, osmotic, acid and ethanol stress, as well as contributing to virulence. Recently, light therapies have been proposed as additional means for the control of contamination, with applications focused on the food and medical device industries. Several variations of light treatment are currently being investigated, among them UV light therapy and photodynamic inactivation therapies. While these therapies are not limited solely to the management of *L. monocytogenes*, the discovery of the blue light photoreceptor Lmo0799 upstream of the *Listerial* σ^B activation cascade has resulted in increased interest in the effects of light on *L. monocytogenes*. This chapter reviews the mechanisms by which light is known to stimulate the σ^B activation cascade, and light treatments currently being investigated as suitable regimes for management of contamination. A detailed understanding of the mechanisms by which light influences virulence and the stress response in *L. monocytogenes*, as well as the mechanisms by which light causes cellular damage, is key for the successful development of new light-based control measures.

Chapter 6 – The idea of primary colonization of surfaces by *Pseudomonas* as a requirement or a decisive step for *Listeria monocytogenes* (Lm) integration in biofilms (Sasahara & Zottola, 1993) has been of primary influence in food industry sanitation. The assumption has been that *L. monocytogenes* is not capable of sufficient EPS production to create biofilms on its own and that *Pseudomonas* (*fluorescens, putida, fragi* or other species) are clearly superior in this respect and can provide a generous matrix able to host unobtrusive guests such as *Listeria*. For further insight into this classic 20-year-old topic, mixed biofilms with several strains of Lm and *P. fluorescens* (Pf) were obtained at room and refrigeration temperatures to be studied with confocal microscopy. Differential labeling of both species allowed identification of their respective location in biofilm structure. Both types of cells were separately quantified by plating in specific media. Co-culture benefits for Lm were confirmed in biofilms, particularly at 4°C, although not in planktonic cells. Besides, *Listeria* cells appeared preferentially located in the deeper, more anaerobic layers, with *Pseudomonas* cells placed on top of them. It is well known that bacteria in biofilms are better protected against various antimicrobial agents than planktonic cells; the protection is increased for cells in mixed biofilms and even more so when the cells are located in their deeper layers. The observed distribution pattern may guide the development of new and more effective cleaning and disinfection strategies to be used in food processing associated surfaces.

Chapter 7 – The Atlantic salmon (*Salmosalar L.*) is the most significant aquaculture species in Europe, both in terms of production biomass and economic value, with Norway, followed by Scotland and Ireland as the three major European producers.

The market for "processed" salmon subjected to one or more treatments sensibly changing the original characteristics of the products, such as salting, marinating and smoking is constantly and quickly growing. Nowadays, thanks to modern production technologies and the large scale retail trade, the smoked salmon market has gained great appreciation across Europe.

This product, normally cold-smoked and vacuum packed, with a refrigerated shelf-life of 3-8 weeks, is considered a risk product for human listeriosis: several studies demonstrated that is an excellent substrate for *Listeria monocytogenes*. The processing plants environment is the major source of finished-product contamination for cold-smoked salmon. It is possible that considerable contamination occurs during processing, in particular during evisceration, fish handling and packaging as a result of inadequate hygiene.

The growth of this microorganism in smoked salmon has been reported most of all during the storage period suggesting that the stress applied during the technological processes may extend the lag phase or reduce the growth rate, but is not able to affect the vitality of the cells. In the following, through an up-to-date review of (*personal and non*) published data, the main hygienic aspects of processing and marketing will be discussed.

Chapter 8 – *Listeria monocytogenes* is one of the most important foodborne pathogens because of the high mortality rate of this pathogen infection. Thus, its presence in food along the food chain is intensively controlled. The detection of Listeria in food samples is traditionally performed by culture-based methods using selective enrichment and plating media. The isolated strains should then be confirmed by biochemical characterization. This method, described in the ISO 11290-1, is the gold standard but is time consuming and may not be suitable to test food with short shelf life or in the case of outbreaks. Faster alternative methods based on different technologies have been developed and validated, they are antibodies-based (ELISA, ELFA) or acid nucleic based (RNA or DNA hybridization, PCR, RT-PCR, qPCR) technologies. In this chapter, the standard method for detecting *L. monocytogenes* in food and feed products as well as the different alternative technologies currently available are described in detail. The advantages and drawbacks of each technology are presented as well as a time scale comparison between the different technologies.

Chapter 9 – *Listeria monocytogenes* is an ubiquitous foodborne pathogen responsible for 'listeriosis' which is one of the most severe foodborne diseases in humans. Although the disease has an overall low incidence, it has a very high mortality rate of 20-30%. A diverse range of food products have been implicated with sporadic and epidemic incidents of listeriosis linked to the consumption of ready-to-eat foods, dairy products, seafood, cooked or frozen meat, pork and fresh produce. The pathogen is a major threat to the food industry since it can survive the harsh conditions encountered in food processing environments including high salinity, acidity and refrigeration temperatures. More importantly, it has the ability to attach to abiotic surfaces and form biofilms that can frequently contaminate food products and the production environment. Such biofilms tend to be more resistant to antimicrobial treatments. This has generated the need to develop reliable control strategies for *L. monocytogenes* (existing either as planktonic or sessile communities) in processing environments. With the recent emergence of bacteria resistant to antimicrobials and the inclination of consumers towards products derived from natural sources, it is important to explore alternatives for disinfection. This review will focus on the prevalence and control of *L. monocytogenes* in food processing environments. In particular, the role biofilms play in persistence of the pathogen in production environments and the use of essential oils as suitable disinfectant agents will be highlighted.

Chapter 10 – *Listeria monocytogenes* is a facultative intracellular bacterium that causes life-threatening illnesses including meningitis, stillbirth, and abortion in humans. This pathogen is ubiquitously distributed in nature, and in the most cases, transmission of this pathogen to humans are achieved via the intake of contaminated foods. In order to trace the sources and routes of infection, it is a first-line to understand the prevalence of this pathogen in a series of foods. Here the authors review the prevalence of this pathogen in a number of ready-to-eat (RTE) foods that have been concerned as one of the most high risk foods for listerial infection. In addition, they also summarize the characteristics of *L. monocytogenes* isolates from different sources, especially in terms of their antimicrobial resistance (for clinical aspects) and growth kinetics in food environments (supporting to construct microbial risk modeling). Finally, the authors discuss future perspective of *L. monocytogenes* study from a viewpoint of food hygiene.

Chapter 11 – *Listeria monocytogenes* is an emerging Gram-positive pathogenic microorganism widely distributed in the environment, food processing areas and in foods. This microorganism has been isolated from several ready-to-eat (RTE) products from meat, poultry, seafood, dairy, fruits and vegetables origin, thus representing, high risk foods for transmitting listeriosis to susceptible individuals. In the last decades, this food-borne disease has increased due mainly to the changes in food habits of consumers, and new practices of handling, preparation and commercialization of retails and food industries. Novel technologies for the RTE products preservation have recently been developed, among them: the use of natural antimicrobials, edible films and coatings, irradiation, UV-C light, pulsed light, pulsed electric fields, high-pressure processing and modified atmosphere packaging for ensuring the safety and quality of these products. However, hurdle technologies are preferred to control *L. monocytogenes* populations and to decrease the impact on the sensory attributes in these kinds of products. This chapter provides an overview of novel technologies applied to RTE products to controlling *L. monocytogenes.*

Chapter 12 – Soil is one of the ecological niches of *Listeria monocytogenes*, where these bacteria are equal members of the microbial community, although the mechanisms of their adaptation and forms of existence in soil ecosystems are studied insufficiently. Up to now, scientists considered the influence of soil on the reproduction of *L. monocytogenes* without regard for the soil genesis and soil chemical and physical properties. It is obvious that the type of soil and different soil properties play a certain part in the reproduction of pathogenic microorganisms, because soils differ sharply in their humus content, concentrations of nutrients, pH, particle-size composition, and microbial associations.

In this relation, the physical and chemical properties of brown forest, brown podzolic, tidal marsh, maritime meadow and urban soils that affect the reproduction of *L. monocytogenes* were studied. The results of the experiments suggest that a high content of exchangeable bases, domination of the fraction of humic acids bound with clay minerals and the fraction of fulvic acids in humus, and high soil temperature stimulate the reproduction of the studied bacteria in soils. High humus content, a predominance of humates, and a low soil temperature have an inhibiting effect on bacteria.

Influence of mineral fertilizers on the reproduction of *L. monocytogenes* in soil ecosystems under different temperatures (4-6^{0}C and 20-22^{0}C) were studied. Application of the fertilizers favored preservation and active reproduction of *L. monocytogenes* in brown podzolic and brown forest soils. Duration of application of the mineral fertilizers positively correlated with preservation and reproduction of *L. monocytogenes* in different types of soil.

The biological activity of volatile metabolites of volatile metabolites of saprophytic soil bacteria and germinating seeds of plant against *L. monocytogenes* was studied. It was shown that volatile metabolites are transfer factors and can be the sole carbon and energy source for *L. monocytogenes*. Different character of inter specific relationships between bacteria, influencing their propagation, can be observed on the metabolic level. Volatile compounds produced by microorganisms are capable to act as both intra- and interspecific regulators of microbial communities. In this connection the propagation of *L. monocytogenes* inhabiting soil may be stimulated or inhibited by the metabolic products of soil microorganisms. Methanol is the main substance affecting their growth and reproduction.

Chapter 13 – Listeria pathogenicity island 1 (LIPI-1) is located between *prs* and *orfX* loci. It consists of six genes that possess central role in the infection cycle of *L. monocytogenes* and is divided into three transcriptional units; the *hly* monocistron is considered as the center, downstream of which *mpl*, *actA* and *plcB* are found and upstream of which *plcA* and *prfA* are located. LIPI-1 sequence analysis has been used to improve our understanding regarding the phylogeny and evolution of both pathogenic and nonpathogenic *Listeria* spp. Moreover, a significant amount of data has been published, especially over the last decade, regarding their regulation and the effect of environmental stimuli, such as growth conditions and food processing interventions on their transcription rate. In this chapter all information relevant to LIPI-1 structure and function is integrated and critically reviewed.

Chapter 14 – Human foodborne listeriosis has been mainly linked to the consumption of poultry, beef and dairy products. However, the number of studies reporting incidence of *L. monocytogenes* in ready-to-eat fruits and vegetables has been dramatically increased over the last decade; prevalence has been reported to range from 0.04% to 36.8% to a variety of products, such as carrots, cabbage, cauliflower, celery, cucumber, various herbs, lettuce, rocket, strawberries etc. At the same time listeriosis outbreaks epidemiologically associated with the consumption of such products have also increased, in both number and severity, with the 2011 US multistate outbreak linked to whole cantaloupes being the most characteristic. In this chapter prevalence of *L. monocytogenes* in unprocessed and minimally processed fruits and vegetables as well as listeriosis outbreaks linked to the consumption of such products are addressed.

In: *Listeria monocytogenes*
Editor: Edmund C. Hambrick

ISBN: 978-1-63117-054-6

Chapter 1

NATURAL APPROACHES FOR CONTROLLING *LISTERIA MONOCYTOGENES*

Abhinav Upadhyay and Kumar Venkitanarayanan*
Department of Animal Science, University of Connecticut, Storrs, CT, US

ABSTRACT

Listeria monocytogenes has emerged as one of the major foodborne pathogens, characterized by high hospitalization and case fatality rates in humans. The ubiquitous distribution of *L. monocytogenes* in the environment along with its ability to form biofilms results in its frequent persistence in food processing and packaging facilities, thereby contaminating a variety of foods, especially ready-to-eat meat products. The drug of choice for the treatment of listeriosis in humans has been antibiotic, however, there have been reports of development of antibiotic resistance in *L. monocytogenes*, with isolation of multidrug resistant strains from cattle carcasses and meat. This increasing antibiotic resistance in *L. monocytogenes* and growing concerns over the use of synthetic chemicals in the food industry has ignited an interest in exploring the potential of various natural approaches as an alternative strategy to prevent food contamination and control listeriosis in humans.

This chapter discusses the feasibility, effectiveness and advantages of alternative approaches using plant derived antimicrobials and probiotic bacteria in controlling *L. monocytogenes* in food processing environments, high-risk foods, and humans.

1. INTRODUCTION

Despite significant advancements in food science and technology, foodborne illnesses due to microbial contamination continue to be a serious public health concern. Annually, an estimated 48-million foodborne illnesses, 128,000 hospitalizations and nearly 3000 deaths are

* Corresponding author, Department of Animal Science, University of Connecticut, 3636, Horsebarn Hill Rd Ext., Unit 4040, Storrs Mansfield, CT 06269, USA. Phone: (860)-486-0947, Fax: (860)-486-4375.
E-mail: kumar.venkitanarayanan@uconn.edu.

caused by foodborne pathogens in the United States (Scallan et al., 2011). The annual health care costs in the US due to foodborne diseases are estimated to be as high as $77.7 billion (Scharff, 2012). *Listeria monocytogenes* has emerged as one of the major foodborne pathogens resulting in nearly 19% of all deaths due to foodborne infections in the US (Scallan et al., 2011). First recognized as a foodborne pathogen in the 1980s (Schlech et al., 1983), *L. monocytogenes* has been implicated in several large multistate outbreaks resulting in high hospitalization (90%) and case fatality rates (20-30%) in infected individuals (Cartwright, 2012; Rocourt and Brosch, 1992; Schuchat et al., 1991). Susceptible population to listeriosis includes the elderly, infants (Munoz et al., 2012), immuno-compromised patients (Martins et al., 2010) and pregnant women (Lamont, et al., 2011). In addition, recent outbreak investigations have demonstrated that *L. monocytogenes* can cause non-invasive gastrointestinal infections with fever and flu like manifestations in apparently healthy individuals (Aureli et al., 2000; Ooi and Lorber, 2005).

The widespread distribution of *L. monocytogenes* in diverse environments such as soil, water, decaying vegetation, silage, sewage (Watkins and Sleath, 1981), and farm effluents (Beuchat, 1996; Steele and Odumeru, 2004), along with its ability to tolerate desiccation, temperature fluctuations, and variations in pH and osmolarity, leads to frequent contamination of food processing areas (Chaturongakul et al., 2008; Donnelly, 2001). In particular, serotypes 1/2a, 1/2b, and 1/2c *L. monocytogenes* are better adapted to food processing plant environment, and have been frequently isolated from these niches (Jay et al., 1996; Norwood and Gilmour, 2000). In food processing facilities, *L. monocytogenes* persists by forming biofilms (Fatemi and Frank, 1999) on a variety of food processing and equipment surfaces, including stainless steel, polycarbonate surface, plastic, teflon and rubber (Borucki et al., 2003; Chavant et al., 2002; Frank and Koffi, 1990; Moretro and Langsrud, 2004). Biofilms protect the embedded bacteria from the action of sanitizers (Borucki et al., 2003; Folsom and Frank, 2007) and adverse external conditions (Blackman and Frank, 1996), thereby serving as a continuous source of microbial contamination to food products and causing food spoilage and human infections (Aarnisalo et al., 2007).

The major food products implicated in listeriosis include ready-to-eat (RTE) meat products (USFSIS, 2004), soft cheese, unpasteurized milk (Latorre et al., 2011) and fresh produce such as bean sprouts, celery (CDC 2012b, FDA 2010) and cantaloupes (Scallan et al., 2011). The bacterium has also been isolated from potatoes, cabbage, and cucumber (Beuchat, 1996; Farber and Peterkin, 1991; Heisick et al., 1995; Heisick et al., 1989). The psychrotrophic and halotolerant nature enables *L. monocytogenes* to proliferate on RTE meat products during their long-term cold storage (Brakat, 1999; Farber and Peterkin, 1991).

The currently employed therapy for the treatment of human listeriosis includes a combination of ampicillin and aminoglycoside such as gentamicin (Dutta et al., 2009; Poros-Gluchowska and Markiewicz, 2003; Temple and Nahata, 2000); however, there have been reports of development of antibiotic resistance in *L. monocytogenes* upon exposure to these antibiotics (Cekmez et al., 2012; Krawczk-Balska et al., 2012). Multidrug resistant strains of *L. monocytogenes* have been isolated from cattle carcasses (Wieczorek, et al., 2012) and meat (Pesavento et al., 2010). This increasing antibiotic resistance in *L. monocytogenes* and growing concerns over the use of synthetic chemical disinfectants in food industry have led to renewed interest in exploring the potential of various natural approaches as an alternative strategy to prevent food contamination by *L. monocytogenes* and combat foodborne listeriosis in humans.

2. Alternative Approaches

Since ancient times, plant extracts have been used as food preservatives, flavor enhancers and dietary supplements to prevent food spoilage, and maintain human health. In addition, fermented foods containing beneficial microbiota have been a part of traditional diets for their health benefits. The antimicrobial activity of several plant-derived antimicrobials (PDAs) has been documented (Ahmad and Beg, 2001; Bhatt and Negi, 2012; Kubo et al., 1993; Negi et al., 1999, 2003a, 2003b, 2005, 2010; Silva et al., 1996; Zeng et al., 2012). A great majority of these compounds are secondary metabolites, and are produced as a result of reciprocal interactions between plants, microbes, and animals (Reichling, 2010). These secondary metabolites could be species or genera specific in their action, and do not primarily contribute to major metabolic processes in plants, but potentiate their ability to survive local environments (Harborne, 1993) and defend the plants against microorganisms such as bacteria, fungi and viruses (Kennedy and Wightman, 2011). The major benefit of using PDAs as food preservatives is that they do not exert deleterious effects usually associated with synthetic chemicals (VanWyk and Gericke, 2000). Moreover, there is evidence that the antimicrobial action of PDAs does not induce resistance in pathogens (Ali et al., 2005; Ohno et al., 2003). The major groups of PDAs include polyphenols, flavonoids, alkaloids, lectins, and tannins (Cowan, 1999; Geissman, et al., 1963). Some of the PDAs that possess significant antimicrobial activity include *trans*-cinnamaldehyde, eugenol, thymol, carvacrol, β-resorcyclic acid, and caprylic acid. All the aforementioned compounds have been shown to exert antimicrobial effects against both gram-positive and -negative bacteria, primarily by damaging the cell wall and membrane integrity, thereby leading to leakage of cellular contents and cell death (Burt, 2004). In addition, a majority of PDAs are generally regarded as safe (GRAS) compounds with low mammalian cytotoxicity and quick biodegradability in soil and water (Isman, 2000), thus making them safe and environmental friendly as natural disinfectants.

The choice of PDAs for application in different foods to enhance food safety depends upon a variety of factors. The antimicrobial efficacy of PDAs is determined by physiochemical properties such as pH, pKa, hydrophobicity, solubility in aqueous solutions, and stability (Negi, 2012; Stratford and Eklund, 2003). The presence of fat (Cava-Roda et al., 2010), sugars (Gutierrez et al., 2008a), and proteins (Cerrutti and Alzamora, 1996; Hyldgaard et al., 2012; Kyung, 2011) modulate the antimicrobial efficacy of essential oils in foods. In addition, the type of spoilage microorganisms and extrinsic factors such as temperature, water activity and atmospheric composition exert a significant impact on the antimicrobial property of phytochemicals (Gould, 1989). Thus, besides the antimicrobial efficacy of a plant compound, the choice of phytochemicals depends upon the target food product (Owen and Palombo, 2007).

The Food and Agriculture organization (FAO) and World Health Organization (WHO) define probiotics as "live microorganisms which when administered in adequate amounts confer a health benefit on the host". Probiotic bacteria exert multiple benefits to the host, such as nutrient digestion and assimilation (Sonnenburg et al., 2005; Yatsunenko et al., 2012), potentiating host immune function (Olszak, T. et al., 2012), and protection against enteric pathogens (Candela et al., 2008; Fukuda et al., 2011). Although several species of bacteria have been identified, the majority of probiotic bacteria supplemented in human diet belong to

Lactobacilli and Bifidobacteria. Interestingly, recent studies have shown that PDAs that are highly bactericidal towards enteric pathogens exert low antimicrobial effect against commensal gut microbiota (Hawrelak et al., 2009; Di Pasqua et al., 2005). Since PDAs and probiotics exert their antimicrobial effects by different mechanisms (Shipradeep et al., 2012), a combinatorial approach using PDAs and probiotics could be more effective in controlling *L. monocytogenes*, however studies eliciting their synergistic interactions are lacking.

3. Control of *Listeria monocytogenes* in Food Processing Environment

The widespread presence of *L. monocytogenes* in the environment potentially leads to reoccurring introduction of the pathogen in food processing areas (Tompkin, 2002). Therefore, by controlling the establishment of biofilms and persistence of *L. monocytogenes* in such environments, it is possible to reduce the risk of food contamination and subsequent human infections. Implementation of appropriate standards, good manufacturing and hygiene practices, implementation of hazard analysis and critical control point are important for controlling the persistence of *L. monocytogenes* in food processing facilities (Tompkin et al., 1999). In this regard, both preventing the formation of *L. monocytogenes* biofilm as well as eradication of mature biofilm are critical, since biofilms contribute to the environmental persistence of *L. monocytogenes*. Chemical disinfectants commonly used for processing plant cleaning and sanitation include surfactants and alkali/acid chemicals such as quaternary ammonium compounds, hypochlorite (Krysinski et al., 1992), chlorine, peracetic acid, and peroctanoic acid (Fatemi and Frank, 1999). In addition, hydrogen peroxide and ozone treatments are employed to control *L. monocytogenes* biofilms in processing plants (Robbins et al., 2005). These treatments disintegrate food residues by decreasing surface tension, emulsifying fats, and denaturing proteins (Forsythe & Hayes, 1998; Maukonen et al., 2003; Mosteller & Bishop, 1993; Simoes et al., 2010); however, they are found to be not very effective in completely inactivating *L. monocytogenes* biofilms, especially in the presence of organic matter and low temperatures (Heir et al., 2004; Holah et al., 2002; Pan et al., 2006; Romanova et al., 2002). Moreover, hardness of water and chemical inhibitors negatively affect the disinfection efficiency (Bremer et al., 2002; Cloete et al., 1998; Kuda et al., 2008; Simoes et al., 2010). Consequently, the efficacy of alternate approaches such as plant derived antimicrobials, competitive exclusion bacteria and/or their antimicrobial metabolites have been investigated for controlling *L. monocytogenes* in food-processing environment.

3.1. Plant Derived Antimicrobials as Natural Disinfectants

The use of PDAs as disinfectants is a viable alternative approach to control *L. monocytogenes* planktonic cells and biofilms in food-processing environment. A recent study by Upadhyay et al. (2013b) investigated the efficacy of several PDAs, including trans-cinnamaldehyde, carvacrol, thymol and eugenol in eliminating *L. monocytogenes* biofilms developed on plastic and steel surfaces at 37, 25 and 4°C. The plant compounds at concentrations ranging from 0.1 to 0.4% were found to be very effective in rapidly inactivating mature biofilms of *L. monocytogenes* at all three temperatures. In another study, Desai and co-workers (2012) found that thyme oil, oregano oil and carvacrol at 0.1 to 0.5% concentrations, were effective in inactivating *L. monocytogenes* biofilms developed from 21 *L. monocytogenes* strains representing 13 serotypes. Plant-derived antimicrobials have also been used with surfactants to improve the disinfection efficacy. The use of PDAs in combination with surfactants reduces the low water solubility of essential oils, thereby increasing the effective compound concentration and antimicrobial activity. Conesa and coworkers investigated the efficacy of micellar encapsulated eugenol and carvacrol in reducing *L. monocytogenes* colony biofilms (Perez-Conesa et al., 2006) and *L. monocytogenes* biofilms on steel chips (Perez-Conesa et al., 2011). These researchers encapsulated the essential oils in micellar nonionic surfactants and observed that the two antimicrobials were very effective in inactivating *L. monocytogenes* biofilms, and decreased viable bacterial counts by 3.5 to 4.08 log CFU/cm^2 within 20 min of exposure. The aforementioned studies primarily focused on inactivating pre-formed mature biofilms, however with increasing advancement in the understanding of molecular processes involved in biofilm formation, new methodologies are targeted at interrupting critical cell-to-cell signaling for enhancing biofilm control. These include bacterial quorum sensing antagonists (Dunstall et al., 2005; Rasmussen et al., 2005a), two-component signal transduction inhibitors (Worthington et al., 2012) and antibiofilm chemical modulators (Worthington et al., 2012) such as eukaryotic protein kinase inhibitors (Wenderska et al., 2011). The selection of such compounds is based on their ability to interact with a wide range of molecular targets critical for biofilm formation in bacteria. Plant compounds have been reported to inhibit quorum sensing in human pathogens (Adonizio et al., 2006; Vattem et al., 2007). Several quorum-sensing inhibitory phytochemicals, including polyphenols have been found effective in reducing biofilm formation in pathogenic bacteria (Cragg et al., 1997; Huber et al., 2003) such as *Burkholderia cenocepacia* (Riedel et al., 2006), *Cronobacter sakazakii* (Amalaradjou and Venkitanarayanan, 2011) and *Pseudomonas aeruginosa* (Sarabhai et al., 2013). Our laboratory recently reported that subinhibitory (concentrations not inhibiting bacterial growth and viability) concentrations of trans-cinnamaldehyde, carvacrol, thymol and eugenol down-regulated the expression of genes responsible for bacterial motility, quorum sensing and initial surface attachment in *L. monocytogenes* (Upadhyay et al., 2013b). These low concentrations of plant compounds also reduced exopolysaccharide production that contributes to biofilm mass and strength.

3.2. Competitive Exclusion Bacteria for Environmental Pathogen Control

Physical interactions and chemical cross talk between bacteria of different species with the production of antagonistic metabolites is known to modulate bacterial colonization and biofilm formation in a niche (Carpentier & Chassing, 2004; Kives et al., 2005; Rossland et al., 2005; Simoes et al., 2010; Tait and Sutherland, 1998; Valle et al., 2006). Previously, Zhao and co-workers (2004) found that microorganisms obtained from floor drains of food processing facilities with no history of detectable *L. monocytogenes* exhibited significant antilisterial activity, and were able to inhibit biofilm formation by *L. monocytogenes.* The inhibitory bacteria isolated were identified as *Enterococcus durans*, *Lactococcus lactis* subsp. *lactis* and *L. plantarum.* Similarly, in a study by Leriche and Carpentier (2000),

Staphylococcus sciuri biofilms were found to limit the adhesion and growth of *L. monocytogenes* on stainless steel surface. Although the exact mechanism of antibiofilm action induced by probiotics remains unclear, it appears to be through competition, exclusion and displacement from adhesion surfaces, and nutrient scarcity (Woo and Ahn, 2013). The ability of probiotic bacteria to synthesize and secrete anti-adhesive chemicals has been observed by several researchers (Desai & Banat, 1997; Nitschke & Costa, 2007; Rodrigues et al., 2004; van Hamme et al., 2006). This includes antagonism against major foodborne pathogens such as *Staphylococcus aureus* (Rodrigues et al., 2004), *Salmonella* (Mireles et al., 2001) and *L. monocytogenes* (Garcia-Almendarez et al., 2008; Schobitz et al., 2014). In addition, bacterial metabolites such as nisin and reuterin have been well documented for their antibiofilm potential against various microorganisms, including *L. monocytogenes* (Bower et al., 1995; Chi-Zhang et al., 2004; Dufour et al., 2004; Mahdavi et al., 2007). Although the use of PDAs and probiotics as bio-disinfectants is promising, further studies are needed to investigate the safety, feasibility and effectiveness of such approaches in commercial settings, specifically relating to their effects on the shelf life, sensory attributes and consumer acceptance of the final food products (Mosteller & Bishop, 1993; Wirtanen et al., 2000).

4. Control of *Listeria Monocytogenes* in Foods

Approximately 99% of all *L. monocytogenes* infections occur via contaminated food (Swaminathan et al., 2007). Due to the high case fatality rates (20-30%) associated with listeriosis, the United States Department of Agriculture Food Safety and Inspection Service (USDA-FSIS) has adopted a zero tolerance policy for the presence of *L. monocytogenes* on RTE foods (Gerba et al., 1996; U.S Food Safety and Inspection Service, 1989), which implies that *L. monocytogenes* detection ($\geq$ 1 log CFU/25 g of sample) on RTE foods initiates a recall of the product leading to significant economic losses. Additionally, the USDA-FSIS alternatives 1 and 2 of the *L. monocytogenes* regulations mandate that all RTE meat-processing plants apply post-processing treatments of food products, which may include the use of antimicrobials to inactivate or suppress *L. monocytogenes* growth (Anonymous, 2003 and U.S. Food Safety and Inspection Service, 2004). Traditionally, various standard food preservation techniques, including refrigeration, desiccation, pasteurization, and synthetic antimicrobials for the control of Listeria in foods have been investigated with varying degree of success (Davidson et al., 2001). Primarily, they are targeted at enhancing food safety and shelf life without compromising the sensory and nutritional attributes of food products (Campos et al., 2011; Gandhi and Chikindas, 2007). However, in light of the recent reports on foodborne listeriosis outbreaks (CDC, 2011; CDC, 2012a,b; CDC, 2013), the contamination of food products with *L. monocytogenes* still remains as a significant challenge that needs to be adequately addressed. Modern methods of food preservation such as activated antimicrobial films, irradiation, modified atmosphere packaging have limited application owing to their deleterious effects on the organoleptic properties and consumer acceptability of foods (Negi, 2012). With increasing consumer demands for minimally processed, microbiologically safe, and organic foods, there is significant focus on the use of natural preservatives to enhance the shelf-life and microbiological safety of food products.

4.1. Plant-derived Antimicrobials for Reducing Listeria Contamination in Foods

Plant-derived antimicrobials are viable candidates to improve the safety and shelf-life of food products. In addition to the significant antimicrobial effects exerted by these compounds, the major benefits of using PDAs in foods include reduced application of potentially harmful synthetic chemical preservatives, increase in flavor, nutritional, and medicinal value of foods. A majority of PDAs are GRAS status compounds, and are approved for addition in foods by the US FDA.

4.1.1. Essential Oils

Essential oils are volatile, aromatic, oils extracted from various parts of plant, such as flowers, bark, leaves and fruits (Deans and Ritchie, 1987; Sanchez et al., 2010) by distillation (Negi, 2012) or supercritical fluid extraction (Bakkali et al., 2008; Pereira and Meireles, 2010). Individual active components present in essential oils are also synthesized chemically. Essential oils contain a variety of compounds such as terpenes, phenols, aldehydes or esters, which are classified as GRAS substances. The antimicrobial action of essential oils or their active components involves multiple mechanisms with several cellular targets (Burt, 2004). Thus, the development of bacterial resistance against these natural interventions is a formidable challenge for pathogens. Essential oils from cinnamon, clove, oregano, thyme, nutmeg, bay, and coriander exhibit a high degree of antimicrobial activity against *L. monocytogenes* (Barbosa et al., 2009; Djenane et al., 2011; Firouzi et al., 2007; Gutierrez et al., 2008a,b; Singh et al., 2003, Smith-palmer et al., 1998; Zhang et al., 2009). In a recent study from our laboratory, post processing antimicrobial washing of frankfurters with trans-cinnamaldehyde, carvacrol and thymol was found to significantly reduce *L. monocytogenes* contamination on frankfurter surface (Upadhyay et al., 2013a). Clove and cinnamon extracts have also been found to exert significant antilisterial effect in other food systems such as chicken meat (Hoque et al., 2008) and cheese (Vrinda Menon and Garg, 2002; Smith-Palmer et al., 2001). In addition to their antimicrobial properties, essential oils such as wintergreen or clove oil containing high a concentration of phenols have been found to exhibit significant anti-inflammatory, antioxidant (Bhat et al., 2011; Hamed et al., 2012) and cardioprotective benefits to the host (Smith, 2012).

4.1.2. Plant Extracts

Plant extracts have potential application in foods as preservatives and flavor enhancers. The antibacterial activity of plant extracts is primarily due to bacterial membrane disruption and cell content leakage (Otake et al., 1991). The extracts of cinnamon bark and fruit have been reported to possess antimicrobial activity (Agnihotri and Vaidya, 1996). Yuste and Fung (2002) reported a significant reduction in the number of *L. monocytogenes* in apple juice supplemented with 0.1-0.3% cinnamon. Similarly, plant extracts have been found effective in reducing *L. monocytogenes* in meat products (Hao et al., 1998; Ward et al., 1998). Water soluble extracts of arrowroot tea extract (Kim and Fung, 2004) at 6% (w/w) concentration was found to significantly reduce *L. monocytogenes* in ground beef samples, however the extracts of green and jasmine tea exerted no antimicrobial effect (Kim et al., 2004). A combination of oregano and cranberry extract significantly reduced *L. monocytogenes* at pH

6.0, however the antilisterial effect was lesser at neutral pH. Likewise, Ruiz and coworkers (2009) observed that rosemary extract was more effective in reducing *L. monocytogenes* counts in turkey and ham meat when used in combination with nisin. The antimicrobial efficacy of plant extracts is generally lesser than the individual antimicrobial components, probably due to the use of crude extracts and the presence of compounds such as sugars that protect the bacteria (Kapoor et al., 2007; Parvathy et al., 2009).

4.3. Organic Acids

Organic acids such as acetic, lactic, and citric acid have been commonly used for food preservation. The antimicrobial effect of organic acids is exerted primarily through pH-induced stress, acidification of cytoplasm, reduced enzymatic activity, metabolism and cell injury (Samelis and Sofos, 2003). In addition, extrinsic stress factors such as heat and low pH enhance the antimicrobials efficacy of organic acids. Organic acids are widely applied in preventing microbial contamination and extending the shelf-life of ready-to-eat meat products (Campos et al., 2011; Georgaras et al., 2006; Murphy et al., 2006). Currently, sodium or potassium lactate (2%) in combination with 0.05-0.15% sodium diacetate is commonly used in the food industry as preservative in meat due to their strong synergistic antimicrobial effect (Abou-Zeid et al., 2007) and minimal effect on organoleptic quality of the food (Abou-Zeid et al., 2007; Porto-Feit et al., 2011).

An increasing body of evidence has demonstrated the potential of application of antimicrobials on food surface to inactivate foodborne pathogens (Garcia et al., 2007; Mattson et al., 2011; Upadhyay et al., 2013a), since bacteria are primarily present at the product surface in cases of post-processing contamination. Moreover, since the amount of antimicrobial necessary for surface application is less, the deleterious effect on organoleptic properties of foods are minimal. For example, a dip treatment of frankfurters inoculated with *L. monocytogenes* in a solution containing 2% acetic acid, 1% lactic acid, or 0.1% benzoic acid followed by steam treatment for 1.5 s inhibited growth of the pathogen for 14 weeks at 7°C (Murphy et al., 2006). Other combinations of organic acids that exhibit significant antilisterial activity on foods include acetic acid and monocaprylin on frankfurters (Garcia et al., 2007), propionate with lactate on pork (Porto-Feit et al., 2011), and lactate in combination with diacetate on sausage (Georgaras et al., 2006) and ham (Stopforth et al., 2010).

A newer approach to increase the antimicrobial efficacy of organic acids is using them as constituents of edible films and gels. The incorporation of antimicrobials in edible coating increases the contact time of the antimicrobial with meat surface (Siracusa et al., 1992), thereby potentially increasing their efficacy. Lactic and acetic acids incorporated in calcium alginate gels were found to be more effective in inactivating *L. monocytogenes* than acid treatment alone (Siracusa et al., 1992).

4.4. Probiotic Bacteria as Biopreservatives in Food

Probiotic microbes such as Lactic acid bacteria and their antimicrobial metabolites are a viable and safe (GRAS-status) approach to inhibit spoilage and pathogenic microorganisms in foods (Holzapfel et al., 1995; Ross et al., 2002). The use of probiotic bacteria as biocides in foods depends upon the viability and stability of probiotic and their beneficial properties during the product's shelf life (Mattila-Sandholm et al., 2002). The various metabolites produced by these bacteria such as bacteriocins, diacetyl, hydrogen peroxide and organic acids are used for preserving food products without significantly affecting the organoleptic characteristics of foods (Galvez et al., 2007).

4.4.1. Bacteriocins

Bacteriocins are ribosomally synthesized, antimicrobial peptides mostly produced by food-grade microbes (Balciunas et al., 2013; Galvez et al., 2007). They are promising candidates as bioprotective agents by virtue of their safety, nontoxic nature, acid heat tolerance, coupled with a wide antimicrobial spectrum against spoilage and disease causing foodborne pathogens. Most bacteriocins exert their antimicrobial effects by forming a pore in the cell membrane, thereby disrupting the membrane potential leading to cell death. The bacteriogenic strains can either be used as starter cultures or as protective co-culture with a starter culture in fermented or non-fermented foods (Galvez et al., 2007). Due to this reason, bacteriocins are being used in several applications, including biopreservation and shelf-life extension of food products (Balciunas et al., 2013).

Nisin, a class I bacteriocin, first discovered in 1928 as a metabolite of *Lactococcus lactis* (Roger, 1928), is a USFDA-approved bio-preservative. The efficacy of nisin in inactivating *L. monocytogenes* has been demonstrated in many foods, especially cheese (Davies et al., 1997), sausages (Hampikyan and Ugur, 2007), and pork (Kouakou et al., 2008). The form of nisin commonly available commercially is Nisaplin (Danisco), which consists of 2.5% nisin in sodium chloride and non-fat dried milk (McAuliffe and Jordan, 2012). In addition, the class IIa group that includes pediocin-like peptides, in particular Pediocin PA-1 has received substantial attention due to their significant antimicrobial efficacy (Rodriguez et al., 2002) specifically against *L. monocytogenes* (Katla et al., 2001, 2002; Naghmouchi et al., 2006). The stability of pediocin PA-1 in foods such as cheese, frankfurters, and sausage has also been demonstrated (Nieto-Lozano et al., 2010). Other bacteriocins with significant efficacy against *L. monocytogenes* in foods include divergicin M35 (Tahiri et al., 2004, 2009), enterocin AS-48 (Ananou et al., 2005, Molinos et al., 2008), lacticin 3147 (Morgan et al., 1999, 2001) and piscicosin CS526 (Azuma et al., 2007). These bacteriocins have been successfully used in various food systems to reduce *L. monocytogenes* contamination. For example, enterocin AS-48 was found to be effective in reducing *L. monocytogenes* counts in sausage (Ananou et al., 2005), soy-based desserts (Martinez et al., 2009) and processed vegetables (Molinos et al., 2005).

Recent studies have investigated the efficacy of bacteriocins in combination with emerging technologies such as antimicrobial packaging for increasing shelf life and food safety (Blanco et al., 2008, 2012; Ercolini et al., 2006; Mauriello et al., 2004). Alginate films containing enterocin at 2000 activity units/cm^2 was effective in controlling *L. monocytogenes* in vacuum packaged, cooked ham during refrigerated storage (Marcos et al., 2007). Similarly, the combination of chitosan and divergicin M35 (class IIa bacteriocins) exhibited an additive effect against *L. monocytogenes* (Benabbou et al., 2009). The use of antimicrobial film is especially applicable for solid food products such as frankfurters, where the surface contaminants come in close contact with the antimicrobial films.

Foods can either be supplemented with bacteriocins produced ex situ, or by inoculation of bacteriocin producing cultures (Stiles, 1996). However, stability of the strain, ability to produce sufficient concentrations of bacteriocins, economic feasibility, and deleterious effects of bacteriocins on food properties are some of the challenges in the application of bacteriocins in foods (Fallico et al., 2010). In addition, various extrinsic factors such as interaction with food chemicals and components, inactivation, or precipitation of proteins in the food matrix need to be considered as well while selecting bacteriocins for commercial application in foods. Due to these reasons, there are strict regulatory requirements for the use of bacteriocins in foods.

5. Control of *Listeria Monocytogenes* Virulence

L. monocytogenes infection in humans is characterized by febrile gastroenteritis, bacteremia, and sepsis followed by meningioencephalitis, placentitis, abortion and death in susceptible individuals (Jaradat and Bhunia, 2003; Racz et al., 1972). Since crossing the intestinal barrier is the first critical step for *L. monocytogenes* infection, reducing *L. monocytogenes* adhesion, colonization and subsequent invasion of intestinal epithelial cells could potentially control listeriosis in humans. Various virulence determinants in *L. monocytogenes* critical for its attachment, colonization, and invasion of host intestinal epithelium, subsequent intracellular proliferation and systemic spread have been identified (Boland et al., 2001). The major factors include internalins (Lingnau et al., 1995; Southwick and Purich, 1996), listeriolysin O (LLO) (Greiffenberg et al., 1998; Smith-Palmer et al., 2002), phospholipases (Grundling et al., 2003; Marquis and Hager, 2000) and actin polymerization protein (ActA) (Alvarez-dominguez et al., 1997; Kocks et al., 1992).

Antibiotics are used to treat listerial infections due to the high severity and mortality rate of the disease. β-lactam antibiotics such as penicillin G and ampicillin, either alone or in combination with gentamicin are the current drugs of choice for treating listeriosis in humans (Krawczyk-Balska et al., 2012; Temple and Nahata, 2000). However, the indiscriminate and excessive use of antibiotics in humans and farm animals has lead to the development of antibiotic resistance in *L. monocytogenes.* Recent investigations have reported the development of multidrug resistance and reduced susceptibility to effective antibiotics in *L. monocytogenes* strains isolated from cattle carcass, meat and food processing environments (Kovacevic et al., 2013). Thus, there is a renewed interest in exploring the potential of natural approaches, including PDAs and probiotics as an alternative strategy to combat listeriosis in humans.

5.1. Plant-derived Antimicrobials as Food Supplements

Changes in dietary components have been associated with fluctuations in the composition of gut microbial population and diversity (Ley et al., 2006; Duncan et al., 2007), which in turn modulate host's metabolic functions (Brown et al., 2012) and susceptibility to gastro-intestinal bacterial infections (Ghosh et al., 2011). Recent research has revealed that low concentrations of plant essential oils are able to reduce bacterial pathogenicity by modulating gene transcription (Goh et al., 2002; Tsui et al., 2004) and virulence protein production in many foodborne pathogens (Azizkhani et al., 2013; de Souza et al., 2010; De Wit et al., 1979;

Gonzalez-Fandos et al., 1994; Li et al., 2011; Parsaeimehr et al., 2010; Qiu et al., 2010a,b, 2011a,b, 2012), including *L. monocytogenes* (Smith-Palmer et al., 2002). In a recent study from our laboratory, sub-inhibitory concentrations of three plant compounds, namely *trans*-cinnamaldehyde, carvacrol and thymol significantly reduced *L. monocytogenes* adhesion and invasion of human intestinal cells *in vitro* (Upadhyay et al., 2012). These compounds also significantly decreased *L. monocytogenes* motility, listeriolysin production and lecithinase activity. Real-time quantitative PCR data revealed that the plant compounds down-regulated the expression of *L. monocytogenes* virulence genes (Upadhyay et al., 2012). Carvacrol and thymol have also been found to exert a similar effect on virulence attributes of other foodborne pathogens, including *Salmonella* Enteritidis (Upadhyaya et al., 2013) and *Staphylococcus aureus* (Qiu et al., 2010; Smith-Palmer et al., 2004). Several researches have investigated the potential of plant-derived compounds as quorum sensing inhibitors (Koh et al., 2013; Persson et al., 2005; Rasmussen et al., 2005a,b), since quorum sensing partially regulates the expression of virulence determinants in pathogenic bacteria. This strategy is unlikely to contribute to development of multidrug resistant pathogens due to the absence of selection pressure on bacteria (Koh et al., 2013). Nakayama and coworkers (2009) identified that ambuic acid has the potential to inhibit quorum sensing in *L. innocua* by inhibiting signal peptide biosynthesis. In our investigation (Upadhyay et al., 2013), plant-derived antimicrobials such as trans-cinnamaldehyde, carvacrol, thymol and eugenol significantly reduced the expression of Agr quorum sensing peptide genes (*agrA, agrB, agrC*) critical for pathogen invasion of intestinal epithelial cells and virulence in host (Riedel et al. 2009).

5.2. Probiotic Bacteria as Antilisterial Supplements

The human intestinal tract is colonized by large and complex communities of bacterial species that include as many as 10^{12} cells per 1g of fecal mass in an average human being (Hattori and Taylor, 2008; Savage et al., 1977). The gut microbiota interacts with the host intestinal tissue to perform various biological processes (Dethlefsen et al., 2007), including nutrition, immunohomeostasis and defense against microbes (Sekirov et al., 2010). With advances in high throughput sequencing, metagenomics, and development of gnotobiotic animals, the ability to explore the variations in gut microbiota composition and their effect on human health has markedly enhanced (Gordon et al., 2011; Khor et al., 2011). Thus improving gut microbial diversity and abundance could facilitate gut health (Dominguez-Bello and Blaser, 2008) and prevent foodborne infections. Bifidobacterium and Lactobacillus, two genera of the human gut microflora, are among the well-characterized candidates proposed to possess potential probiotic effects (Guerin-Danan et al., 1998). Toure and coworkers (2003) screened and characterized isolates of Bifidobacteria with significant antilisterial potency from newborn infants. Moroni and coworkers (2006) found that bacteriocin-producing bifidobacterial isolates (*Bifidobacterium thermacidophilum, B. thermophilum*) from humans (Toure et al., 2003) reduced adhesion and invasion of *L. monocytogenes* on human intestinal cells (Caco2, HT-29) *in vitro*. Visualization of adhesion sites by fluorescent *in situ* hybridization revealed the adherence of probiotic bacteria and *L. monocytogenes* in close proximity. A similar reduction in the adhesion and invasion efficacy of *L. monocytogenes* in the presence of *Lactobacillus* and *Bifidobacterium* was observed by Corrs and co-workers (2007). In addition, these authors demonstrated that interactions between probiotic bacteria and epithelial cells significantly affect the host immunological response. Similar probiotic-mediated modulation of host intestinal cell immune response to prevent enterohaemorrhagic *E. coli* O157:H7 infection has been observed by Jandu and co-workers (2009). These authors demonstrated that the presence of probiotic *L. helveticus* R0052 reduced pathogen-mediated disruption of IFN-gamma-STAT-1 signaling.

The antilisterial efficacy of probiotic bacteria has been recently demonstrated in gnotobiotic mice (Bambirra et al., 2007; dos Santos et al., 2011). Archambaus and group (2012) studied the effect of two *Lactobacillus* species, *L. paracasei* CNCM I-3689 and *L. casei* BL23 on *L. monocytogenes* and orally acquired listeria infection in gnotobiotic mouse. The probiotic bacteria decreased *L. monocytogenes* counts in intestinal tissue and reduced systemic spread of the pathogen to various organs, including spleen. Moreover, whole genome intestinal transcriptomics analysis revealed a modulation in host gene expression, primarily IFN-stimulated genes and remodulation of *L. monocytogenes* transcriptome.

Another approach for using probiotic bacteria as a prophylactic or therapeutic strategy to control listeriosis in humans is based on molecular mimicry of host receptor or targeted clearance of the pathogen. Previously, *E. coli* strain expressing glycosyltransferase genes and producing a chimeric LPS ending in a mimic of ganglioside GM1 exhibited cholera toxin binding property against *Vibrio cholerae* (Focareta et al., 2006). In another study, *L. paracasei* expression of Listeria adhesion protein was tested for its efficacy to prevent *L. monocytogenes* adhesion, invasion and translocation in a Caco-2 cell culture system (Koo et al., 2012). These investigators observed that the recombinant probiotic was more effective than normal probiotic strain in reducing *L. monocytogenes* infection *in vitro.*

Conclusion and Future Directions

The use of plant-derived antimicrobials and probiotics is a promising strategy for controlling *L. monocytogenes* in food processing environment, foods and humans due to the diverse biological effects exerted by these interventions. However, the organoleptic impact of plant compounds in food products limit their usage as natural antimicrobials. Similarly, the use of probiotics to control foodborne pathogen necessitates the need for proper characterization of the probiotics to better understand their properties before using as antimicrobial agents. Further studies are also needed in exploring efficient synergistic interactions between different plant compounds and probiotics for enhanced antimicrobial efficacy and broader application range. Moreover, a better understanding of the mode of action and interaction between plant compounds, probiotics, and bacteria is critical for devising efficient control measures against *L. monocytogenes.* Metagenomic analysis is an efficient tool to explore the nature of such interactions. The impact of plant compounds and probiotics on the epigenome of host cells and its effect on bacterial infection is another unexplored area that might lead to the development of natural therapeutics against *L. monocytogenes* and other enteric pathogens.

REFERENCES

Aarnisalo, K. Lunde'n, J. Korkeala, H. & Wirtanen, G. (2007). Susceptibility of *Listeria monocytogenes* strains to disinfectants and chlorinated alkaline cleaners at cold temperatures. *LWT – Food Science and Technology*, 40, 1041–1048.

Abou-Zeid, K. A. Yoon, K. S. Oscar, T. P. Schwarz, J. G. Hashem F. M. & Whiting, R. C. (2007). Survival and Growth of *Listeria monocytogenes* in broth as a function of temperature, pH, and potassium lactate and sodium diacetate concentrations. *Journal of Food Protection*, 70, 2620-2625.

Adonizio, A. L. Downum, K. Bennett, B. C. & Mathee, K. (2006). Antiquorum sensing activity of medicinal plants in southern Florida. *Journal of Ethnopharmacology*, 105, 427–435.

Agnihotri, S. & Vaidya, A. D. B. (1996). A novel approach to study antibacterial properties of volatile components of selected Indian medicinal herbs. *Indian Journal of Experimental Biology*, 34, 712–715.

Ahmad, I. & Beg, A. Z. (2001). Antimicrobial and phytochemical studies on 45 Indian medicinal plants against multi-drug resistant human pathogens. *Journal of Ethnopharmacology* 74, 113–123.

Ali S. M. Khan, A. A. Ahmed, I. Musaddiq, M. Ahmed, K. S. Polasa, H. Rao, L. Habibullah, C. M. Sechi, L. A. & Ahmed, N. (2005). Antimicrobial activities of eugenol and cinnamaldehyde against the human gastric pathogen *Helicobacter pylori*. *Annals of Clinical Microbiology and Antimicrobials*, 4, 1-7.

Alvarez-dominguez, C. Vazquez-boland, J. Carrasco-marin, E. Lopezmato, P. & Leyvacobia, F. (1997). Host cell heparan sulfate proteoglycans mediate attachment and entry of *Listeria monocytogenes*, and the Listerial surface protein ActA is involved in heparan sulfate receptor recognition. *Infection and Immunity*, 65, 78–88.

Amalaradjou, M. A. R. & Venkitanarayanan, K. (2011). Effect of trans-Cinnamaldehyde on inhibition and inactivation of *Cronobacter sakazakii* biofilm on abiotic surfaces. *Journal of Food Protection* 74, 200-208.

Ananou, S. Garriga, M. Hugas, M. Maqueda, M. Martinez-Bueno, M. Galvez, A. & Valdivia, E. (2005). Control of *Listeria monocytogenes* in model sausages by enterocin AS-48. *International Journal of Food Microbiology*, 103, 179-90.

Anonymous (2003). Control of *Listeria monocytogenes* in ready-to-eat meat and poultry products; final rule. *Federal Register*, 68, 109, Washington, D.C.

Aureli, P. Fiorucci, G. C. Caroli, D. Marchiaro, G. Novara, O. Leone, L. & Salmaso, S. (2000). An outbreak of febrile gastroenteritis associated with corn contaminated by *Listeria monocytogenes*. *The New England Journal of Medicine*, 342, 1236-41.

Azizkhani, M. Misaghi, A. Basti, A. A. Gandomi, H. & Hosseini, H. (2013). Effects of *Zataria multiflora* Boiss. essential oil on growth and gene expression of enterotoxins A, C and E in *Staphylococcus aureus* ATCC 29213. *International Journal of Food Microbiology*, 163, 159-65.

Azuma, T. Bagenda, D. K. Yamamoto, T. Kawai, Y. & Yamazaki, K. (2007). Inhibition of *Listeria monocytogenes* by freeze-dried piscicocin CS526 fermentate in food. *Letters in Applied Microbiology*, 44, 138-44.

Backhed, F. Ding, H. Wang, T. Hooper, L. V. Koh, G. Y. Nagy, A. Semenkovich, C. F. & Gordon, J. I. (2004). The gut microbiota as an environmental factor that regulates fat storage. *Proceedings of the National Academy of Sciences USA*, 101, 15718–15723.

Backhed, F., Manchester, J. K., Semenkovich, C. F. & Gordon, J. I. (2007). Mechanisms underlying the resistance to diet-induced obesity in germ-free mice. *Proceedings of the National Academy of Sciences USA*, 104, 979–984.

Bakkali, F. Averbeck, S. Averbeck, D. & Idaomar, M. (2008). Biological effects of essential oils – A review. *Food and Chemical Toxicology*. 46, 446-475.

Balciunas, E. M. Martinez, F. A. C. Todorov, S. D. Franco, B. D. G. M. Converti, A. & Oliveira. R. P. S. (2013). Novel biotechnological applications of bacteriocins: A review. *Food Control*, 32, 134-142.

Bambirra, Lima, K. G. Franco, B. D. Cara, D. C. Nardi, R. M. Barbosa, F. H. & Nicoli, J. R. (2007). Protective effect of *Lactobacillus sakei* 2a against experimental challenge with *Listeria monocytogenes* in gnotobiotic mice. *Letters in Applied Microbiology*, 45, 663-667.

Barbosa, L. N. Fernandes, A. A, Ushimaru, P. I. de Silva, P. I. & Fernandes, A. Jr. (2009). Essential oils against foodborne pathogens and spoilage bacteria in minced meat. *Foodborne pathogen and Disease*, 6, 725-728.

Benabbou, R. Zihler, A. Desbiens, M. Kheadr, E. Subirade, M. & Fliss, L. (2009). Inhibition of *Listeria monocytogenes* by a combination of chitosan and divergicin M35. *Canadian Journal of Microbiology*, 55, 347-355.

Beuchat, L. R. (1996). *Listeria monocytogenes*: incidence in vegetables. *Food Control*, 7, 223-228.

Beuchat, L. R. Farber, J. M. Garrett, E. H. Harris, L. J. Parish, M. E. Suslow, T. V. & Busta, F. F. (2001). Standardization of a method to determine the efficacy of sanitizers in inactivating human pathogenic microorganisms on raw fruits and vegetables. *Journal of Food Protection*, 64, 1079–1084.

Bhat , R. Alias, A. K. & Paliyath, G. (2011). Essential oils and other plant extracts as food preservatives. *Progress in Food Preservation*, New Jersey, John Wiley & Sons. 539-580.

Bhatt, P. & Negi, P.S. (2012). Antioxidant and antibacterial activities in the leaf extracts of Indian Borage (*Plectranthus amboinicus*). *Food and Nutrition Sciences*, 3, 146-152.

Blackman, I. C. and Frank, J. F. (1996). Growth of *Listeria monocytogenes* as a biofilm on various food-processing surfaces. *Journal of Food Protection*, 59, 827-831.

Blanco Massani, M. Fernandez, M. R. Ariosti, A. Eisenberg, P. & Vignolo, G. (2008). Development and characterization of an active polyethylene film containing *lactobacillus curvatus* CRL 705 bacteriocins. Food Additives and contaminants. Chemistry, Analysis, Control, Exposure and Risk Assessment, 25, 1424-1430.

Boland, J. A. V. Kuhn, M. Berche, P. Chakraborty, T. Bernal, G. D. Goebel, W. Bruno Zorn, B. G. Wehland, J. Kreft, J. (2001). Listeria pathogenesis and molecular virulence determinants. *Clinical Microbiology Reviews*, 14, 584-640.

Borucki, M. K. Peppin, J. D. White, D. Loge, F. & Call, D.R. (2003). Variation in biofilm formation among strains of *Listeria monocytogenes. Applied and Environmental Microbiology*, 69, 7336-7342.

Bower, C. K. McGuire, J. & Daeschel, M. A. (1995). Suppression of *Listeria monocytogenes* colonization following adsorption of nisin onto silica surfaces. *Applied and Environmental Microbiology*, 61, 992-997.

Brakat, R. K. & Harris, L. J. (1999). Growth of *Listeria monocytogenes* and *Yersinia enterocolitica* on cooked modified-atmosphere-packaged poultry in the presence and absence of a naturally occurring microbiota. *Applied and Environmental Microbiology*, 65, 342–345.

Bremer, P. J. Monk, I. & Butler, R. (2002). Inactivation of *Listeria monocytogenes/* flavobacterium spp. biofilms using chlorine: impact of substrate, pH, time and concentration. *Letters in Applied Microbiology*, 35, 321–325.

Brown, K. DeCoffe, D. Molcan, E. & Gibson D. L. (2012). Diet-Induced Dysbiosis of the Intestinal Microbiota and the Effects on Immunity and Disease. *Nutrients*, 4, 1095-1119.

Burt, S. (2004). Essential oils: their antibacterial properties and potential applications in foods—a review. *International Journal of Food Microbiology*, 94, 223–253.

CAC Drafting Group. 1999. Management of *Listeria monocytogenes* in foods. Draft document prepared by the Codex Drafting Group and submitted in preparation for the meeting of the Codex Committee on Food Hygiene to be held in 1999.

Campos, C. A. Castro, M. P. Gliemmo, M. F. & Schelegueda, L. I. (2011). Use of natural antimicrobials for the control of *Listeria monocytogenes* in foods. Science against microbial pathogens: communicating current research and technological advances, 2, 1112-1123.

Campos, C. A. Gerschenson, L. N. & Flores, S. K. (2010). Development of edible films and coatings with antimicrobial activity. *Food and Bioprocess Technology*, 6, 849-875.

Candela, M. Perna, F. Carnevali, P. Vitali, B. Ciati, R. Gionchetti, P. Rizzello, F. Campieri, M. & Brigidi, P. (2008). Interaction of probiotic *Lactobacillus* and *Bifidobacterium* strains with human intestinal epithelial cells: adhesion properties, competition against enteropathogens and modulation of IL-8 production. *International Journal of Food Microbiology*, 125, 286–292.

Canillac, N. & Mourey, A. (2004). Effects of several environmental factors on the anti-*Listeria monocytogenes* activity of an essential oil of *Picea excels*. *International Journal of Food Microbiology*, 92, 95– 103.

Carpentier, B. & Chassing, D. (2004). Interactions in biofilms between *Listeria monocytogenes* and resident microorganisms from food industry premises. *International Journal of Food Microbiology*, 97, 111–122.

Cartwright, E. J. Jackson, K. A. Shacara, D. J. Graves, L. M. Silk, B. J. & Mahon, B. E. (2013). Listeriosis outbreaks and associated food vehicles, United States, 1998-2008. *Emerging Infectious Disease,* 19, 1-10.

Cava-Roda, R. M. Taboada-Rodríguez, A. Valverde-Franco, M. T. & Marín- Iniesta, F. (2010). Antimicrobial activity of vanillin and mixtures with cinnamon and clove essential oils in controlling *Listeria monocytogenes* and *Escherichia coli* O157:H7 in milk. *Food and Bioprocess Technology*, 5, 2120-2131.

CDC, Centers for Disease control and prevention (2012a). Listeriosis Linked to Imported Frescolina Marte Brand Ricotta Salata Cheese.

CDC, Centers for Disease control and prevention, (2011). Listeriosis Linked to Whole Cantaloupes from Jensen Farms, Colorado.

CDC, Centers for Disease control and prevention, (2012b). Foodborne outbreak online database (FOOD). Centers for Disease control and prevention, Atlanta, GA.

CDC, Centers for Disease control and prevention, (2013). Multistate Outbreak of Listeriosis Linked to Crave Brothers Farmstead Cheeses (Final Update).

Cekmez, F. Tayman, C. Saglam, C. Cetinkaya, M. Bedir, O. Gnal, A. Tun, T. & Sarici, S. U. (2012). Well-known but rare pathogen in neonates: *Listeria monocytogenes*. *European Review for Medical and Pharmacological Sciences*, 4, 58-61.

Cerrutti, P. & Alzamora, S. M. (1996). Inhibitory effects of vanillin on some food spoilage yeasts in laboratory media and fruit purées. *International Journal of Food Microbiology*, 29, 379–386.

Chaturongakul, S. Raengpradub, S. Wiedmann, M. & Boor, K.J. (2008). Modulation of stress and virulence in *Listeria monocytogenes*. *Trends Microbiology*, 16, 388-396.

Chavant, P. Martinie, B. Meylheuc, T. Bellon-Fontaine, M. N. & Hebraud, M. (2002). *Listeria monocytogenes* LO28: surface physicochemical properties and ability to form biofilms at different temperatures and growth phases. *Applied and Environmental Microbiology*, 68, 728-737.

Chi-Zhang, Y. Yam, K. L. and Chikindas, M. L. (2004). Effective Control of *Listeria monocytogenes* by combination of nisin formulated and slowly released into a broth system. *International Journal of Food Microbiology*, 90, 15–22.

Cloete, T. E. Jacobs, L. & Brozel, V. S. (1998). The chemical control of biofouling in industrial water systems. *Biodegradation*, 9, 23–37.

Corrs, S. C. Gahan, C. G. & Hill, C. (2007). Impact of selected Lactobacillus and Bifidobacterium species on *Listeria monocytogenes* infection and the mucosal immune response. *FEMS Immunology and Medical Microbiology*, 50, 380-8.

Cowan, M. M. (1999). Plant products as antimicrobial agents. *Clinical Microbiology Review* 12, 564–582.

Cox, L. J. Kleiss, T. Cordier, J. L. Cordellana, C. Konkel, P. Pedrazzini, C. Beumer, R. & Siebenga, A. (1989). *Listeria* spp. in food processing, non-food and domestic environments. *Food Microbiology*, 1989, 6, 49–61.

Cox, L. J. Kleiss, T. Cordier, J. L. Cordellana, C. Konkel, P. Pedrazzini, C. Beumer, R. & Siebenga. A. (1989). *Listeria* spp. in food processing, non-food and domestic environments. *Food Microbiology*, 6, 49–61.

Cragg, G. M. Newman, D. J. & Snader K. M (1997). Natural products in drug discovery and development. *Journal of Natural Products* 60, 52–60.

Davidson, P. M. (2001). Chemical preservatives and natural antimicrobial compounds, In: Doyle, M. P., Beuchat, L. R., Montville, T. J. (Eds.), Food Microbiology: Fundamental and Frontiers, 2nd ed. ASM Press, Washington DC, pp. 593–627.

Davies, E. A. Bevis, H. E. & Delves-Broughton, J. (2003). The use of the bacteriocin, nisin, as a preservative in ricotta-type cheeses to control the food-borne pathogen *Listeria monocytogenes*. *Letters in Applied Microbiology*, 24, 343-346.

de Souza, E. L. de Barros, J. C. de Oliveira, C. E. & da Conceição, M. L. (2010). Influence of *Origanum vulgare* L. essential oil on enterotoxin production, membrane permeability and surface characteristics of *Staphylococcus aureus*. *International Journal of Food Microbiology*, 137, 308-11.

De Wit J. C. Notermans S. H. W. Gorin W, & Kampelmacher E. H. (1979). Effect of garlic oil or onion oil on toxin production by Clostridium botulinum in meat slurry. *Journal of Food Protection*, 42, 222–224.

Deans, S. G. & Ritchie, G. (1987). Antimicrobial properties of plant essential oils. *International Journal of Food Microbiology* 5, 165-180.

Desai, J. & Banat, I. M. (1997). Microbial production of surfactants and their commercial potential. *Microbiology and Molecular Biology Reviews*, 61, 47-64.

Desai, M.A. Soni, K.A. Nannapaneni, R. Schilling, M.W. & Silva, J.L. (2012). Reduction of *Listeria monocytogenes* biofilms on stainless steel and polystyrene surfaces by essential oils. Journal of Food Protection 75, 1332-1337.

Dethlefsen, L. McFall-Ngai, M. & Relman, D.A. (2007). An ecological and evolutionary perspective on human–microbe mutualism and disease. Nature, 449, 811-818.

Dethlefsen, L. McFall-Ngai, M. & Relman, D.A. (2007). An ecological and evolutionary perspective on human-microbe mutualism and disease. Nature, 449, 811-818.

Di Pasqua, R. De Feo, V. Villani, F. & Mauriello, G. (2005). *In vitro* antimicrobial activity of essential oils from Mediterranean Apiaceae, Verbenaceae and Lamiaceae against foodborne pathogens and spoilage bacteria. Annals of Microbiology, 55, 139–143.

Djenane, D. Yangüela, J. Montañés, L. Djerbal, M. & Roncaléset, P. (2011). Antimicrobial activity of *Pistacia lentiscus* and *Satureja montana* essential oils against *Listeria monocytogenes* CECT 935 using laboratory media: Efficacy and synergistic potential in minced beef. Food Control, 22, 1046-1053.

Dominguez-Bello, M. G. & Blaser, M. J. (2008). Do you have a probiotic in your future? *Microbes and Infection*, 10, 1072–1076.

Donnelly, C. W. (2001). *Listeria monocytogenes*: a continuing challenge. *Nutrition Reviews*, 59, 183-194.

Dos Santos, L. M. Santos, M. M. de Souza Silva, H. P. Arantes, R. M. Nicholi, J. R. & Vieira, L. Q. (2011). Monoassociation with probiotic *Lactobacillus delbrueckii* UFV-H2b20 stimulates the immune system and protects germfree mice against *Listeria monocytogenes* infection. *Medical Microbiology and Immunology*, 200, 29-38.

Dufour, M. Simmonds, R. S. & Bremer, P. J. (2004). Development of a laboratory scale clean-in-place system to test the effectiveness of ''natural'' antimicrobials against dairy biofilms. *Journal of Food Protection*, 67, 1438–1443.

Duncan, S. H. Belenguer, A. Holtrop, G. Johnstone, A. M. Flint, H. J. & Lobley, G. E. (2007). Reduced dietary intake of carbohydrates by obese subjects results in decreased concentrations of butyrate and butyrate-producing bacteria in feces. *Applied and Environmental Microbiology*, 73, 1073–1078.

Dunstall, G. Rowe, M. T. Wisdom, G. B. & Kilpatrick, D. (2005). Effect of quorum sensing agents on the growth kinetics of *Pseudomonas* spp. of raw milk origin. *Journal of Dairy Research*, 72, 276–280.

Dutta, N. K. Mazumdar, K. & Park, J. H. (2009). *In vitro* synergistic effect of gentamicin with the anti-inflammatory agent diclofenac against *Listeria monocytogenes. Letters in Applied Microbiology*, 48, 783-785.

Ercolini, D. Storia, A. Villani, F. & Mauriello, G. (2006). Effect of a bacteriocin-activated polythene film on *Listeria monocytogenes* as evaluated by viable staining and epifluorescence microscopy. *Journal of Applied Microbiology*, 100, 765-772.

Fallico, V. McAuliffe, O. Fitzgerald, G. Hill, and C. Ross, R. P. In: Protective cultures, antimicrobial metabolites and bacteriophages for food and beverage biopreservation. C Lacroix, Ed. Woodhead Publishing 2010; 4:pp. 100-28.

Farber, J. M. & Peterkin, P. I. (1991). *Listeria monocytogenes,* a food-borne pathogen. *Microbiology Reviews*, 55, 476–511.

Fatemi, P. & Frank, J. F. (1999). Inactivation of *Listeria monocytogenes/Pseudomonas fluorescens* biofilms by peracid sanitizers. *Journal of Food Protection*, 62, 761–765.

FDA, Food and Drug Administration, (2010). DSHS orders Sangar produce to close, recall products. Food and Drug Administration, Silver Spring, MD. http://www.fda.gov/safety/recalls/2010/ucm230709.htm.

Firouzi, R. Shekarforoush, S.S. Nazer, A. H. K. Borumand, Z. & Jooyandeh, A. R. (2007). Effects of Essential Oils of Oregano and Nutmeg on Growth and Survival of *Yersinia enterocolitica* and *Listeria monocytogenes* in Barbecued Chicken. *Journal of Food Protection,* 70, 2626-2630.

Focareta, A. Paton, J.C. Morona, R. Cook, J. & Paton, A. W. (2006). A recombinant probiotic for treatment and prevention of cholera. *Gastroenterology*, 130, 1688-95.

Folsom, J. P. & Frank, J. F. (2007). Proteomic analysis of a hypochlorous acid-tolerant *Listeria monocytogenes* cultural variant exhibiting enhanced biofilm production. *Journal of Food Protection*, 5, 1129-1136.

Forsythe, S. J., and Hayes, P. R. (1998). Food hygiene, microbiology and HACCP (3rd ed.). Aspen Publishers.

Frank, J., & Koffi, R. (1990). Surface-adherence growth of *Listeria monocytogenes* is associated with increased resistance to surfactant sanitizers and heat. *Journal of Food Protection* 53, 550-554.

Frank, J. F. & Koffi, R. A. (1990). Surface adherent growth of *Listeria monocytogenes* in associated with increased resistance to surfactant sanitizers and heat. *Journal of Food Protection* 53, 550-554.

Fukuda, S. Toh, H. Hase, K. Oshima, K. Nakanishi, Y. Yoshimura, K. Tobe, T. Clarke, J. M. Topping, D. L. Suzuki, T. Taylor, T. D. Itoh, K. Kikuchi, J. Morita, H. Hattori, M. & Ohno, H. (2011). Bifidobacteria can protect from enteropathogenic infection through production of acetate. *Nature*, 469, 543–547.

Galvez, A. Abriouel, H. Lopez, R. L. & Ben Omar, N. (2007). Bacteriocin based strategies for food biopreservation. *International journal of food microbiology*,120, 51-70.

Gandhi, M. & Chikindas, M. L. (2007). Listeria: A foodborne pathogen that knows how to survive. *International Journal of Food Microbiology*, 113, 1-15.

Garcia-Almendarez, B. E., Cann, I. K. O., Martin, S. E., Guerrero-Legarreta, I., &

Geissman, T. A. (1963). Flavonoid compounds, tannins, lignins and related compounds. In: Florkin, M., Stotz, E. H. (Eds.), Pyrrole Pigments, Isoprenoid Compounds and Phenolic Plant Constituents. Elsevier, New York, p. 265.

Georgaras, I. Panagiotis, N. Skandamis, N. Belk, K. E. Scanga, J. A. Kendall, P. A. Smith, G. C. & Sofos, J. N. (2006). Post-processing application of chemical solutions for control of *Listeria monocytogenes*, cultured under different conditions on commercial smoked sausage formulated with and without lactate-sodium diacetate. *Food Microbiology*, 23, 762-771.

Gerba, C. P. Rose, J. B. & Haas, C. N. (1996). Sensitive populations: Who is at the greatest risk. *International Journal of Food Microbiology*, 30, 113-123.

Ghosh, S. Dai, C. Brown, K. Rajendiran, E. Makarenko, S. Baker, J. Ma, C. Halder, S. Montero, M. Ionescu, V. A. Klegeris, A. Vallance, B. A. & Gibson, D. L. (2011). Colonic microbiota alters host susceptibility to infectious colitis by modulating inflammation, redox status, and ion transporter gene expression. *American Journal of Physiology Gastrointestinal Liver Physiology*, 301, G39–G49.

Gill, A. O. Delaquis, P. Russo, P. & Holley, R. A. (2002). Evaluation of antilisterial action of cilantro oil on vacuum packed ham. *International Journal of Food Microbiology*, 73, 83-92.

Goh, E. B. Yim, G. Tsui, W. McClure, J. Surette, M. G. & Davies, J. (2002). Transcriptional modulation of bacterial gene expression by subinhibitory concentrations of antibiotics. *Proceedings of the National Academy of Sciences of the United States of America*. 99, 17025- 17030.

Gonzalez-Fandos, E. Garc´ıa-Lo´pez, M. L. Sierra, M. L. & Otero, A. (1994). Staphylococcal growth and enterotoxins (A–D) and thermonuclease synthesis in the presence of dehydrated garlic. *Journal of Applied Bacteriology*, 77, 549 – 552.

Gordon, J. I. & Klaenhammer, T. R. (2011). A rendezvous with our microbes. *Proceedings of the National Academy Sciences of the United States of America*. 108, 4513–4515.

Gould, G. W. (1989). Introduction. In: Gould, G. W. (Ed.), Mechanisms of Action of Food Preservation Procedures. Elsevier Applied Science, London, pp. 1–42.

Greiffenberg, L. Goebel W. Kim, K. S. Weiglein, I. Bubert, A. Engelbrecht, F. Stins, M. Kuhn, M. (1998). Interaction of *Listeria monocytogenes* with human brain microvascular endothelial cells: InlB-dependent invasion, long-term intracellular growth, and spread from macrophages to endothelial cells. *Infection and Immunity*, 66, 5260-5267.

Grundling, A. Gonzalez, M. D. Higgins, D. E. (2003). Requirement of the *Listeria monocytogenes* broad-range phospholipase PC-PLC during infection of human epithelial cells. *Journal of Bacteriology*, 185, 6295-6307.

Guerin-Danan, C. Chabanet, C. Pedone, C. Popot, F. Vaissade, P. Bouley, C. Szylit, O. & Andrieux, C. (1998). Milk fermented with yogurt cultures and *Lactobacillus casei* compared with yogurt and gelled milk: influence on intestinal microflora in healthy infants. *American Journal of Clinical Nutrition*, 67, 111–117.

Gutierrez, J. Barry-Ryan, C. & Bourke, P. (2008a). The antimicrobial efficacy of plant essential oil combinations and interactions with food ingredients. *International Journal of Food Microbiology*, 124, 91–97.

Gutierrez, J. Barry-Ryan, C. & Bourke, P. (2009). Antimicrobial activity of plant essential oils using food model media: Efficacy, synergistic potential and interactions with food components. *Food Microbiology*, 26, 142-150.

Gutierrez, J. Rodriguez, G. Barry-Ryan, C. & Bourke, P. (2008b). Efficacy of Plant Essential Oils against Foodborne Pathogens and Spoilage Bacteria Associated with Ready-to-Eat Vegetables: Antimicrobial and Sensory Screening. *Journal of Food Protection*, 71, 1846-1854.

Hamed, S. F. Sadek, Z. & Edris, A. (2012). Antioxidant and antimicrobial activities of clove bud essential oil and eugenol nanoparticles in alcohol-free microemulsion. *Journal of Oleo Science*, 61, 641-648.

Hampikyan, H. & Ugur, M. (2007). The effect of nisin on *L. monocytogenes* in Turkish fermented sausages (sucuks). *Meat Science*, 76, 327-32.

Hao, Y. Y. Brackett, R. E. Doyle, M. P. (1998). Inhibition of *Listeria monocytogenes* and *Aeromonas hydrophila* by plant extracts in refrigerated cooked beef. *Journal of Food Protection* 61, 307–312.

Harborne, J. R. (1993). Introduction to ecological biochemistry. 4th ed. London: Elsevier.

Hattori, M. & Taylor, T. D. (2009). The Human Intestinal Microbiome: A New Frontier of Human Biology. *DNA Research*, 16, 1-12.

Hawrelak, J. A. Cattley, T. & Myers, S. P. (2009). Essential oils in the treatment of intestinal dysbiosis: a preliminary *in vitro* study. *Alternative Medicine Review*, 14, 380–384.

Heir, E. Lindstedt, B. A. Rotterud, O. J. Vardund, T. Kapperud, G. Nesbakken, T. (2004). Molecular epidemiology and disinfectant susceptibility of *Listeria monocytogenes* from meat processing plants and human infections. *International Journal of Food Microbiology*, 96, 85-96.

Heisick, J. E. Wagner, D. E. Nierman, M. L. & Peeler, J. T. (1989). Listeria spp. found on fresh market produce. *Applied and Environmental Microbiology*, 55,1925-1927.

Heisick, J. E. Rosas-Marty, L. I. & Tatini, S. R. (1995). Enumeration of viable *Listeria* species and *Listeria monocytogenes* in foods. *Journal of Food Protection*, 58, 733-736.

Holah, J. T. Taylor, J. H. Dawson, D. J. Hall, K. E. (2002). Biocide use in the food industry and the disinfectant resistance of persistent strains of *Listeria monocytogenes* and *Escherichia coli*. *Journal of Applied Microbiology*, 92, 111S-120S.

Holley, R. A. & Patel, D. (2005). Improvement in shelf-life and safety of perishable foods by plant essential oils and smoke antimicrobials. *Food Microbiology*, 22, 273–292.

Holzapfel, W. H. Geisen, R. & Schillinger, U. (1995). Biological preservation of foods with reference to protective cultures, bacteriocins and food- grade enzymes. *International Journal of Food Microbiology*, 24, 343-62.

Hoque, M. M. Bari, M. L. Juneja, V. K. & Kawamoto, S. (2008). Antimicrobial activity of cloves and cinnamon extracts against food borne pathogens and spoilage bacteria, and inactivation of *Listeria monocytogenes* in ground chicken meat with their essential oils. Report of National Food Research Institute, 72, 9-21.

Hu. Y. Gall, K. Ho, A. Ivanek, R. Gröhn, Y. T. & Wiedmann, M. (2006). Daily variability of *Listeria* contamination patterns in a cold-smoked salmon processing operation. *Journal of Food Protection*, 69, 2123–2133.

Huber, B. Eberl, L. Feucht, W. & Polster, J. (2003). Influence of polyphenols on bacterial biofilm formation and quorum-sensing. *Z. Naturforsch.* 58, 879–884.

Hyldgaard, M. Mygind, T. & Meyer, R. L. (2012). Essential oils in food preservation: mode of action, synergies, and interactions with food matrix components. *Frontiers in Microbiology*, 3, 12.

International Commission on Microbiological Specifications for Foods. (1994). Choice of sampling plan and criteria for *Listeria monocytogenes. International Journal of Food Microbiology*, 22, 89–96.

International Commission on Microbiological Specifications for Foods. (1996). Annex to Codex document on establishment of sampling plans for *L. monocytogenes* in foods in international trade. Submitted by ICMSF to the Codex Food Hygiene Committee, September 1996.

Isman, M. B. (2000). Plant essential oils for pest and disease management. *Crop Protection*, 19, 603– 608.

Jandu, N. Zeng, Z. J. Johnson-henry, K. C. & Sherman, P. M. (2009). Probiotics prevent enterohaemorrhagic *Escherichia coli* O157:H7- mediated inhibition of interferon-gamma-induced tyrosine phosphorylation of STAT-1. *Microbioogy*, 155, 531-40.

Jaradat, Z. W. & Bhunia A. K. (2003). Adhesion, invasion, and translocation characteristics of *Listeria monocytogenes* serotypes in Caco-2 cell and mouse models. *Applied and Environmental Microbiology*, 69, 3640-3645.

Jay, J. M. (1996). Prevalence of *Listeria* spp. in meat and poultry products. *Food Control*, 7, 209-214.

Kapoor, N. Narain, U. and Misra, K. (2007). Bioactive conjugates of curcumin having ester, peptide, thiol and disulphide links. *Journal of Scientific and Industrial Research*, 66, 647–650.

Katla, T. Moretro, T. Aasen, I. M. Holck, A. Axelsson, L. & Naterstad, K. (2001). Inhibition of *Listeria monocytogenes* in cold smoked salmon by addition of sakacin P and/or live *Lactobacillus sakei* cultures. *Food Microbiology*, 18, 431-39.

Katla, T. Moretro, T. Sveen, I. Aasen, I. M. Axelsson, L. Rorvik, L. M. & Naterstad, K. (2002). Inhibition of *Listeria monocytogenes* in chicken cold cuts by addition of sakacin P and sakacin P-producing *Lactobacillus sakei*. *Journal of Applied Microbiology*, 93, 191-196.

Kennedy, D. O. & Wightman, E. L. (2011). Herbal extracts and phytochemicals: plant secondary metabolites and the enhancement of human brain function. *Advances in Nutrition*, 2, 32-50.

Khor, B. Gardet, A. & Xavier, R. J. (2011). Genetics and pathogenesis of inflammatory bowel disease. *Nature*, 474, 307–317.

Kim, S. Fung, D. Y. (2004). Antibacterial effect of water-soluble arrowroot (*Puerariae radix*) tea extracts on foodborne pathogens in ground beef and mushroom soup. *Journal of Food Protection* 67, 1953–1956.

Kim, S. Ruengwilysup, C. Fung, D. Y. (2004). Antibacterial effect of water-soluble tea extracts on foodborne pathogens in laboratory medium and in a food model. *Journal of Food Protection*, 67, 2608–2612.

Kives, J. Guadarrama, D. Orgaz, B. Rivera-Sem, A. Vazquez, J. & SanJose, C. (2005). Interactions in biofilms of *Lactococcus lactis* ssp. cremoris and *Pseudomonas fluorescens* cultured in cold UHT milk. *Journal of Dairy Science*, 88, 4165–4171.

Kocks, C. Gouin, E. Tabouret, M. Berche, P. Ohayon, H. & Cossart, P. (1992). *L. monocytogenes*-induced actin assembly requires the actA gene product, a surface protein. *Cell*, 68, 521-531.

Koh, C. L. Sam, C. K. Yin, W. F. Tan, L. Y. Krishnan, T. Chong, Y. M. & Chan, K. G. (2013). Plant-derived natural products as sources of anti-quorum sensing compounds. *Sensors*, 13, 6217-28.

Koh, C. L. Sam, C. K. Yin, W. F. Tan, L. Y. Krishnan, T. Chong, Y. M. & Chan, K. G. (2013). Plant-derived natural products as sources of anti-quorum sensing compounds. *Sensors*, 13, 6217-6228.

Koo, O. K. Amalaradjou, M. A. R. & Bhunia, A. K. (2012). Recombinant probiotic expressing Listeria adhesion protein attenuates *Listeria monocytogenes* virulence in vitro. *PLoS One*, 7,29277.

Kouakou P, Ghalfi H, Destin J, Dubois-Dauphin R, Evrard P, & Thonart P. (2008). Enhancing the antilisterial effect of *Lactobacillus curvatus* CWBI-B28 in pork meat and cocultures by limiting bacteriocin degradation. *Meat Science*, 80, 640–648.

Kovacevic, J. Sagert, J. Wozniak, A. Gilmour, M. W. & Allen, K. J. (2013). Antimicrobial resistance and co-selection phenomenon in *Listeria* spp. recovered from food and food production environments. *Food Microbiology*, 34, 319-27.

Kovacevic, J. Sagert, J. Wozniak, A. Gilmour, M. W. & Allen, K. J. (2013). Antimicrobial resistance and co-selection phenomenon in *Listeria* spp. recovered from food and food production environments. *Food Microbiology*, 34, 319-327.

Krawczyk-Balska, A. Marchlewicz, J. Dudek, D. Wasiak, K. & Samluk, A. (2012). Identification of a ferritin-like protein of *Listeria monocytogenes* as a mediator of β-lactam tolerance and innate resistance to cephalosporins. *BMC Microbiology*, 12, 278.

Krysinski, E. P. Brown, L. J. & Marchisello, T. J. (1992). Effect of cleaners and sanitizers on *Listeria monocytogenes* attached to product contact surfaces. *Journal of Food Protection* 55, 246-251.

Kubo, L. Muroi, H. & Himejima, M. (1993). Structure–antibacterial activity relationships of anacardic acids. *Journal of Agricultural and Food Chemistry* 41, 1016–1019.

Kuda, T. Yano, T. & Kuda, M. T. (2008). Resistances to benzalkonium chloride of bacteria dried with food elements on stainless steel surface. *LWT – Food Science and Technology*, 41, 988–993.

Kyung, K. H. (2011). Antimicrobial properties of Allium species. *Current Opinion in Biotechnology*, 23, 142-147.

Lamont, R. F. Sobel, J. Mazaki-Tovi, S. Kusanovic, J. P. Vaisbuch, E. Kim. S. K. Uldbjerg N. & Romero, R. (2011). Listeriosis in human pregnancy: a systematic review. *Journal of Perinatal Medicine*, 39, 227-36.

Latorre, A. A. Pradhan, A. K. Van Kessel, J. A. Karns, J. S. Boor, K. J. Rice, D. H. Mangione, K. J. Grohn, Y. T. & Schukken, Y. H. (2011). Quantitative risk assessment of listeriosis due to consumption of raw milk. *Journal of Food Protection*, 74, 1268–1281.

Leriche, V. & Carpentier, B. (2000). Limitation of adhesion and growth of *Listeria monocytogenes* on stainless steel surfaces by *Staphylococcus sciuri* biofilms. *Journal of Applied Microbiology*, 88, 594–605.

Ley, R. E. Turnbaugh, P. J. Klein, S. & Gordon, J. I. (2006). Microbial ecology: human gut microbes associated with obesity. *Nature* 444, 1022–1023.

Li, J., Dong, J., Qiu, J. Z., Wang, J. F., Luo, M. J., Li, H. E., Leng, B. F., Ren, W. Z., & Deng, X.M. (2011). Peppermint oil decreases the production of virulence-associated exoproteins by *Staphylococcus aureus*. *Molecules*, 16, 1643-54.

Lingnau, A. Domann, E. Hudel, M. Bock, M. Nichterlein, T. Wehland, J. & Chakraborty, T. (1995). Expression of the *Listeria monocytogenes* EGD *inlA* and *inlB* genes, whose products mediate bacterial entry into tissue culture cell lines, by PrfA-dependent and independent mechanisms. *Infection and Immunity*, 63, 3896–3903.

Garcia M., Amalaradjou, M. A. R. Nair, M. K. M. Annamalai, T. Surendranath, S.. Lee, S Hoagland, T. Dzurec, D. Faustman, C. & Venkitanarayanan, K. (2007). Inactivation of *Listeria monocytogenes* on frankfurters by monocaprylin alone or in combination with acetic acid. *Journal of Food Protection*, 70, 1594-1599.

Mahdavi, M. Jalali, M. & Kermanshahi, R. K. (2007). The effect of nisin on biofilm forming foodborne bacteria using microtiter plate method. *Research in Pharmaceutical Sciences*, 2, 113–118.

Marcos, B. Aymerich, T. Monfort, J. M. & Garriga, M. (2007). Use of antimicrobial biodegradable packaging to control *Listeria monocytogenes* during storage of cooked ham. *International Journal of Food Microbiology*, 120, 152-158.

Marquis, H. & Hager, E. J. (2000). pH-regulated activation and release of a bacteria-associated phospholipase C during intracellular infection by *Listeria monocytogenes*. *Molecular Microbiology*, 35, 289-298.

Martinez Viedma, P. Abriouel, H. Ben Omar, N. Lucas Lopez, R. Valdivia, E. & Galvez, A. (2009). Assay of enterocin AS-48 for inhibition of foodborne pathogens in desserts. *Journal of Food Protection*, 72, 1654-1659.

Martins, I. S. Faria, F. C. Miguel, M. A. Dias, M. P. Cardoso, F. L. Magalhães, A. C. Mascarenhas, L.A. Nouér, S.A. Barbosa, A.V. Vallim, D.C. Hofer, E. Rabello, R.F. Riley, L. W. & Moreira, B. M. (2010). A cluster of *Listeria monocytogenes* infections in hospitalized adults. *American Journal of Infection Control*, 38, e31-36.

Mattila-Sandholm, T. Myllärinen, P. Crittenden, P. R. Mogensen, G. Fondén, R. & Saarela M. (2002). Technological challenges for future probiotic foods. *International Dairy Journal*, 12, 173-182.

Maukonen, J. Matto , J. Wirtanen, G. Raaska, L. Mattila-Sandholm, T. & Saarela, M. (2003). Methodologies for the characterization of microbes in industrial environments: a review. *Journal of Industrial Microbiology and Biotechnology*, 30, 327–356.

Mauriello, G. Ercolini, D. La Storia, A. Casaburi, A. & Villani, F. (2004). Development of polyethene films for food packaging activated with an antilisterial bacteriocins from *Lactobacillus curvatus* 32Y. *Journal of Applied Microbiology*, 97, 314-322.

McAuliffe, O. & Jordan, K. N. (2012). Biotechnological approaches for control of *Listeria monocytogenes* in foods. *Current Biotechnology*, 1, 267-280.

Mead, P. S. Dunne, E. F. Graves, L. Wiedmann, M. Patrick, M. Hunter, S. Salehi, E. Mostashari, F. Craig, A. Mshar, P. Bannerman, T. Sauders, B.D. Hayes, P. Dewitt, W. Sparling, P. Griffin, P. Morse, D. Slutsker, L. & Swaminathan. B. (2006). Nationwide outbreak of listeriosis due to contaminated meat. *Epidemiology and Infection*, 134, 744–751.

Mireles, J. R. Toguchi, A. & Harshey, R. M. (2001). *Salmonella enterica* serovar Typhimurium swarming mutants with altered biofilm-forming abilities: surfactin inhibits biofilm formation. *Journal of Bacteriology*, 183, 5848–5854.

Molinos, A. C. Abriouel, H. Ben Omar, N. Lucas, R. Valdivia, E. & Galvez, A. (2008). Inactivation of *Listeria monocytogenes* in raw fruits by enterocin AS-48. *Journal of Food Protection*, 71, 2460-7.

Molinos, A. C. Abriouel, H. Ben Omar, N. Valdivia, E. Lopez, R. L. Maqueda, M. Canamero, M. M. & Galvez, A. (2005). Effect of immersion solutions containing enterocin AS-48 on *Listeria monocytogenes* in vegetable foods. *Applied and Environmental Microbiology*, 71, 7781-7.

Moretro, T. & Langsrud, S. (2004). *Listeria monocytogenes*: biofilm formation and persistence in food-processing environments. *Biofilms*, 1, 107-121.

Morgan, S. M. Galvin, M. Kelly, J. Ross, R. P. & Hill, C. (1999). Development of a lacticin 3147-enriched whey powder with inhibitory activity against foodborne pathogens. *Journal of Food Protection*, 62, 1011-1016.

Morgan, S. M. Galvin, M. Ross, R. P. & Hill, C. (2001). Evaluation of a spray- dried lacticin 3147 powder for the control of *Listeria monocytogenes* and *Bacillus cereus* in a range of food systems. *Letters in Applied Microbiology*, 33, 387-91.

Moroni, O. Kheadr, E. Boutin, Y. Lacroix, C. & Fliss, I. (2006). Inactivation of adhesion and invasion of food-borne *Listeria monocytogenes* by bacteriocin-producing

Bifidobacterium strains of human origin. *Applied and Environmental Microbiology*, 72, 6894-901.

Mosteller, T. M. & Bishop, J. R. (1993). Sanitizer efficacy against attached bacteria in milk biofilm. *Journal of Food Protection*, 56, 34–41.

Munoz, M. Guevara, L. Palop, A. Tabera, J. & Fernández, P.S. (2009). Determination of the effect of plant essential oils obtained by supercritical fluid extraction on the growth and viability of *Listeria monocytogenes* in broth and food systems using flow cytometry. *LWT - Food Science and Technology*, 42, 220-227.

Munoz, P. Rojas, L. Bunsow, E. Saez, E. Sánchez-Cambronero, L. Alcalá, L. Rodríguez-Creixems, M. & Bouza, E. (2012). Listeriosis: An emerging public health problem especially among the elderly. *The Journal of Infection*, 64, 19-33.

Murphy. R. Y. Hanson, R. E. Johnson, N. R. Chappa, K. and Berrang, M. E. (2006). Combining organic acid treatment with steam pasteurization to eliminate *Listeria monocytogenes* on fully cooked frankfurters. *Journal of Food Protection*, 69, 47-52.

Naghmouchi, K. Drider, D. Kheadr, E,. Lacroix , C. Prévost, H. & Fliss, I. (2006). Multiple characterizations of *Listeria monocytogenes* sensitive and insensitive variants to divergicin M35, a new pediocin-like bacteriocin. *Journal of Applied Microbiology*. 10, 29–39.

Negi, P. S. (2012). Plant extracts for the control of bacterial growth: Efficacy, stability and safety issues for food application. *International Journal of Food Microbiology*, 156, 7-17.

Negi, P. S. Anandharamakrishnan, C. & Jayaprakasha, G. K. (2003a). Antibacterial activity of *Aristolochia bracteata* root extracts. *Journal of Medicinal Food* 6, 401–403.

Negi, P. S. Chauhan, A. S. Sadia, G. A. Rohinishree, Y. S. & Ramteke, R. S. (2005). Antioxidant and antibacterial activities of various Seabuckthorn (*Hippophae rhamnoides* L.) seed extracts. *Food Chemistry*, 92, 119–124.

Negi, P. S. Jayaprakasha, G. K. & Jena, B. S. (2003b). Antioxidant and antimutagenic activities of pomegranate peel extract. *Food Chemistry*, 80, 393–397.

Negi, P. S. Jayaprakasha, G. K. & Jena, B. S. (2010). Evaluation of antioxidant and antimutagenic activities of the extracts from the fruit rinds of *Garcinia cowa*. *International Journal of Food Properties*, 13, 1256–1265.

Negi, P. S. Jayaprakasha, G. K. Rao, L. J. M. & Sakariah, K. K. (1999). Antibacterial activity of turmeric oil—a byproduct from curcumin manufacture. *Journal of Agricultural and Food Chemistry*, 47, 4297–4300.

Nguyen, U. T. Wenderska, I. B. Chong, M. A. Koteva, K. Wright, G. D. & Burrows, L. L. (2012). Small-molecule modulators of *Listeria monocytogenes* biofilm development. *Applied and Environment Microbiology*, 78, 1454-65.

Nieto-Lozano, J. C., Reguera-Useros. J. I., Peláez-Martínez, M. C., Sacristán-Pérez-Minayo, G., Gutiérrez-Fernández, A. J., Hardisson, T. A. (2010). The effect of the pediocin PA-1 produced by *Pediococcus acidilactici* against *Listeria monocytogenes* and *Clostridium perfringens* in Spanish dry-fermented sausages and frankfurters. *Food Control*, 21, 679–685.

Nitschke, M. & Costa, S. G. V. A. O. (2007). Biosurfactants in food industry. *Trends in Food Science and Technology*, 18, 252–259.

Norwood, D. E. & Gilmour, A. (2000). The growth and resistance to sodium hypochlorite of *Listeria monocytogenes* in a steady-state multispecies biofilm. *Journal of Applied Microbiology*, 88, 512-20.

Ohno, T. Kita, M. Yamaoka, Y. Imamura, S. Yamamoto, T. Mitsufuji, S. Kodama, T. Kashima, K. & Imanishi, J. (2003). Antimicrobial activity of essential oils against *Helicobacter pylori. Helicobacter*, 8, 207–215.

Olszak, T. An, D. Zeissig, S. Vera, M. P. Richter, J. Franke, A. Glickman, J. N. Siebert, R. Baron, R. M. Kasper, D. L. & Blumberg, R. S. (2012). Microbial exposure during early life has persistent effects on natural killer T cell function. *Science*, 336, 489–493.

Ooi, S. T. & Lorber, B. (2005). Gastroenteritis due to *Listeria monocytogenes. Clinical Infectious Disease*, 40, 1327-32.

Otake, S. Makimura, M. Kuroki, T. Nishihara, Y. & Hirasawa, M. (1991). Anticaries effects of polyphenolic compounds from Japanese green tea. *Caries Research*, 25, 438–443.

Owen, R. J. & Palombo, E. A. (2007). Anti-listerial activity of ethanolic extracts of medicinal plants, *Eremophila alternifolia* and *Eremophila duttonii*, in food homogenates and milk. *Food Control*, 18, 387–390.

Pan, Y. K. Breidt, & Jr. Kathariou, S. (2006). Resistance of *Listeria monocytogenes* biofilms to sanitizing agent in a simulated food-processing environment. *Applied and Environmental Microbiology*, 72, 7711-7717.

Parsaeimehr, M. Basti, A. A. Radmehr, B. Misaaghi, A. Abbasifar, A. Karim, G. Rokni, N. Motlagh, M. S. Gandomi, H. Noori, N. & Khanjari, A. (2010). Effect of *Zataria multiflora* Boiss. essential oil, nisin, and their combination on the production of enterotoxin C and alpha-hemolysin by *Staphylococcus aureus. Foodborne Pathogen and Disease*, 7, 299-305.

Parvathy, K. S. Negi, P. S. Srinivas, P. (2009). Antioxidant, antimutagenic and antibacterial activities of curcumin-β-diglucoside. *Food Chemistry*, 115, 265–271.

Pereira C. G. & Meireles M. A. A. (2010). Supercritical fluid extraction of bioactive compounds: fundamentals, applications and economic perspectives. *Food Bioprocess Technology* 3, 340–372.

Perez-Conesa, D. Mclandsborough, L. & Weiss, J. (2006). Inhibition and inactivation of *Listeria monocytogenes* and *Escherichia coli* O157:H7 colony biofilms by micellar-encapsulated eugenol and carvacrol. *Journal of Food Protection*, 69, 2947e2954.

Perez-Conesa, D., Cao, J., Chen, L., McLandsborough, L., & Weiss, J., 2011. Inactivation of *Listeria monocytogenes* and *Escherichia coli* O157:H7 biofilms by micelle- encapsulated eugenol and carvacrol. *Journal of Food Protection* 74, 55-62.

Persson, T. Hansen, T. H. Rasmussen, T. B. Skinderso, M. E. Givskov, M. & Nielsen, J. (2005). Rational design and synthesis of new quorum-sensing inhibitors derived from acylated homoserine lactones and natural products from garlic. *Organic Biomolecular Chemistry*, 3, 253–262.

Pesavento, G. Ducci, B. Nieri, D. Comado, A. & Nostro, L. (2010). Prevalence and antibiotic susceptibility of *Listeria* spp. isolated from raw meat and retail foods. *Food Control*, 5, 708-713.

Poros-Gluchowska, J. & Markiewicz, Z. (2003). Antimicrobial resistance of *Listeria monocytogenes. Acta Microbiologica Polonica*, 52, 113-129.

Porto-Feit, A. C. S. Campano, S. G. Call, J. E. Shoyer, B. A. Gartner, K. Tufft, I. Oser, A. Lee, J. and Luchansky, J. B. (2011). Validation of food-grade salts of organic acids as ingredients to control *Listeria monocytogenes* on pork scrapple during extended refrigerated storage. *Journal of Food Protection*. 74, 394-402.

Qiu, J. Feng, H. Lu, J. Xiang, H. Wang, D. Dong, J. Wang, J. Wang, X. Liu, J. & Deng, X. (2010a). Eugenol reduces the expression of virulence-related exoproteins in *Staphylococcus aureus. Applied and Environmental Microbiology*, 76, 5846-5851.

Qiu, J. Wang, J. Luo, H. Du, X. Li, H. Luo, M. Dong, J. Chen, Z. & Deng, X. (2011a). The effects of subinhibitory concentrations of costus oil on virulence factor production in *Staphylococcus aureus*. *Journal of Applied Microbiology*, 110, 330-40.

Qiu, J. Zhang, X. Luo, M. Li, H. Dong, J. Wang, J. Leng, B. Wang, X. Feng, H. Ren, W. & Deng, X. (2011b). Subinhibitory concentrations of perilla oil affect the expression of secreted virulence factor genes in *Staphylococcus aureus. PLoS One*, 6, e16160.

Qiu, J., Li, H., Su, H., Dong, J., Luo, M., Wang, J., Leng, B., Deng, Y., Liu. J., & Deng X. (2012). Chemical composition of fennel essential oil and its impact on *Staphylococcus aureus* exotoxin production. *World Journal of Microbiology and Biotechnology*, 28, 1399-1405.

Qiu, J., Wang, D., Xiang, H., Feng, H., Jiang, Y., Xia, L., Dong, J., Lu, J., Yu, L., & Deng, X. (2010b). Subinhibitory concentrations of thymol reduce enterotoxins A and B and α-hemolysin production in *Staphylococcus aureus* isolates. *PLoS One*, 5, e9736.

Racz, P. Tenner, K. & Mero, E. (1972). Experimental Listeria enteritis. I. An electron microscopic study of the epithelial phase in experimental Listeria infection. *Laboratory Investigation* 26, 694–700.

Rasmussen, T. B. Bjarnsholt, T. Skindersoe, M. E. Hentzer, M. Kristoffersen, P. Kote, M. Nielsen, J. Eberl, L. & Givskov, M. (2005a). Screening for quorum-sensing inhibitors (QSI) by use of a novel genetic system, the QSI selector. *Journal of Bacteriology*, 187, 1799–1814.

Rasmussen, T. B. Skindersoe, M. E. Bjarnsholt, T. Phipps, R. K. Christensen, K. B. Jensen, P. O. Andersen, J.B. Koch, B. Larsen, T.O. Hentzer, M. Eberl, L. Hoiby, N. & Givskov, M. (2005b). Identity and effects of quorum-sensing inhibitors produced by Penicillium species. *Microbiology*, 151, 1325–1340.

Regalado, C. (2008). Effect of *Lactococcus lactis* UQ2 and its bacteriocin on *Listeria monocytogenes* biofilms. Food Control, 19, 670–680. *Listeria monocytogenes* biofilms. *Food Control*, 19, 670–680.

Reichling, J. (2010) Plant-Microbe Interactions and Secondary Metabolites with Antibacterial, Antifungal and Antiviral Properties, In Annual Plant Reviews Volume 39: Functions and Biotechnology of Plant Secondary Metabolites, Second edition (ed M. Wink), Wiley-Blackwell, Oxford, UK.

Riedel, C. U. Monk, I. R. Casey, P. G. Waidmann, M. S. Gahan, C. G. & Hill, C. (2009). AgrD-dependent quorum sensing affects biofilm formation, invasion, virulence and global gene expression profiles in *Listeria monocytogenes. Molecular Microbiology*, 71, 1177-89.

Riedel, K. Kothe, M. Kramer, B. Saeb, W. Gotschlich, A. Ammendola, A. & Eberl, L. (2006). Computer-aided design of agents that inhibit the cep quorum-sensing system of *Burkholderia cenocepacia*. *Antimicrobial Agents and Chemotherapy*, 50, 318–323.

Robbins, J. B. Fisher, C. W. Moltz, A. G. and Martin, S. E. (2005). Elimination of *Listeria monocytogenes* biofilms with ozone, chlorine, and hydrogen peroxide. *Journal of Food Protection*, 68, 494–498.

Rocourt, J. R. and Brosch, R. (1992). Human listeriosis: 1990. Document WHO/HPP/FOS/92.4. World Health Organization, Geneva, Switzerland.

Rodrigues, L. van der Mei, H. Teixeira, J.A, & Oliveira, R. (2004). Biosurfactant from Lactococcus lactis 53 inhibits microbial adhesion on silicone rubber. *Applied and Environmental Microbiology*, 66, 306-11.

Rodriguez, J. M. Martinez, M. I. and Kok, J. (2002). Pediocin PA-1, a wide-spectrum bacteriocins from lactic acid bacteria. *Critical Reviews in Food Science and Nutrition*, 42, 91-121.

Roger, L. A. (1928). The inhibiting effect of *Streptococcus lactis* on *Lactobacillus burgaricus*. *Journal of Bacteriology*, 16, 321-5.

Romanova, N. Farvin, S. & Griffiths, M. W. (2002). Sensitivity of *Listeria monocytogenes* to sanitizers used in the meat processing industry. *Applied and Environmental Microbiology*, 68, 6405-6409.

Ross, R. P. Morgan, S. & Hill, C. (2002). Preservation and fermentation: past, present and future. *International Journal of Food Microbiology*, 79, 3-16.

Rossland, E. Langsrud, T. Granum, P. E. and Sørhaug, T. (2005). Production of antimicrobial metabolites by strains of Lactobacillus or Lactococcus co-cultured with *Bacillus cereus* in milk. *International Journal of Food Microbiology*, 98, 193–200.

Rossland, E., Langsrud, T., Granum, P. E., & Sørhaug, T. (2005). Production of antimicrobial metabolites by strains of Lactobacillus or Lactococcus co-cultured with Bacillus cereus in milk. *International Journal of Food Microbiology*, 98, 193–200.

Ruiz, A. Williams, S. K. Djeri, N. Hinton Jr. A. Rodrick, G. E. (2009). Nisin, rosemary, and ethylenediaminetetraacetic acid affect the growth of *Listeria monocytogenes* on ready-to-eat turkey ham stored at four degrees Celsius for sixty-three days. *Poultry Science*, 88, 1765–1772.

Samelis J, and Sofos J. N. Organic acids. In: Roller S, ed. *Natural antimicrobials for the minimal processing of foods*. Cambridge:Woodhead Publishing Limited and CRC Press LLC; 2003.

Samelis, J. & Metaxopoulos, J. (1999). Incidence and principal sources of Listeria spp. and *L. monocytogenes* contamination in processed meats and a meat processing plant. *Food Microbiology*, 16, 465–477.

Sanchez, E. García S. Heredia N. (2010). Extracts of edible and medicinal plants damage membranes of *Vibrio cholerae*. *Applied Environmental Microbiology* 76, 6888–6894.

Sánchez, E. García, S. & Heredia, N. (2010). Extracts of edible and medicinal plants damage membranes of *Vibrio cholerae*. *Applied and Environmental Microbiology*, 76, 6888–6894.

Sarabhai, S. Sharma, P. & Capalash, N. (2013). Ellagic acid derivatives from *Terminalia chebula* Retz. downregulate the expression of quorum sensing genes to attenuate *Pseudomonas aeruginosa* PAO1 virulence. *PLoS One*, 8, e53441.

Savage, D. C. (1977). Microbial ecology of the gastrointestinal tract. *Annual Reviews in Microbiology*, 31, 107-133.

Scallan, E. Hoekstra, R. M. Angulo, F. J. Tauxe, R. V. Widdowson, M. A. Roy, S. L. Jones, J. L. & Griffin, P. M. (2011). Foodborne illness acquired in the United States— major pathogens. *Emerging Infectious Diseases,* 17, 7–15.

Scharff, R. L. (2012). Economic burden from health losses due to foodborne illness in the United States. *Journal of Food Protection*, 75, 123-31.

Schlech, W. F. 3rd, Lavigne, P. M. Bortolussi, R. A. Allen, A. C. Haldane, E. V. Wort, A. J. Hightower, A. W. Johnson, S. E. King, S. Nicholls, E. S. & Broome, C. V. (1983). Epidemic listeriosis –evidence for transmission by food. *New England Journal of Medicine* 318, 203–206.

Schobitz, R. Gonzalez, C. Katherine, V. Mariela, H. Yanina, N. & Richardo, F. (2014). A biocontroller to eliminate *Listeria monocytogenes* from the food processing environment. *Food Control*, 36, 217-223.

Schuchat, A. Swaminathan, B. & Broome, C. V. (1991). Epidemiology of human listeriosis. *Clinical Microbiology Reviews*, 4, 169–183.

Sekirov, I. Russell, S. L. Antunes, L. C. & Finlay, B. B. (2010). Gut microbiota in health and disease. *Physiological Reviews*, 90, 859–904.

Serra, A. T. Matias, A. A. Nunes, A. V. M. Leitao, M. C. Brito, D. Bronze, R. Silva, S. Pires, A. Crespo, M. T. Romao, M. V. S. & Duarte, C. M. (2008). *In vitro* evaluation of olive- and grape-based natural extracts as potential preservatives for food. *Innovative Food Science and Emerging Technologies,* 9, 311–319.

Shipradeep, Karmakar, S. Khare, R. S. Ojha, S. Kundu, K. & Kundu, S. (2012). Development of probiotic candidate in combination with essential oils from medicinal plant and their effect on enteric pathogens: A review. *Gastroenterology Research Practice*, 457150, 1-6.

Silva, O. Duarte, A. Cabrita, J. Pimentel, M. Diniz, A. & Gomes, E. (1996). Antimicrobial activity of Guinea-Bissau traditional remedies. *Journal of Ethnopharmacology*, 50, 55–59.

Simoes, M. Simoes, L. & Vieira, M. J. (2010). A review of current and emergent biofilm control strategies. *LWT- Food Science and Technology* 43, 573-583.

Singh, A. Singh, R. K. Bhunia A. K. & Singh, N. (2003). Efficacy of plant essential oils as antimicrobial agents against *Listeria monocytogenes* in hotdogs. *LWT - Food Science and Technology*, 36, 787-794.

Siracusa, G. R. & Dickson, J. S. (1992). Inhibition of *Listeria monocytogenes* on beef tissue by application of organic acids immobilized in a calcium alginate. *Journal of Food Science*, 57, 293-296.

Smith-Palmer, A. Stewart, J. & Fyfe, L. (2001). The potential application of plant essential oils as natural food preservatives in soft cheese. *Food Microbiology*, 18, 463-470.

Smith-Palmer, A. Stewart, J. & Fyfe. L. (1998). Antimicrobial properties of plant essential oils and essences against five important foodborne pathogens. *Letters in Applied Microbiology*, 26, 118-122.

Smith-palmer, A. Stewartt, J. & Fyfe, L. (2002). Inhibition of listeriolysin O and phosphatidylcholine-specific production in *Listeria monocytogenes* by subinhibitory concentrations of plant essential oils. *Journal of Medical Microbiology*, 51, 567-74.

Smith-palmer, A. Stewartt, J. and Fyfe, L. (2004). Influence of subinhibitory concentrations of plant essential oils on the production of enterotoxins A and B and alpha-toxin by *Staphylococcus aureus*. *Journal of Medical Microbiology*, 53, 1023-1027.

Smith, C. E. 2012. Plant Oils and Cardiometabolic Risk Factors: The Role of Genetics. *Current Nutrition reports*, 1, 161-168.

Sonnenburg, J. L. Xu, J. Leip, D. D. Chen, C. H. Westover, B. P. Weatherford, J. Buhler, J. D. & Gordon, J. I. (2005). Glycan foraging in vivo by an intestine-adapted bacterial symbiont. *Science*, 307, 1955–1959.

Southwick, F.S. & Purich, D.L. (1996). Intracellular pathogenesis of listeriosis. *The New England Journal of Medicine*, 334, 770-776.

Steele, M. & Odumeru, J. (2004). Irrigation water as source of foodborne pathogens on fruits and vegetables. *Journal of Food Protection*, 67, 2839-2849.

Stiles, M. E. (1996). Biopreservation by lactic acid bacteria. *Antonie van Leeuwenhoek*. 70, 331–345.

Stopforth, J. D. Visser, D. Zumbrink, R., van Dick, L. & Bontenbal, E. W. (2010). Control of *Listeria monocytogenes* on cooked ham by formulation with a lactate-diacetate blend and surface treatment with lauroc arginate. *Journal of Food Protection*, 73, 552-555.

Stratford, M. & Eklund, T. (2003). Organic acids and esters. In: Russell, N. J., Gould, G. W. (Eds.), Food Preservatives. Kluwer Academic/Plenum Publishers, London, pp. 48–84.

Swaminathan, B. & Gerner-Smidt, P. (2007). The epidemiology of human listeriosis. *Microbes and Infections,* 9, 1236-1243.

T. E. Mattson, Johny, A. K. Amalaradjou, M. A. R. More, K. Schreiber, D. T. Patel, J. and Venkitanarayanan. K. (2011). Inactivation of Salmonella spp. on tomatoes by plant molecules. *International Journal of Food Microbiology*, 144, 464-468.

Tahiri I, Desbiens M, Benech R, Kheadr E, Lacroix C, Thibault S, Ouellet D, & Fliss I. (2004). Purification, characterization and amino acid sequencing of divergicin M35: a novel class IIa bacteriocin produced by *Carnobacterium divergens* M35. *International Journal of Food Microbiology*, 97, 123-36.

Tahiri, I. Desbiens, M. Benech, R. Kheadr, E. Lacroix, C. Thibault, S. Quellet, D. & Fliss, I. (2004). Purification, characterization and amino acid sequencing of divergicin M35: a novel class IIa bacteriocin produced by *Carnobacterium divergens* M35. *International Journal of Food Microbiology*, 97, 123-36.

Tahiri, I. Desbiens, M. Kheadr, E. Lacroix, C. & Fliss, I. (2009). Comparison of different application strategies of divergicin M35 for inactivation of *Listeria monocytogenes* in cold-smoked wild salmon. *Food Microbiology*, 26, 783-93.

Tahiri, I. Desbiens, M. Kheadr, E. Lacroix, C. & Fliss, I. (2009). Comparison of different application strategies of divergicin M35 for inactivation of *Listeria monocytogenes* in cold-smoked wild salmon. *Food Microbiology*, 26, 783-93.

Tait, K. & Sutherland, I. W. (1998). Antagonistic interactions amongst bacteriocin- producing enteric bacteria in dual species biofilms. *Journal of Applied Microbiology*, 93, 345–352.

Temple, M. E. & Nahata, M. C. (2000). *Treatment of Listeriosis. The Annals of Pharmacotherapy*, 34, 656-661.

Tompkin, R. B. Christiansen, L. N. Shaparis, A. B. Baker, R. L. & Schroeder, J. M. (1992). Control of *Listeria monocytogenes* in processed meats. *Food Australia*, 44, 370–376.

Tompkin, R. B. Scott, V. N. Bernard, D. T. Sveum, W. H. & Gombas. K.S. (1999). Guidelines to prevent post-processing contamination from *Listeria monocytogenes*. *Dairy Food and Environmental Sanitation*, 19, 551–562.

Tompkin, R. B. (2002). Control of *Listeria monocytogenes* in the food processing environment. *Journal of Food Protection*, 65, 709-725.

Toure, R. Kheadr, E. Lacroix, C. Moroni, O. & Fliss, I. (2003). Production of antibacterial substances by Bifidobacterial isolates from infant stool active against *Listeria monocytogenes*. *Journal of Applied Microbiology*, 95, 1058-1069.

Tsui, W. H. Yim, G. Wang, H. H. McClure, J. E. Surette, M. G. & Davies, J. (2004). Dual effects of MLS antibiotics: transcriptional modulation and interactions on the ribosome. *Chemistry and Biology* 11, 1307-1316.

Turnbaugh, P. J. Ley, R. E. Mahowald, M. A. Magrini, V. Mardis, E. R. & Gordon J. I. (2006). An obesity-associated gut microbiome with increased capacity for energy harvest. *Nature*, 444, 1027–1031.

U.S. Food Safety and Inspection Service, (1989). Revised policy for controlling Listeria monocytogenes. Food Safety and Inspection Service, U.S. Dept. of Agriculture, Washington, D.C. Federal Register, 54, 22345.

U.S. Food Safety and Inspection Service, 2004. Compliance guidelines to control Listeria monocytogenes in post-lethality exposed ready-to-eat meat and poultry products. Available at: http://www.fsis.usda.gov/lm_rule_compliance_guidlines_may_2006.Pdf.

Unnerstad, H. Bannerman, E. Bille, J. Danielsson-Tham, M. L. Waak, E. & Tham, W. (1996). Prolonged contamination of a dairy with *Listeria monocytogenes*. *Netherlands Milk and Dairy Journal*, 50, 493–499.

Upadhyay, A. Johny-Kollanoor, A. Amalaradjou, M. A. R. Ananda Baskaran, S. & Venkitanarayanan, K. (2012). Plant-derived antimicrobials reduce *Listeria monocytogenes* virulence factors in vitro and down-regulate expressioin of virulence genes. *International Journal of Food Microbiology* 157, 88-94.

Upadhyay, A. Upadhyaya, I. Johny-Kollanoor, A. & Venkitanarayanan, K. (2013b). Antibiofilm effect of plant derived antimicrobials on *Listeria monocytogenes*. *Food Microbiology*, 36, 79-89.

Upadhyay, A. Upadhyaya, I. Johny-Kollanoor, A. Ananda Baskaran, S. Mooyottu, S. Karuathil, D. & Venkitanarayanan, K. (2013a). Inactivation of *Listeria monocytogenes* on frankfurters by plant-derived antimicrobials alone or in combination with hydrogen peroxide. International Journal of Food Microbiology, 163, 114-118.

Upadhyaya, I. Upadhyay, A. Kollanoor-Johny, A. Darre, M. J. & Venkitanarayanan, K. (2013). Effect if plant derived antimicrobials on *Salmonella* Enteritidis Adhesion to and invasion of Primary chicken epithelial cells in vitro and virulence gene expression. *International Journal of Molecular Sciences*, 14, 10608-10625.

Valle, J. Re, D. S. Henry, N. Fontaine, T. Balestrino, D. Latour-Lambert, P. & Ghigo, J. M. (2006). Broad-spectrum biofilm inhibition by a secreted bacterial polysaccharide. *Proceedings of the National Academy of Sciences USA*, 103, 12558–12563.

Van Hamme J. D. Singh, A. & Ward, O. P. (2006). Surfactants in microbiology and biotechnology: Part 1. Physiological aspects. *Biotechnological advances*, 24, 604-620.

Van Wyk B. E, & Gericke, N. (2000). People's Plants. (Edition 1) Briza Publications, Arcadia, Pretoria, South Africa.

Vattem, D. A. Mihalik, K. Crixell, S. H. & McClean, R. J. C. (2007). Dietary phytochemicals as quorum sensing inhibitors. *Fitoterapia*, 78, 302–310.

Vilar, M. J. Yus, E. Sanjuan, M. L. Dieguez, F. J. & Rodriguez-Otero, J. L. (2007). Prevalence of and risk factors for *Listeria* species on dairy farms. *Journal of Dairy Science*, 90, 5083–5088.

Vrinda Menon, K. & Garg, S.R. (2001). Inhibitory action of clove on *Listeria monocytogenes* in meat and cheese. *Food Microbiology*, 39, 432-434.

Waak, E. Tham, W. & Danielsson-Tham, M. L. (2002). Prevalence and fingerprinting of *Listeria monocytogenes* strains isolated from raw whole milk in farm tanks and in dairy plant receiving tanks. *Applied and Environmental Microbiology*, 68, 3366–3370.

Ward, S. M. Delaquis, P. J. Holley, R. A. Mazza, G. (1998). Inhibition of spoilage and pathogenic bacteria on agar and pre-cooked roast beef by volatile horseradish distillates. *Food Research International*, 31, 19–26.

Watkins, J. & Sleath, K. P. (1981). Isolation and enumeration of *Listeria monocytogenes* from sewage, sewage sludge and river water. *Journal of Applied Microbiology*, 50, 1-9.

Wenderska, I. B. Chong, M. McNulty, J. Wright, G. D. & Burrows, L. L. (2011). Palmitoyl-DL-carnitine is a multitarget inhibitor of *Pseudomonas aeruginosa* biofilm development. *ChemBioChem: A European Journal of Chemical Biology* 12, 2759–2766.

Wieczorek, K. Dmowska, K. & Osek, J. (2012). Prevalence, characterization and antimicrobial resistance of *Listeria monocytogenes* isolated from bovine hides and carcasses. *Applied and Environmental Microbiology*, 78, 2043-2045.

Wirtanen, G. Saarela, M. and Mattila-Sandholm, T. (2000). Biofilms – impact on hygiene in food industries. In J. D. Bryers (Ed.), *Biofilms II: Process analysis and applications* 327–372.

Woo, J. and Ahn, J. (2013). Probiotic-mediated competition, exclusion and displacement in biofilm formation by food-borne pathogens. *Letters in Applied Microbiology*, 56, 307-313.

Worthington, R. J. Richards, J. J. & Melander, C. (2012). Small molecule control of bacterial biofilms. *Organic and Biomolecular Chemistry*, 10, 7457-74.

Yatsunenko, T. Rey, F. E. Manary, M. J. Trehan, I. Dominguez-Bello, M. G. Contreras, M. Magris, M. Hidalgo, G. Baldassano, R. N. Anokhin, A. P. Heath, A. C. Warner, B. Reeder, J. Kuczynski, J. Caporaso, J. G. Lozupone, C. A. Lauber, C. Clemente, J. C. Knights, D. Knight, R. and Gordon, J. I. (2012). Human gut microbiome viewed across age and geography. *Nature*, 486, 222–227.

Yuste, J. & Fung, D. Y. C. (2002). Inactivation of Listeria monocytogenes Scott A 49594 in apple juice supplemented with cinnamon. *Journal of Food Protection* 65, 1663–1666.

Zeng, W. C. He, Q. Sun, Q. Zhong, K. Gao, H. (2012). Antibacterial activity of water- soluble extract from pine needles of *Cedrus deodara. International Journal of Food Microbiology*, 153, 78–84.

Zhang, H. Kong, B. Xiong, Y. L. and Sun, X. (2009). Antimicrobial activities of spice extracts against pathogenic and spoilage bacteria in modified atmosphere packaged fresh pork and vacuum packaged ham slices stored at 4°C. *Meat Science*, 81, 686-692.

Zhao, T. Doyle, M. P. and Zhao, P. (2004). Control of *Listeria monocytogenes* in a biofilm by competitive-exclusion microorganisms. *Applied and Environmental Microbiology*, 70, 3996–4003.

In: *Listeria monocytogenes*
Editor: Edmund C. Hambrick
ISBN: 978-1-63117-054-6

Chapter 2

LISTERIA MONOCYTOGENES IN READY-TO-EAT FOODS AND INTERVENTION STRATEGIES

Byong Kwon Yoo and Cheng-An Hwang
Residue Chemistry and Predictive Microbiology Research Unit,
Eastern Regional Research Center, Agricultural Research Service,
U.S. Department of Agriculture, Wyndmoor, PA, US

ABSTRACT

Listeria monocytogenes is a foodborne pathogen capable of causing listeriosis, a severe illness that has a high fatality rate. It has been a significant food safety concern for decades due to its ubiquitous presence in the environment, ability to grow at refrigeration temperature, and resistance to general cleaning and sanitizing procedures by forming biofilms. Food products that have been linked to the outbreaks of listeriosis include meat, poultry, seafood, dairy products, vegetables, and fruits. Since they are normally consumed without prior cooking, ready-to-eat (RTE) foods, such as frankfurters, deli meats, salads, and smoked seafood, pose a higher risk of causing listeriosis than other types of food products. This chapter will review the food safety aspect of *L. monocytogenes* in RTE foods including its prevalence, causes of contamination, food safety regulations, and current and developing intervention technologies to control *L. monocytogenes* in RTE foods.

INTRODUCTION

Listeria monocytogenes is a Gram positive foodborne pathogen and causative agent of listeriosis, which emerged as a major foodborne illness in the early 1980's, in spite of its discovery dated back to 80 years ago [1]. *L. monocytogenes* is transmitted through contaminated food in most cases [2]. Listeriosis has become a primary concern to public health authorities because it has serious clinical outcomes in pregnant women, children, elderly, and individuals with compromised immunity [3, 4]. When pregnant women are infected with *L. monocytogenes*, they experience a mild influenza-like symptom or asymptomatic infection, but later they can encounter preterm delivery, fetal loss, invasive neonatal infection, and infant death [5]. Foodborne transmission of listeriosis was first recognized after an outbreak in Canada in 1981 which was associated with consumption of contaminated coleslaw [2]. Pasteurized milk was associated with the first recognized foodborne outbreak of listeriosis in the United States in 1983 [6]. The Centers for Disease Control and Prevention (CDC) estimated that approximately 1600 illnesses and 260 deaths due to listeriosis occurred annually in the United States in 2011 [7]. The average annual incidence in the United States was estimated to be 2.9 cases per million populations for 2009-2011 [8]. The risk of listeriosis is greater when the food is contaminated with high levels of the organism; however, the infectious dose of the pathogen also depends on strain virulence, type of foods, and host susceptibility [9].

Characteristics

Classified as the sub-branch of *Clostridium*, the genus *Listeria* has low G+C content (36-42%), consistent with the other members such as *Staphylococcus*, *Streptococcus*, and *Lactobacillus* in the corresponding phylogenetic position [10]. Among 6 species of *Listeria*, only *L. monocytogenes* and *L. ivanovii* are regarded as pathogenic. A biochemical test confirming the ability of acid production from xylose may be used to distinguish *L. monocytogenes* from other *Listeria* species [11], and the CAMP (Christie Atkins Munch-Petersen) test [12] checking the occurrence of hemolysis is used to identify pathogenic *L. monocytogenes* from other non-pathogenic species. Molecular typing methods such as multilocus enzyme electrophoresis, pulsed field gel electrophoresis (PFGE), and ribotyping have been developed to further characterize subtypes of *L. monocytogenes* as well as conventional serotyping or phage typing [13, 14, 15]. Out of 13 serotypes causing human illnesses, serotypes 1/2a, 1/2b, and 4b cover 95% of isolates from patients suffering from listeriosis [16]. *L. monocytogenes* is distinctive in many ways from other pathogenic bacteria. It exists extensively in the natural environment including surface water, soil, sewage water, plants, and animal feces. Since it can survive under environmental stresses such as low water activity (e.g., high NaCl concentrations), low pH, refrigerated temperature as low as 3 to 4°C, and low oxygen, once it enters the food processing plant, its persistent survival in food processing equipment is a major concern for the food industry. The infectious dose depends on the immunological status of individual. In general, more than 100 colony forming unit (CFU)/g of *L. monocytogenes* are required to cause epidemic foodborne outbreaks, but a contamination level of < 0.3 CFU/g in processed frankfurter meats was responsible for a multistate outbreak of listeriosis in the United States in 1998 [17]. However, there are limited infectious data available to develop a dose-response model due to high fatality upon pathogen infection which makes the estimation of infectious dose less reliable [18, 19].

Surveillance

L. monocytogenes is found in foods in many parts of the world [20], ranging from 0 to 16% in RTE foods in the United States, 14.2 to 30.9% in raw fish and meats in Denmark [21], and 12 to 25% in retail meats in Japan [22]. Because of the widespread distribution of *L. monocytogenes* in foods and in the food processing environment and the potential to be highly fatal to a certain group of people when they are infected, the United States sets a "zero tolerance" for this pathogen in RTE foods as a guidance in spite of the low incidence rate [23]. PulseNet, a U.S. national molecular subtyping network for enteric bacterial disease surveillance, was established in 1998 [5]. When isolates of the pathogen from patients are sent to state public health laboratories for standardized PFGE analysis, the PFGE patterns are uploaded to a central PulseNet database for national comparison. If more or equal to 2 patterns of isolates are uploaded within a 120-day period, investigation is initiated whether the rate is higher than the historical background [24]. Following a Council of State and Territorial Epidemiologists position statement, the CDC launched its *Listeria* initiative in 2004 to collect surveillance data including case-control studies from listeriosis patients and comparative strain typing by local health departments upon the occurrence of outbreaks. Twenty four confirmed listeriosis outbreaks were reported during 1998-2008, resulting in 359 illnesses, 215 hospitalizations, and 38 deaths [5]. RTE meats caused more early outbreaks, and RTE produce and cheese products were associated with outbreaks later in the study period. Surveillance data also suggested that RTE foods such as frankfurters, deli meats, poultry products, unpasteurized milk and soft cheeses which are stored under refrigerated temperature for a long duration pose a higher risk than the raw foods that need to be cooked before they are consumed. Notably, the elderly people (age $\geq$ 65) and Hispanics (40.3 and 50.5 million, respectively, based on 2010 Census estimates) are more susceptible to the risk of listeriosis in the United States [25].

FOOD VEHICLES

Both processed and raw foods are involved in the occurrence of listeriosis, and RTE foods such as deli meats, poultry products, processed seafoods, dairy products, fresh-cut fruits and leafy greens, have become a major food group that linked to the illnesses. The presence of *L. monocytogenes* in foods is likely to be associated with poor hygiene of food manufacturers or indigenous presence of the pathogen [26], even though the route of contamination varies in each food category. The existence of *L. monocytogenes* in cooked deli meats and poultry products seems to be the result of post-contamination, which occurs when they are further sliced into different shapes and repackaged [27]. The absence of competitive microflora may facilitate proliferation of *L. monocytogenes* in smoked fish products during storage, based on the incidence report in Sweden [28]. Dietary habit of consuming raw seafood products without cooking may contribute to the illnesses by the pathogen [29]. When dairy products are manufactured from contaminated milk, *L. monocytogenes* often survives the manufacturing processing and is present in finished products [30]. The prevalence rate of *L. monocytogenes* in dairy products is generally lower than that in RTE deli meats and seafood products [31]. Recently, produce has become a major food implicated in the occurrence of listeriosis. The consumption of raw and minimally processed produce has increased dramatically mainly driven by the pursuit of healthy diet by the consumers [32]. These products are prone to *L. monocytogenes* contamination during post-harvest handling [33], and they generally do not receive treatments during processing to eliminate the contamination. Therefore, to the consumers, raw and minimally processed produce pose a significant risk when they are contaminated with *L. monocytogenes.*

BIOFILMS

Biofilm consists of bacterial cells surrounded by an extracellular polysaccharide (EPS) matrix that attaches to a solid surface. The matrix harbors the bacterial cells and protects the cells from cleaning and sanitizing agents and adverse environmental conditions. The formation of biofilm in food processing environment is a major microbiological safety and quality concern in food production. Biofilm-forming bacteria include *Pseudomonas, Campylobacter, Salmonella, Escherichia coli,* and *Listeria,* and they are in mixed combination or one species dominates over others [34]. *L. monocytogenes* may form biofilms in all areas of food processing environments such as floors, plant walls, tanks, gaskets, conveyor belts, pipes, freezers, smoke houses, air ducts, and drains [35]. Attachment of bacteria to the surface of foods or food utensils may reduce the efficacy of antimicrobial interventions [36] and biofilm-forming *L. monocytogenes* may survive in food plants up to several years [27]. However, *L. monocytogenes* is unable to form thick biofilms densely populated with 10^9–10^{12} CFU/cm^2, like other common biofilm-forming bacteria, but it can reach 10^4–10^7 CFU/cm^2 [37]. In biofilms, *L. monocytogenes* is protected from a variety of environmental factors, such as ultraviolet rays, acids, desiccation, and antimicrobials, and it tolerates higher concentrations of sanitizers [38, 39]. The ability of *L. monocytogenes* to persist in the food industry is closely related to the inability of elimination of this bacterium from the facility as well as its ability to survive at low temperatures [38, 40]. Little is known about the EPS composition of *L. monocytogenes* and EPS are mainly produced by *L. monocytogenes* strains that form greater amount of biofilm [41]. Extracellular DNA is necessary for both initial attachment and early biofilm formation as a central component in *L. monocytogenes* biofilm matrix, and N-acetylglucosamine is also present interacting with DNA to preserve the adhesion capacity [42]. The biofilm structure can vary depending on the cultivation system used by *L. monocytogenes*. Under static conditions, *L. monocytogenes* biofilm generally consists of a homogeneous layer of rod cells and microcolonies. By contrast, *L. monocytogenes* biofilms grown under continuous flow conditions consist of ball-shaped microcolonies that are surrounded by a network of knitted chains composed of elongated cells [43].

INTERVENTION STRATEGIES

The lethality treatment received by processed RTE foods generally eliminates *L. monocytogenes*; however products can be re-contaminated by exposure after the lethality treatment during peeling, slicing, repackaging, and other procedures [44]. RTE meat and poultry processing plants must include control programs for *L. monocytogenes* in their Hazard Analysis Critical Control Point (HACCP) plans, Sanitation Standard Operating Procedures (SSOP), or prerequisite programs to prevent its growth and proliferation in the plant environment and equipment, thus preventing the cross-contamination of RTE products [45].

The regulation in Title 9 of the Code of Federal Regulations [46] states that establishments must comply with at least one of the three alternatives to provide an intervention if an RTE product is produced and exposed to the processing environment after the primary lethality process. If an establishment chooses Alternative 1, it must use a post-lethality treatment that reduces or eliminates microorganisms on the product and an antimicrobial agent or process that suppresses or limits the growth of *L. monocytogenes*. If an establishment chooses Alternative 2, it can use either the post-lethality treatment of product (Alternative 2a) or an antimicrobial agent or process that suppresses or limits growth (Alternative 2b). If an establishment selects Alternative 1 or Alternative 2, it must have also a sanitation program that addresses the testing of food contact surfaces. For Alternative 3, an establishment can choose to use sanitation as its intervention strategy. Minimum routine sampling frequencies are required based on the Alternative and daily production volume size. Alternative 1 will receive a Food Contact Surface (FCS) testing per production line every 6 months. Alternative 2 will take four times of FCS testing per production line a year. Alternative 3, except for deli and hotdog products, will have a FCS testing per production line every month. Alternative 3 for deli and hotdog products will receive three different levels of FCS testing frequency based on the ranges of daily production volume (lbs): every month for very small (1-6000); every two weeks for small (6001-50000); and every week for large (>50001).

Several intervention strategies have been adopted to prevent and control *L. monocytogenes* in RTE food products. The strategies include post-package decontamination technology such as high pressure processing and irradiation, pre/post-package processing such as surface pasteurization by steam or hot water, and formulating products with antimicrobial additives such as lactates, diacetates, and organic acids [47]. Diacetates are frequently used as stand-alone antimicrobials in meat products due to the beneficial properties they confer on the quality of meat when they are applied at appropriate concentrations [48]. Recent literature has shown that sodium diacetate at 6% is effective in inhibiting the growth of *L. monocytogenes* up to 90 d in frankfurters based on predictive models [49]. Research has focused on the combined application of lactates and diacetates for their synergistic inhibitory effect against the growth of pathogenic organisms in meat and poultry products. Combination of 2.5% sodium lactate and 0.1% sodium diacetate prevented outgrowth of *L. monocytogenes* in turkey slurries for about 42 days at 4°C, but 0.1% sodium diacetate alone was not effective [50]. Experiments conducted in small RTE meat processors in three U.S. states revealed that inclusion of lactate-diacetate combination in product formulation was most effective to control *L. monocytogenes* in RTE meat products [51]. The use of potassium lactate and the combination of potassium lactate/sodium diacetate was investigated as potential inhibitors for *L. monocytogenes* in modified atmosphere packaged sliced ham [52]. The presence of potassium lactate or the combination of potassium lactate/sodium diacetate in a modified atmosphere package sliced ham inhibited the growth of *L. monocytogenes* during the normal shelf life of the product while stored at 4°C and at abusive temperature of 8°C. Organic acids, such as acetic acid and lactic acid, are more inhibitory against *L. monocytogenes* on RTE meat products when applied as dipping solutions than lactoferrin used in formulation or surface treatments in frankfurters [53]. The effect of 2% (v/v) malic acid on washing *L. monocytogenes* attached to poultry leg skins showed 1.7 log reductions without adverse sensory effect [54].

Post Processing Treatments

The U.S. Federal regulations require RTE meat and poultry processors to control *L. monocytogenes* by using intervention measures that achieve at least 1 log reduction in post processing contamination [55]. Processors using intervention measures that achieve more than 2 log reduction deserve less frequent microbial testing in the plant. Post-lethality treatments such as steam pasteurization, hot water pasteurization, radiant heating, and high-pressure processing have been developed to prevent or eliminate post-processing contamination by *L. monocytogenes* [45]. Ultra violet treatment can be used either as a post-lethality treatment or antimicrobial agent or process depending on whether it eliminates, reduces, or suppresses growth of *L. monocytogenes* [45]. Post-cook hot-water pasteurization and steam pasteurization resulted in a 7 log CFU/g reduction of *L. monocytogenes* populations in inoculated vacuum packaged fully cooked sliced chicken [56]. Radiant heating was evaluated as a thermal process for surface pasteurization of RTE meat products (turkey bologna, roast beef, corned beef, and ham) for the reduction of *L. monocytogenes* [57]. With the use of radiant heat for pre-package surface pasteurization, 1.3 to 3.5 log reductions of *L. monocytogenes* populations were achieved with treatment times of 60 to 120 s at air temperatures of 246 to 399°C. Post-package submersion heating was evaluated as a method for reducing *L. monocytogenes* populations on the surfaces of contaminated RTE deli-style whole or formed turkey, ham, and roast beef. When the RTE deli-meats were processed at 90.6, 93.3, or 96.1°C for 2-10 min, a 2-4 log CFU/cm^2 reduction of *L. monocytogenes* was achieved [58]. Overall, studies on post-lethality treatments showed reductions of *L. monocytogenes* from 1 to 7 log CFU/g or cm^2 depending on product type, duration of treatment, temperature, and pressure of treatment [45]. Higher log reductions were obtained when both pre-packaging and post-packaging surface pasteurizations were applied and when post-lethality pasteurization was combined with the use of antimicrobial agents.

Sanitation and Hygiene Practices

Due to the prevalence of *L. monocytogenes*, RTE foods may be contaminated with the pathogen through raw materials, processing environment, and at retail [26]. Manufacturing processes for RTE foods such as thermal treatments, fermentation, drying, and high pressure process have been designed to eliminate expected levels of the pathogen to render the products free of *L. monocytogenes*. To avoid the exposure of the processed RTE products to *L. monocytogenes* in the processing environment, good manufacturing, hygienic, and sanitation practices are required. Education is also a key for workers to understand the difference between safe and unsafe practices and how they can contribute to prevention and control strategies [59].

Novel Interventions

Some conventional intervention approaches for non-*Listeria* pathogens are not highly effective and often result in less than 2 log reductions of *L. monocytogenes*. For example, sodium nitrite which is commonly used in cured meat products is not effective against *L. monocytogenes* [60]. The demand for natural and minimally processed RTE foods has grown dramatically over the past decade [61]. To improve the food safety of these products, several GRAS (generally recognized as safe) natural preservatives have been tested on RTE products to identify effective anti-listerial additives without adversely affecting the product quality. These natural preservatives include essential oils, plant derived antimicrobials, cinnamon powder, and leaf extract. Oregano essential oil was able to reduce *L. monocytogenes* in minced beef by 2.5 log CFU/g when it was combined with 0.5% (v/v) caprylic acid during 10 days of storage time at 3°C [62]. Plant derived antimicrobials such as β-resorcylic acid (BR), carvacrol (CR), and trans-cinnamaldehyde (TC) have been evaluated as post-processing dip treatments on frankfurters. Frankfurters were surface-inoculated with a 5-strain mixture of *L. monocytogenes* at 6 log CFU and treated with dip solutions consisting of 0.75% TC, 0.75% CR, or 1.5% BR alone or in combination with hydrogen peroxide (0.1%). The treated frankfurters showed more than 2 log reductions on day 0 compared to controls during the 70-day refrigerated storage [63]. Pulp of tucuma Amazonian fruit (Amazon *Astrocaryum aculeatum*) was investigated to extract antimicrobial compounds against *L. monocytogenes*, *Bacillus cereus*, and *Enterococcus faecalis*. The extract was found to contain growth inhibitory compounds that interrupt the oxidative metabolism of these microorganisms [64]. *Annoa squamosa* leaf extract, which was conventionally used as a cure for diarrhea in Thailand, showed an inhibitory effect against *L. monocytogenes*, though it was not effective against *B. cereus* and *Campylobacter jejuni* [65]. Bacteriocin-producing cultures could be incorporated directly into food or in the form of purified preparations, such as nisin produced by *Lactococcus lactis* that is permitted for use as a food additive in many countries.

The main strategies to prevent and control the formation of biofilms on food contact surfaces include (a) inhibition of initial adherence and colonization on the surface, (b) interference with the molecular signals that modulate the growth of biofilms, and (c) disintegration of the EPS [66]. The control of biofilms of *L. monocytogenes* in food processing plants is done mainly by cleaning and sanitizing of equipment, with compounds such as acetic acid, lactic acid, sodium hypochlorite, quaternary ammonium, and hydrogen peroxide [67]. It is recognized that the conventional methods of cleaning and sanitizing are more efficient against bacterial cells recently deposited on the surface than on sessile cells, mainly due to the protection conferred by the EPS in biofilms [68]. Due to the variability in the composition of the EPS, the use of enzymes or enzyme mixtures from different sources can be an effective tool for biofilm eradication. Enzymes produced by fungi, acting on pectin, xylan, alginate, and arabic gum, are aids in removing bacterial biofilms [69]. The use of antimicrobial peptides (bacteriocins) or bacteriocin-producing strains may also contribute to control the initial adhesion and biofilm formation by *L. monocytogenes* on abiotic surfaces [70]. Similarly, natural compounds, such as essential oils and culinary herb extracts, could be used to prevent *L. monocytogenes* biofilm formation on food contact surfaces [71, 72], while other researchers are also investigating the use of small molecules with known modes of action (e.g., eukaryotic protein kinase inhibitors) to identify possible targets in *L. monocytogenes* to prevent its biofilm formation [73].

Ultra high pressure processing in dairy products [74], intense pulsed light in eggs [75], pulsed electric field in liquid whole eggs [76], X-ray in fresh produce [77], electron beam in meat products [78], microwave heating in seafood products [79], electrolyzed oxidizing water in fish products [80], and cold plasma in various RTE foods [81] are emerging processing technologies which are developed to enhance the safety while maintaining the fresh-like quality of food. Introducing hurdles that reduce the numbers of *Listeria* present in RTE foods and the potential for any remaining *L. monocytogenes* to grow during storage are required for prolonged shelf life of RTE products. One example of an effective hurdle is the use of nanoparticles to encapsulate antimicrobials that release the antimicrobials at controlled timing. Studies have shown that nisin (0.25 mg/ml) encapsulated in soy lecithin-based nanoparticles exhibited a bacteriocidal effect in Minas Frescal cheese [82]; poly-(DL-lactide-co-glycolide) (PLGA) nanoparticles with entrapped cinnamon bark extract inhibited antimicrobial activity against *L. monocytogenes* [83]; and Ply500, *Listeria*-specific cell lytic enzyme (*Listeria* bacteriophage endolysin) covalently attached to silica nanoparticles selectively killed *Listeria* on lettuce [84]. More examples include: photo-activated zinc oxide in the presence of visible light to inactivate 7 log CFU/ml of *L. monocytogenes* in 60 min by cell wall disintegration [85], and pectin-based antimicrobial (sodium lactate and sodium diacetate) edible coatings on chilled or frozen roasted turkeys to minimize risks of foodborne listeriosis [86]. Synergistic effects of hurdle technology include ultraviolet with ultrasound [87], nisin with gamma irradiation [88], lactic acid with sodium lauryl sulfate [89], steam and lactic acid [90], benzalkonium chloride and ultrasound [91], and sodium nitrite with high hydrostatic pressure [92]. Consumers demanding safe fresh products without additives, but with long shelf life preparations necessitate improved packaging method called modified atmosphere packaging (MAP) which includes removal or replacing the atmosphere using a specific combination of carbon dioxide, nitrogen, and oxygen [93]. Modified atmospheres with 80% carbon dioxide were effective to control *L. monocytogenes* in sea bream compared to control when stored at 0°C [94]. Compared to RTE mixes of iceberg lettuce and collard greens in modified atmosphere, RTE vegetables in perforated films led to high populations in spite of low inoculation at 10 CFU/g [95]. Modified atmosphere packaging using 20 ppm (w/w) ozonated dry ice showed 5 log reductions of *L. monocytogenes* during bioaerosol contamination on chicken breast [96]. Antimicrobial film packaging using immobilized lysozyme attached to ethylene vinyl alcohol films achieved approximately 1 log reduction in *L. monocytogenes* [97].

More targeted interventions as alternatives to the use of antibiotics or chemical decontamination in food supply systems are emerging. It has been reported that a high concentration of strain-specific bacteriophages in seafood processing plants can successfully decontaminate pathogens and their biofilms on the surface of stainless steel by one hour phage treatments [98] and meat borne bacteria such as *Lactobacillus curvatus* and *Lactobacillus sakei* effectively reduced biofilm formation on stainless steel surfaces [99]. Treatment of competitive exclusion bacteria include *Lactococcus lactis* subsp. *lactis* and *Enterococcus durans* which reduced biofilm formation by *L. monocytogenes* in floor drains in RTE poultry processing plant [100].

Conclusion

Listeriosis outbreaks continue to cause considerable illness and death in spite of recent declines in laboratory-confirmed infections. The presence of *L. monocytogenes* in various RTE products in conjunction with the multiple transmission routes that may be encountered in food processing plants renders control of the organism as critical in RTE food operations. Despite the unique challenges that the food industry faces, a comprehensive food safety system based on the philosophy of HACCP, a series of sound prerequisite programs such as GMP and SSOP, and processing technologies can effectively control *L. monocytogenes* in RTE foods.

REFERENCES

[1] Murray EGD, Webb RA, Swann MBR (1926) A disease of rabbits characterised by a large mononuclear leucocytosis, caused by a hitherto undescribed bacillus *Bacterium monocytogenes* (n.sp.). *J. Pathol.* 29: 407-439.
[2] Schlech WF, 3rd, Lavigne PM, Bortolussi RA, Allen AC, Haldane EV, et al., (1983) Epidemic listeriosis--evidence for transmission by food. *N. Engl. J. Med.* 308: 203-206.
[3] Doganay M (2003) Listeriosis: clinical presentation. *FEMS Immunol. Med. Microbiol.* 35: 173-175.
[4] Brosch R, Brett M, Catimel B, Luchansky JB, Ojeniyi B, et al., (1996) Genomic fingerprinting of 80 strains from the WHO multicenter international typing study of *Listeria monocytogenes* via pulsed-field gel electrophoresis (PFGE). *Int. J. Food Microbiol.* 32: 343-355.
[5] Cartwright EJ, Jackson KA, Johnson SD, Graves LM, Silk BJ, et al., (2013) Listeriosis outbreaks and associated food vehicles, United States, 1998-2008. *Emerg. Infect. Dis.* 19: 1-9; quiz 184.
[6] Fleming DW, Cochi SL, MacDonald KL, Brondum J, Hayes PS, et al., (1985) Pasteurized milk as a vehicle of infection in an outbreak of listeriosis. *N. Engl. J. Med.* 312: 404-407.
[7] Scallan E, Hoekstra RM, Angulo FJ, Tauxe RV, Widdowson MA, et al., (2011) Foodborne illness acquired in the United States--major pathogens. *Emerg. Infect. Dis.* 17: 7-15.
[8] CDC (2013) Vital signs: *Listeria* illnesses, deaths, and outbreaks--United States, 2009-2011. *MMWR Morb. Mortal Wkly Rep.* 62: 448-452.
[9] McLauchlin J, Mitchell RT, Smerdon WJ, Jewell K (2004) *Listeria monocytogenes* and listeriosis: a review of hazard characterisation for use in microbiological risk assessment of foods. *Int. J. Food Microbiol.* 92: 15-33.
[10] Poyart C, Berche P, Trieu-Cuot P (1995) Characterization of superoxide dismutase genes from gram-positive bacteria by polymerase chain reaction using degenerate primers. *FEMS Microbiol. Lett.* 131: 41-45.
[11] Kampfer P, Bottcher S, Dott W, Ruden H (1991) Physiological characterization and identification of *Listeria* species. *Zentralbl. Bakteriol.* 275: 423-435.
[12] Christie K, Atkins NE, Munch-Petersen E (1944) A Note On a Lytic Phenomenon Shown by Group B *Streptococci. Aust. J. Exp. Biol. Med.* 22: 197-200.
[13] Wang Y, Zhao A, Zhu R, Lan R, Jin D, et al., (2012) Genetic diversity and molecular typing of *Listeria monocytogenes* in China. *BMC Microbiol.* 12: 119.

[14] Parisi A, Latorre L, Normanno G, Miccolupo A, Fraccalvieri R, et al., (2010) Amplified Fragment Length Polymorphism and Multi-Locus Sequence Typing for high-resolution genotyping of *Listeria monocytogenes* from foods and the environment. *Food Microbiol.* 27: 101-108.

[15] Rudi K, Katla T, Naterstad K (2003) Multi locus fingerprinting of *Listeria monocytogenes* by sequence-specific labeling of DNA probes combined with array hybridization. *FEMS Microbiol. Lett.* 220: 9-14.

[16] Graves LM, Swaminathan B, Hunter SB, editors (1999) Subtyping *Listeria monocytogenes*. 2nd ed. New York: Marcel Dekker. 279-297 p.

[17] Mead PS, Slutsker L, Dietz V, McCaig LF, Bresee JS, et al., (1999) Food-related illness and death in the United States. *Emerg. Infect. Dis.* 5: 607-625.

[18] Mataragas M, Skandamis PN, Drosinos EH (2008) Risk profiles of pork and poultry meat and risk ratings of various pathogen/product combinations. *Int. J. Food Microbiol.* 126: 1-12.

[19] Smith MA, Takeuchi K, Anderson G, Ware GO, McClure HM, et al., (2008) Dose-response model for *Listeria monocytogenes*-induced stillbirths in nonhuman primates. *Infect. Immun.* 76: 726-731.

[20] Hutchins AD (1996) Assessment of alimentary exposure to *Listeria monocytogenes. Int. J. Food Microbiol.* 30: 71-85.

[21] Norrung B, Andersen JK, Schlundt J (1999) Incidence and control of *Listeria monocytogenes* in foods in Denmark. *Int. J. Food Microbiol.* 53: 195-203.

[22] Inoue S, Nakama A, Arai Y, Kokubo Y, Maruyama T, et al., (2000) Prevalence and contamination levels of *Listeria monocytogenes* in retail foods in Japan. *Int. J. Food Microbiol.* 59: 73-77.

[23] Chen Y, Ross WH, Scott VN, Gombas DE (2003) *Listeria monocytogenes*: low levels equal low risk. *J. Food Prot.* 66: 570-577.

[24] Graves LM, Hunter SB, Ong AR, Schoonmaker-Bopp D, Hise K, et al., (2005) Microbiological aspects of the investigation that traced the 1998 outbreak of listeriosis in the United States to contaminated hot dogs and establishment of molecular subtyping-based surveillance for *Listeria monocytogenes* in the PulseNet network. *J. Clin. Microbiol.* 43: 2350-2355.

[25] Pouillot R, Hoelzer K, Jackson KA, Henao OL, Silk BJ (2012) Relative risk of listeriosis in Foodborne Diseases Active Surveillance Network (FoodNet) sites according to age, pregnancy, and ethnicity. *Clin. Infect. Dis.* 54 Suppl 5: S405-410.

[26] Lianou A, Sofos JN (2007) A review of the incidence and transmission of *Listeria monocytogenes* in ready-to-eat products in retail and food service environments. *J. Food Prot.* 70: 2172-2198.

[27] Tompkin RB (2002) Control of *Listeria monocytogenes* in the food-processing environment. *J. Food Prot.* 65: 709-725.

[28] Loncarevic S, Tham W, Danielsson-Tham ML (1996) Prevalence of *Listeria monocytogenes* and other *Listeria* spp.in smoked and 'gravad' fish. *Acta Vet. Scand.* 37: 13-18.

[29] Handa S, Kimura B, Takahashi H, Koda T, Hisa K, et al., (2005) Incidence of *Listeria monocytogenes* in raw seafood products in Japanese retail stores. *J. Food Prot.* 68: 411-415.

[30] Oliver SP, Jayarao BM, Almeida RA (2005) Foodborne pathogens in milk and the dairy farm environment: food safety and public health implications. *Foodborne Pathog. Dis.* 2: 115-129.

[31] Gombas DE, Chen Y, Clavero RS, Scott VN (2003) Survey of *Listeria monocytogenes* in ready-to-eat foods. *J. Food Prot.* 66: 559-569.

[32] Beuchat LR (1996) *Listeria monocytogenes*: Incidence on vegetables. *Food Control* 7: 223-228.

[33] NACMCF (1999) Microbiological safety evaluations and recommendations on fresh produce. *Food Control* 10: 117-143.

[34] Chmielewski RA, Frank JF (2003) Biofilm Formation and Control in Food Processing Facilities. *Compr. Rev. Food Sci.* 2: 22-32.

[35] Skandamis PN, Stopforth JD, Ashton LV, Geornaras I, Kendall PA, et al., (2009) *Escherichia coli* O157:H7 survival, biofilm formation and acid tolerance under simulated slaughter plant moist and dry conditions. *Food Microbiol.* 26: 112-119.

[36] Hood SK, Zottola EA (1995) Biofilms in food processing. *Food Control* 6: 9-18.

[37] Gram L, Bagge-Ravn D, Ng YY, Gymoese P, Vogel BF (2007) Influence of food soiling matrix on cleaning and disinfection efficiency on surface attached *Listeria monocytogenes*. *Food Control* 18: 1165-1171.

[38] Carpentier B, Cerf O (2011) Review--Persistence of *Listeria monocytogenes* in food industry equipment and premises. *Int. J. Food Microbiol.* 145: 1-8.

[39] Hall-Stoodley L, Costerton JW, Stoodley P (2004) Bacterial biofilms: from the natural environment to infectious diseases. *Nat. Rev. Microbiol.* 2: 95-108.

[40] Chan YC, Wiedmann M (2009) Physiology and genetics of *Listeria monocytogenes* survival and growth at cold temperatures. *Crit. Rev. Food Sci. Nutr.* 49: 237-253.

[41] Borucki MK, Peppin JD, White D, Loge F, Call DR (2003) Variation in biofilm formation among strains of *Listeria monocytogenes*. *Appl. Environ. Microbiol.* 69: 7336-7342.

[42] Harmsen M, Lappann M, Knochel S, Molin S (2010) Role of extracellular DNA during biofilm formation by *Listeria monocytogenes*. *Appl. Environ. Microbiol.* 76: 2271-2279.

[43] Rieu A, Briandet R, Habimana O, Garmyn D, Guzzo J, et al., (2008) *Listeria monocytogenes* EGD-e biofilms: no mushrooms but a network of knitted chains. *Appl. Environ. Microbiol.* 74: 4491-4497.

[44] Sofos JN, Geornaras I (2010) Overview of current meat hygiene and safety risks and summary of recent studies on biofilms, and control of *Escherichia coli* O157:H7 in nonintact, and *Listeria monocytogenes* in ready-to-eat, meat products. *Meat Sci.* 86: 2-14.

[45] FSIS (2012) Compliance guidelines to control *Listeria monocytogenes* in post-lethality exposed ready-to-eat meat and poultry products. Available from: http://www. fsis. usda.gov/shared/PDF/Controlling_LM_RTE_guideline_0912.pdf. Accessed October 21, 2013.

[46] CFR (2010) 9 CFR 430.4-Control of *Listeria monocytogenes* in post-lethality exposed ready-to-eat products. Available from: http://www.access.gpo.gov/nara/cfr/waisidx_10/9cfr430_10.html. Accessed October 17, 2013.

[47] Lappi VR, Thimothe J, Walker J, Bell J, Gall K, et al., (2004) Impact of intervention strategies on *Listeria* contamination patterns in crawfish processing plants: a longitudinal study. *J. Food Prot.* 67: 1163-1169.

[48] Thompson RL, Carpenter CE, Martini S, Broadbent JR (2008) Control of *Listeria monocytogenes* in ready-to-eat meats containing sodium levulinate, sodium lactate, or a combination of sodium lactate and sodium diacetate. *J. Food Sci.* 73: M239-244.

[49] Lu Z, Sebranek JG, Dickson JS, Mendonca AF, Bailey TB (2005) Application of predictive models to estimate *Listeria monocytogenes* growth on frankfurters treated with organic acid salts. *J. Food Prot.* 68: 2326-2332.

[50] Schlyter JH, Glass KA, Loeffelholz J, Degnan AJ, Luchansky JB (1993) The effects of diacetate with nitrite, lactate, or pediocin on the viability of *Listeria monocytogenes* in turkey slurries. *Int. J. Food Microbiol.* 19: 271-281.

[51] Byelashov OA, Simpson CA, Geornaras I, Kendall PA, Scanga JA, et al., (2008) Evaluation of changes in *Listeria monocytogenes* populations on frankfurters at different stages from manufacturing to consumption. *J. Food Sci.* 73: M430-437.

[52] Mellefont LA, Ross T (2007) Effect of potassium lactate and a potassium lactate-sodium diacetate blend on *Listeria monocytogenes* growth in modified atmosphere packaged sliced ham. *J. Food Prot.* 70: 2297-2305.

[53] Barmpalia IM, Geornaras I, Belk KE, Scanga JA, Kendall PA, et al., (2004) Control of *Listeria monocytogenes* on frankfurters with antimicrobials in the formulation and by dipping in organic acid solutions. *J. Food Prot.* 67: 2456-2464.

[54] Gonzalez-Fandos E, Herrera B (2013) Efficacy of malic acid against *Listeria monocytogenes* attached to poultry skin during refrigerated storage. *Poult. Sci.* 92: 1936-1941.

[55] Wallace FM, Call JE, Luchansky JB (2003) Validation of the USDA/ARS package rinse method for recovery of *Listeria monocytogenes* from naturally contaminated, commercially prepared frankfurters. *J. Food Prot.* 66: 1920-1923.

[56] Murphy RY, Duncan LK, Driscoll KH, Beard BL, Berrang MB, et al., (2003) Determination of thermal lethality of *Listeria monocytogenes* in fully cooked chicken breast fillets and strips during postcook in-package pasteurization. *J. Food Prot.* 66: 578-583.

[57] Gande N, Muriana P (2003) Prepackage surface pasteurization of ready-to-eat meats with a radiant heat oven for reduction of *Listeria monocytogenes*. *J. Food Prot.* 66: 1623-1630.

[58] Muriana PM, Quimby W, Davidson CA, Grooms J (2002) Postpackage pasteurization of ready-to-eat deli meats by submersion heating for reduction of *Listeria monocytogenes*. *J. Food Prot.* 65: 963-969.

[59] Luber P, Crerar S, Dufour C, Farber J, Datta A, et al., (2011) Controlling *Listeria monocytogenes* in ready-to-eat foods: Working towards global scientific consensus and harmonization – Recommendations for improved prevention and control. *Food Control* 22: 1535-1549.

[60] Vitas AI, Garcia-Jalon VA (2004) Occurrence of *Listeria monocytogenes* in fresh and processed foods in Navarra (Spain). *Int. J. Food Microbiol.* 90: 349-356.

[61] Sindelar JJ, Cordray JC, Olson DG, Sebranek JG, Love JA (2007) Investigating quality attributes and consumer acceptance of uncured, no-nitrate/nitrite-added commercial hams, bacons, and frankfurters. *J. Food Sci.* 72: S551-559.

[62] Hulankova R, Borilova G, Steinhauserova I (2013) Combined antimicrobial effect of oregano essential oil and caprylic acid in minced beef. *Meat Sci.* 95: 190-194.

[63] Upadhyay A, Upadhyaya I, Kollanoor-Johny A, Ananda Baskaran S, Mooyottu S, et al., (2013) Inactivation of *Listeria monocytogenes* on frankfurters by plant-derived antimicrobials alone or in combination with hydrogen peroxide. *Int. J. Food Microbiol.* 163: 114-118.

[64] Jobim ML, Santos RC, Dos Santos Alves CF, Oliveira RM, Mostardeiro CP, et al., (2013) Antimicrobial activity of Amazon *Astrocaryum aculeatum* extracts and its association to oxidative metabolism. *Microbiol. Res.*

[65] Dholvitayakhun A, Trachoo N, Sakee U, Cushnie TP (2013) Potential applications for *Annona squamosa* leaf extract in the treatment and prevention of foodborne bacterial disease. *Nat. Prod. Commun.* 8: 385-388.

[66] Francolini I, Donelli G (2010) Prevention and control of biofilm-based medical-device-related infections. *FEMS Immunol. Med. Microbiol.* 59: 227-238.

[67] da Silva EP, De Martinis EC (2013) Current knowledge and perspectives on biofilm formation: the case of *Listeria monocytogenes*. *Appl. Microbiol. Biotechnol.* 97: 957-968.

[68] Czaczyk K, Myszka K (2007) Biosynthesis of extracellular polymeric substances (EPS) and its role in microbial biofilm formation. *Pol. J. Environ. Stud.* 16: 799-806.

[69] Orgaz B, Kives J, Pedregosa AM, Monistrol IF, Laborda F, et al., (2006) Bacterial biofilm removal using fungal enzymes. *Enzyme Microb. Tech.* 40: 51-56.

[70] Minei CC, Gomes BC, Ratti RP, D'Angelis CE, De Martinis EC (2008) Influence of peroxyacetic acid and nisin and coculture with Enterococcus faecium on *Listeria monocytogenes* biofilm formation. *J. Food Prot.* 71: 634-638.

[71] Desai MA, Soni KA, Nannapaneni R, Schilling MW, Silva JL (2012) Reduction of *Listeria monocytogenes* biofilms on stainless steel and polystyrene surfaces by essential oils. *J. Food Prot.* 75: 1332-1337.

[72] Sandasi M, Leonard CM, Viljoen AM (2010) The in vitro antibiofilm activity of selected culinary herbs and medicinal plants against *Listeria monocytogenes*. *Lett. Appl. Microbiol.* 50: 30-35.

[73] Nguyen UT, Wenderska IB, Chong MA, Koteva K, Wright GD, et al., (2012) Small-molecule modulators of *Listeria monocytogenes* biofilm development. *Appl. Environ. Microbiol.* 78: 1454-1465.

[74] Delgado FJ, Delgado J, Gonzalez-Crespo J, Cava R, Ramirez R (2013) High-pressure processing of a raw milk cheese improved its food safety maintaining the sensory quality. *Food Sci. Technol. Int.* Epub ahead of print on May 31, 2013, doi: 10.1177/1082013212455349

[75] Dunn J (1996) Pulsed light and pulsed electric field for foods and eggs. *Poult. Sci.* 75: 1133-1136.

[76] Monfort S, Gayán E, Raso J, Condón S, Alvarez I. (2010) Evaluation of pulsed electric fields technology for liquid whole egg pasteurization. *Food Microbiol.* 27: 845-852.

[77] Mahmoud BS (2012) Effect of x-ray treatments on pathogenic bacteria, inherent microbiota, color, and texture on parsley leaves. *Foodborne Pathog. Dis.* 9: 922-927.

[78] Garcia-Marquez I, Cambero MI, Ordonez JA, Cabeza MC (2012) Shelf-life extension and sanitation of fresh pork loin by E-beam treatment. *J. Food Prot.* 75: 2179-2189.

[79] Sheen S, Huang L, Sommers C (2012) Survival of *Listeria monocytogenes*, *Escherichia coli* O157:H7, and *Salmonella* spp. on catfish fillets exposed to microwave heating in a continuous mode. *J. Food Sci.* 77: E209-214.

[80] McCarthy S, Burkhardt W, III. (2012) Efficacy of electrolyzed oxidizing water against *Listeria monocytogenes* and *Morganella morganii* on conveyor belt and raw fish surfaces. *Food Control* 24: 214-219.

[81] Niemira BA (2012) Cold plasma decontamination of foods. *Annu. Rev. Food Sci. Technol.* 3: 125-142.

[82] Malheiros Pda S, Daroit DJ, Brandelli A (2012) Inhibition of *Listeria monocytogenes* in minas frescal cheese by free and nanovesicle-encapsulated nisin. *Braz. J. Microbiol.* 43: 1414-1418.

[83] Hill LE, Taylor TM, Gomes C (2013) Antimicrobial efficacy of poly (DL-lactide-co-glycolide) (PLGA) nanoparticles with entrapped cinnamon bark extract against *Listeria monocytogenes* and *Salmonella* Typhimurium. *J. Food Sci.* 78: N626-632.

[84] Solanki K, Grover N, Downs P, Paskaleva EE, Mehta KK, et al., (2013) Enzyme-based listericidal nanocomposites. *Sci. Rep.* 3: 1584.

[85] Kairyte K, Kadys A, Luksiene Z (2013) Antibacterial and antifungal activity of photoactivated ZnO nanoparticles in suspension. *J. Photochem. Photobiol.* B 128C: 78-84.

[86] Jiang Z, Neetoo H, Chen H (2011) Efficacy of freezing, frozen storage and edible antimicrobial coatings used in combination for control of *Listeria monocytogenes* on roasted turkey stored at chiller temperatures. *Food Microbiol.* 28: 1394-1401.

[87] Birmpa A, Sfika V, Vantarakis A (2013) Ultraviolet light and Ultrasound as non-thermal treatments for the inactivation of microorganisms in fresh ready-to-eat foods. *Int. J. Food Microbiol.*

[88] Ndoti-Nembe A, Vu KD, Doucet N, Lacroix M (2013) Effect of combination of essential oils and bacteriocins on the efficacy of gamma radiation against *Salmonella* Typhimurium and *Listeria monocytogenes*. *Int. J. Radiat. Biol.* 89: 794-800.

[89] Byelashov OA, Kendall PA, Belk KE, Scanga JA, Sofos JN (2008) Control of *Listeria monocytogenes* on vacuum-packaged frankfurters sprayed with lactic acid alone or in combination with sodium lauryl sulfate. *J. Food Prot.* 71: 728-734.

[90] Ban GH, Park SH, Kim SO, Ryu S, Kang DH (2012) Synergistic effect of steam and lactic acid against *Escherichia coli* O157:H7, *Salmonella* Typhimurium, and *Listeria monocytogenes* biofilms on polyvinyl chloride and stainless steel. *Int. J. Food Microbiol.* 157: 218-223.

[91] Torlak E, Sert D (2013) Combined effect of benzalkonium chloride and ultrasound against *Listeria monocytogenes* biofilm on plastic surface. *Lett. Appl. Microbiol.* 57: 220-226.

[92] Myers K, Cannon J, Montoya D, Dickson J, Lonergan S, et al., (2013) Effects of high hydrostatic pressure and varying concentrations of sodium nitrite from traditional and vegetable-based sources on the growth of *Listeria monocytogenes* on ready-to-eat (RTE) sliced ham. *Meat Sci.* 94: 69-76.

[93] Cutter CN (2002) Microbial control by packaging: a review. *Crit. Rev. Food Sci. Nutr.* 42: 151-161.

[94] Provincial L, Guillen E, Gil M, Alonso V, Roncales P, et al., (2013) Survival of *Listeria monocytogenes* and *Salmonella* Enteritidis in sea bream (*Sparus aurata*) fillets packaged under enriched CO_2 modified atmospheres. *Int. J. Food Microbiol.* 162: 213-219.

[95] Sant'Ana AS, Landgraf M, Destro MT, Franco BD (2013) Growth potential of *Salmonella* and *Listeria monocytogenes* in ready-to-eat lettuce and collard greens packaged under modified atmosphere and in perforated film. *J. Food Prot.* 76: 888-891.

[96] Fratamico PM, Juneja V, Annous BA, Rasanayagam V, Sundar M, et al., (2012) Application of ozonated dry ice (ALIGAL Blue Ice) for packaging and transport in the food industry. *J. Food Sci.* 77: M285-291.

[97] Muriel-Galet V, Talbert JN, Hernandez-Munoz P, Gavara R, Goddard JM (2013) Covalent Immobilization of Lysozyme on Ethylene Vinyl Alcohol Films for Nonmigrating Antimicrobial Packaging Applications. *J. Agric. Food Chem.* 61: 6720-6727.

[98] Ganegama Arachchi GJ, Cridge AG, Dias-Wanigasekera BM, Cruz CD, McIntyre L, et al., (2013) Effectiveness of phages in the decontamination of *Listeria monocytogenes* adhered to clean stainless steel, stainless steel coated with fish protein, and as a biofilm. *J. Ind. Microbiol. Biotechnol.* 40: 1105-1116.

[99] Perez Ibarreche M, Castellano P, Vignolo G (2013) Evaluation of anti-*Listeria* meat borne *Lactobacillus* for biofilm formation on selected abiotic surfaces. *Meat Sci.* 96: 295-303.

[100] Zhao T, Podtburg TC, Zhao P, Chen D, Baker DA, et al., (2013) Reduction by competitive bacteria of *Listeria monocytogenes* in biofilms and *Listeria* bacteria in floor drains in a ready-to-eat poultry processing plant. *J. Food Prot.* 76: 601-607.

In: *Listeria monocytogenes*
Editor: Edmund C. Hambrick

ISBN: 978-1-63117-054-6

Chapter 3

CARBOHYDRATE UTILIZATION BY *LISTERIA MONOCYTOGENES* AND ITS INFLUENCE ON VIRULENCE GENE EXPRESSION

Josef Deutscher[1,2,3,*], Francine Moussan Désirée Aké[2,3], Arthur Constant Zébré[2,3], Thanh Nguyen Cao[2,3], Takfarinas Kentache[2,3], Céline Monniot[2,3], Que Mai Ma Pham[2,3], Abdelhamid Mokhtari[2,3], Philippe Joyet[2,3] and Eliane Milohanic[2,3]

[1]Centre National de la Recherche Scientifique, UPR9073 Expression Génétique Microbienne, Institut de Biologie Physico-Chimique, Paris, France

[2]Institut National de la Recherche Agronomique, UMR1319 Microbiologie de l'alimentation au service de la santé humaine (MICALIS), Jouy en Josas, France

[3]AgroParisTech, UMR MICALIS, Jouy en Josas, France

ABSTRACT

The Gram-positive bacterium *Listeria monocytogenes* possesses a large number of transport systems for sugars and sugar derivatives. The uptake of the triol glycerol occurs in an energy-independent process via facilitated diffusion. In contrast, ion-driven transporters, ATP binding cassette (ABC) transporters and the phosphoenolpyruvate carbohydrate phosphotransferase system (PTS) require energy for the carbohydrate uptake reaction. Rhamnose and several sugar-phosphates are taken up via ion-driven permeases. *L. monocytogenes* contains several ABC transporters one of which was found to transport maltose. However, most carbohydrates used by *L. monocytogenes* are transported by the PTS. Among others they include glucose, fructose, mannose, cellobiose, gentiobiose, trehalose, D-arabitol and several β-glucoside type heterosides (salicin, arbutin, esculin, amygdalin). For many PTS the substrate has not yet been determined. Expression of several genes encoding PTS of unknown specificity, such as *lmo1974-1968*, was found to be upregulated during intracellular growth, suggesting that *L. monocytogenes* uses certain carbon sources transported by the PTS for intracellular

* Corresponding author: Josef.Deutscher@grignon.inra.fr.

proliferation. Certain carbohydrates taken up via the PTS strongly repress the expression of virulence genes. Virulence gene expression depends on the transcription activator PrfA (positive regulatory factor A). Its activity was found to be strongly inhibited by the presence of several PTS substrates. The underlying repression mechanism is not yet understood. It differs from common carbon catabolite repression, which in firmicutes is mediated by the PTS protein seryl-phosphorylated HPr bound to the catabolite control protein A. PrfA activity is probably regulated by a PTS component or a metabolic intermediate derived from an efficiently utilized carbohydrate.

INTRODUCTION

Listeria monocytogenes is a microorganism with two different life styles [1]. It is a soil bacterium which during saprophytic growth uses numerous carbon sources mainly derived from decaying plants. This explains the large number of carbohydrate transporters present in this organism as revealed by its genome sequence [2]. But it is also an anthropozoonotic pathogen, i.e., a usually commensal organism for certain animals causing the foodborne disease listeriosis in humans. It contaminates vegetables, dairy products, sausages, fish and other foods and infects humans during its transit across the intestinal tract. Infection can cause septicemia, encephalitis and frequently leads to miscarriage during pregnancy. The mortality rate is around 25% for people infected with listeriosis. The main site of infection is the intestinal epithelium; in *in vitro* experiments the bacterium was found to invade non-phagocytic cells. After crossing the epithelium *L. monocytogenes* invades phagocytic cells, which allows its dissemination via the lymphatic ganglions and the blood stream followed by the infection of various organs, mainly liver and spleen [3]. For *in vitro* host cell invasion *L. monocytogenes* uses two major proteins, InlA and InlB, belonging to the so-called internalin family. InlA interacts with the host cell adhesion factor E-cadherin whereas InlB interacts with the hepatocyte growth factor receptor Met [4]. After internalization *L. monocytogenes* escapes from the vacuoles/phagosomes with the help of the three virulence proteins listeriolysin, phospholipase A and phospholipase B. Intracellular motility and cell-to-cell spread are assured by the polymerization of host actin at one of the two cell poles of the bacterium [5]. The polymerization of actin is initiated by the bacterial actin assembly-inducing protein ActA. Most of the virulence proteins involved in the various infection steps are not present in the closely related non-pathogenic organism *Listeria innocua*.

Expression of the corresponding virulence genes requires the transcription activator PrfA (positive regulatory factor A), a member of the Crp/Fnr transcription regulator family [6], which is also absent from *L. innocua*. PrfA activity is controlled by several environmental signals at the transcriptional and post-transcriptional level. One of the signalling pathways assures that the virulence genes are only expressed at temperatures around 37°C [7]. Another signal is provided by the presence/absence of efficiently metabolized carbon sources transported via the phosphoenolpyruvate (PEP):carbohydrate phosphotransferase system (PTS). The uptake and metabolism of these carbohydrates lead to strong inhibition of PrfA activity without affecting *prfA* gene expression [8, 9]. Although this mechanism leads to the repression of virulence gene expression it differs from the main carbon catabolite repression mechanism operative in firmicutes, because the catabolite control protein A (CcpA) is not involved [10]. In fact, although first described some 20 years ago, the mechanism controlling PrfA activity in response to the presence or absence of an efficiently metabolizable carbon source has not yet been elucidated. The disaccharide cellobiose was found to have the strongest inhibitory effect on PrfA activity followed by glucose. It was therefore concluded that the presence of cellobiose and other PTS sugars might indicate to the bacterium that it thrives in an environment favouring its saprophytic lifestyle and that consequently there was no need to express its virulence genes.

Carbohydrates are taken up by bacteria via four different mechanisms: facilitated diffusion, ion-driven transport, uptake via ATP binding cassette (ABC) transporters and uptake via the PTS. Glucose-6-P and glycerol [11] are two of the non-PTS carbon sources that do not inhibit PrfA activity. Interestingly, three-carbon compounds and glucose-6-P were found to be the main carbon sources for *L. monocytogenes* during intracellular growth [12, 13]. The almost complete absence of PTS substrates in the cytoplasm of the host cells should therefore allow strong expression of the virulence genes. On the contrary, blood contains between 5 and 7 mM glucose; at these concentrations glucose already significantly inhibits PrfA activity [14]. However, PrfA was found to be highly active in blood and virulence genes were therefore strongly expressed [15]. This paradox is not yet understood.

In this chapter we will describe the presently known complement of carbohydrate transport and utilization systems in *L. monocytogenes* strain EGD-e, the first *L. monocytogenes* strain for which the genome sequence has been determined [2]. We will also discuss the most common mechanisms for induction and repression of the corresponding catabolic operons and, finally, summarize the present knowledge about carbohydrate-mediated virulence gene repression in this human pathogen.

CARBOHYDRATE UPTAKE VIA PTS TRANSPORTERS

General Description of the PTS

The PTS is a complex transport system composed of five or six proteins, which are sometimes fused to each other, thus becoming domains of larger proteins. Four of these proteins or domains form a phosphorylation cascade (Figure 1). In the first step, EI autophosphorylates at the Nε-2 position of a conserved histidyl residue by using phosphoenolpyruvate (PEP) as phosphoryl donor. P~EI subsequently transfers the phosphoryl group to the Nδ-1 position of His-15 in HPr. There are only a few HPr or HPr-like proteins in which the phosphorylatable His is not in position 15. EI and HPr are the general PTS proteins and with the exception of enterobacteriaceae, most bacteria possess only one EI and one HPr. This is also the case for listerial species [2] and the *ptsHI* operon, which encodes the proteins HPr and EI, has been studied in *L. monocytogenes* [16] and HPr and its various phosphorylated forms have been characterized [17, 18] . In contrast to the general PTS proteins, bacteria usually possess several sugar-specific EII complexes composed of EIIA, EIIB and EIIC (members of the mannose class PTS also contain an EIID); the *L. monocytogenes* strain EGD-e contains 29 complete and 12 incomplete EII complexes [2, 19], with two of the twelve incomplete PTS being dihydroxyacetone-specific EIIAs, which will be discussed in the chapter “Carbohydrate uptake via facilitated diffusion”. The EIIA components of the various sugar-specific PTS become phoshorylated by P~His-HPr at the Nε-2 position of a conserved histidyl residue. The P~EIIA’s in turn phosphorylate their corresponding sugar-specific EIIB at either a histidyl or more frequently at a cysteyl residue. These phosphoryl transfer reactions are all reversible. In the last step of the PTS phosphorylation cascade, P~EIIB transfers its phosphoryl group to the carbohydrate bound to the membrane-spanning EIIC (or EIIC and EIID for mannose class PTS) (Figure 1). The phosphorylated substrate is subsequently released into the cytoplasm [20, 21]. The PTS therefore catalyzes not only the uptake of its substrates but also their phosphorylation during the transport process.

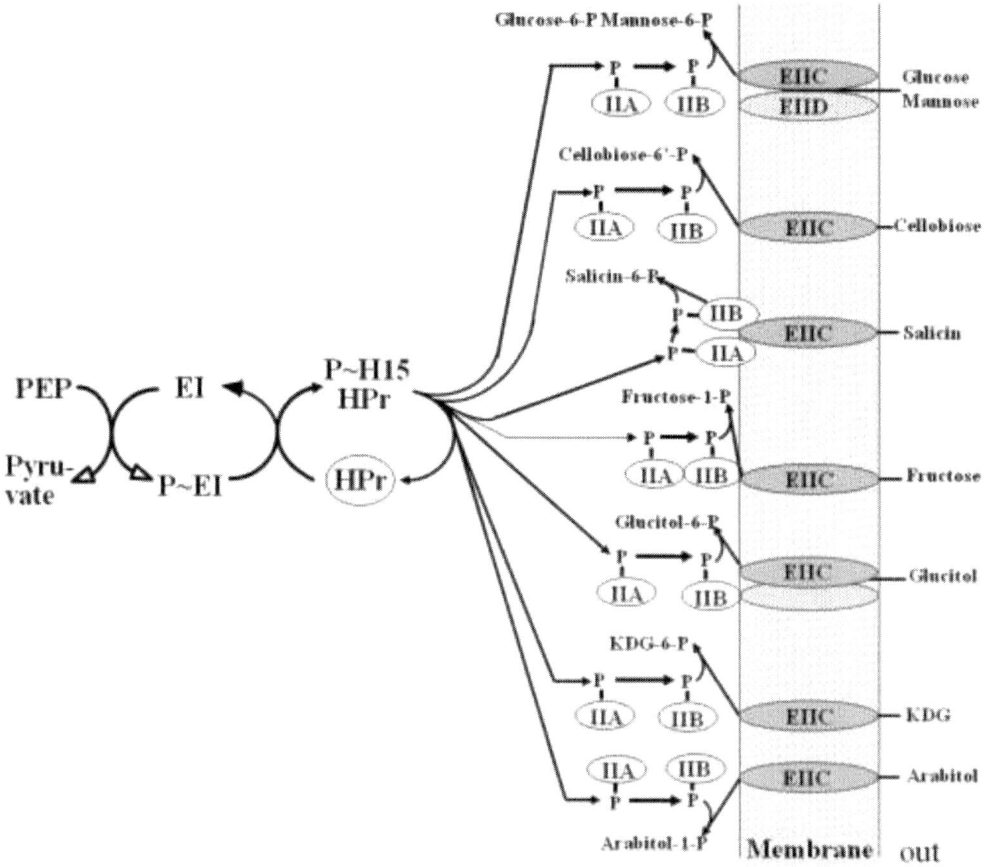

Figure 1. Schematic presentation of the seven different PTS classes. Shown are the following *L. monocytogenes* PTS (from top to the bottom): The mannose class PTS transporting glucose/mannose (ManLMN); the lactose/cellobiose class PTS transporting cellobiose (Lmo2683-2685); the glucose/glucoside class PTS transporting salicin (BvrB, Lmo2787); the fructose/mannitol class PTS presumably transporting fructose (FruA, Lmo2335), a glucitol class PTS presumed to transport glucitol (Lmo0544-0542); an ascorbate class PTS presumably transporting 2-keto-3-deoxygluconate (KDG) (Lmo1973-1971); and finally the galactitol class PTS transporting D-arabitol (Lmo2667-2665). The PTS phosphorylation cascade begins with the two general PTS components EI and HPr. P~His-HPr interacts with the more than 30 EIIAs present in *L. monocytogenes*. Each P~EIIA transfers the phosphoryl group to the EIIB of the same sugar-specificity, although a cross talk between closely related PTS can sometimes be observed. Finally, the P~EIIBs phosphorylate their corresponding carbohydrate bound to the EIIC (or EIIC and EIID for mannose class PTS).

Based on the amino acid sequence of the membrane-integrated EII components, which carry the substrate specificity, as well as of the soluble carbohydrate-specific proteins and the location and surrounding of their phosphorylation sites, seven different classes of PTS can be distinguished in *L. monocytogenes* strain EGD-e (Figure 1) [22, 23]. Members of the mannose class PTS can easily be recognized because they possess two membrane components, EIIC and EIID, with mannose class EIICs being quite different from EIICs of all other PTS classes. In addition, the EIIB components are phosphorylated at a histidyl and not at a cysteyl residue. The EIIA and EIIB components of the glucose/mannose-specific PTS present in most firmicutes [24, 25] and enterobactericeae [26] are usually fused to a single protein, ManL, whereas mannose class PTS with other carbohydrate specificities normally possess distinct EIIA and EIIB components (Figure 1). Lactose/cellobiose and galactitol class PTSs also contain distinct EIIA and EIIB components. These two PTS classes can be distinguished based on the amino acid sequence of the EII components. In addition, EIIA proteins of the lactose/cellobiose class form trimers [27], whereas those of the galactitol class form monomers [28]. Members of the glucose/β-glucoside and fructose/mannitol class PTS resemble each other, but nevertheless can be distinguished based on their amino acid sequence. This also holds for members of the ascorbate class PTS. Finally, PTS belonging to the glucitol class have a quite distinct domain organization with a split EIIC component and EIIB fused to the N-terminus of one of them. The general and specific PTS components have been intensively studied and their phosphorylation sites have been determined. In addition, the crystal and solution structures of EI [29, 30], HPr [31, 32] and numerous sugar-specific EIIAs and EIIBs [33] as well as of several transition complexes for phosphoryl group transfer [34-36] have been determined. The crystal structure of one of the membrane-spanning EIIC has also been determined [37]. As already mentioned, *L. monocytogenes* strain EGD-e contains 29 complete and 12 incomplete EII complexes belonging to the seven different classes and a list of the various PTS has previously been presented in [19]. In the following, we will describe in detail all characterized PTS of *L. monocytogenes* and try to propose potential substrates for the PTS of unknown specificity based on the presumed enzymatic activities of proteins, the genes of which are associated with the various *pts* operons. The predictions should therefore be treated with care and they need of course to be confirmed by experimental data.

Members of the Mannose Class PTS

L. monocytogenes contains two major mannose class PTS (ManLMN or Lmo0096-0098 and MpoABCD or Lmo0784-0781, Mpo stands for **m**annose **p**ermease **o**ne), which both are able to transport glucose and mannose [14, 19]. The two *L. monocytogenes* PTS were initially identified because inactivation of the *manLMN* operon provides high [38] and of the *mpoABCD* genes intermediate immunity [39] against class II bacteriocins. The $EIIC^{Man}$ and $EIID^{Man}$ (ManM and ManN) components were reported to directly interact with several different bacteriocins (class IIa and IIc) by serving as their receptors. In contrast, the components of the PTS^{Mpo} have only an indirect effect. As will be explained later, deletion of *mpoB* or inactivation of the entire operon inhibits the expression of the *manLMN* operon [14, 19]. Only the two membrane-integral PTS components $EIIC^{Man}$ and $EIID^{Man}$, but not

$EIIAB^{Man}$, seem to be necessary for bacteriocin sensitivity [40]. Another study had suggested that only the EIIC protein would be required for bacteriocin sensitivity [41].

While ManLMN transports glucose and mannose with high efficiency, MpoABCD is a low affinity transporter. Consequently, inactivation of ManLMN leads to significantly reduced glucose/mannose utilization [14, 19], whereas inactivation of MpoA has no detectable effect [14]. In fact, the Mpo system turned out to function primarily as glucose sensor. The Mpo components change their phosphorylation state in response to the presence or absence of glucose/mannose and thus control the LevR-like [42] transcription activator ManR, the gene of which (*lmo0785*) is located immediately upstream from the *mpo* operon. ManR belongs to the enhancer binding proteins that interact with RpoN (σ^{54}) [43] and hydrolyze ATP in order to form the activating DNA/protein complex [44, 45]. Indeed, the *mpo* operon is preceded by a RpoN-dependent -12, -24 promoter. However, ManR has no effect on *mpo* operon expression but controls *manLMN* expression. Phosphorylation of ManR by P~MpoB (P~$EIIB^{Mpo}$) at His-871 inactivates the regulator and thus inhibits *man* operon expression. The presence of glucose or mannose leads to dephosphorylation of P~MpoB and therefore also of P~ManR and hence to induction of the *man* operon [14]. In agreement with this concept, deletion of *mpoA*, which also prevents the formation of P~MpoB, leads to constitutive *man* expression. Unexpectedly, deletion of *mpoB*, but not *mpoA*, rendered ManR completely inactive and this mutant utilized glucose even less efficiently than the *man* deletion mutants (PTS^{Man} and PTS^{Mpo} are inactive). In fact, it turned out that in addition to preventing ManR phosphorylation by P~MpoB the transcription activator also needs to interact with unphosphorylated MpoB in order to be active [14]. This interaction has recently been confirmed by yeast two hybrid experiments (M. Ventroux, E. Milohanic, A. Zébré, J. Deutscher and M.F. Noirot-Gros, unpublished results).

L. monocytogenes can slowly transport glucose via other PTS [19], because glucose utilization by a *ptsI* deletion mutant, in which all PTS are inactive, was even less efficient than by the *mpoB* mutant, in which *mpo* and *man* are inactive. The residual glucose uptake observed in the *ptsI* mutant is probably mediated via one or several of the GlcU-like non-PTS permeases Lmo0169, Lmo0176 and Lmo0424, which will be discussed in the paragraph "ion-driven carbohydrate transporters". These transporters have broad substrate specificity and one of them probably also transports mannose, because a *ptsI* mutant is able to ferment not only glucose but also mannose (E. Milohanic and J. Deutscher, unpublished results).

L. monocytogenes possesses two other PTS of the mannose class encoded by the operons *lmo2002-1998* and *lmo0021-0025*. The *lmo2002-1998* operon also contains a gene encoding a protein resembling aminotransferases, suggesting that this PTS might transport an amino sugar. Nothing can be concluded about the substrate specificity of the PTS encoded by the *lmo0021-0025* operon.

Members of the Lactose/cellobiose Class PTS

L. monocytogenes possesses six complete and eight incomplete PTS belonging to the lactose/cellobiose class. Among them, the genes of an incomplete PTS composed of only EIIA and EIIB proteins (Lmo1719 and 1720, respectively) were reported to be induced during growth on cellobiose [46] and this result was confirmed in the study of Stoll and Goebel [19]. The operon contains an additional gene (*lmo1718*) encoding a presumed outer surface protein

(LpoO) with an RGD motif, which is characteristic for proteins involved in cell adhesion. The operon was called the *lpo* operon, and its three genes are induced not only by cellobiose, but also by chitobiose and lactose [46]. Operons of similar composition, but containing also the gene for a cellobiose class EIIC component, are found in several clostridia and bacilli. Expression of the *L. monocytogenes lpo* operon requires a PTS regulation domain (PRD)-containing transcription activator (LevR-like). Its gene, which was originally called *lacR* [46], is located just upstream from *lmo1720*. In a more recent study [19], truncation of the transcription activator gene *lacR*, which was now called *csrA*, was reported to prevent cellobiose-specific regulation of *L. monocytogenes* virulence genes. In fact, it turned out that CsrA is required not only for the expression of the *lmo1720-1718* operon, but also for the expression of an operon encoding an entire cellobiose class PTS, *lmo2683-2685* [19]. This PTS seems to be the major transporter for cellobiose, because its deletion significantly slowed growth on cellobiose as the sole carbon source. In contrast, deletion of the *lmo1719-1720* genes slowed growth on cellobiose only slightly [19]. It therefore came as a surprise that the *lmo1719-1720* deletion mutant exhibits increased expression of the *lmo2683-2685* genes when it was grown in glucose-containing medium.

One or several components of the Lmo2683-2685 PTS seem to function as negative effectors of the expression of the *lmo0027* gene, which encodes an EIICBA of the glucose/glucoside PTS class. Deletion of *lmo2683-2685* or the regulator gene *lmo1721*, which is required for *lmo2683-2685* expression, led to strong expression of *lmo0027* [19]. A similar effect on *lmo0027* expression was observed when either *manR* or the *manR*-controlled genes of the major glucose/mannose PTS, *lmo0096-0098*, were deleted. Lmo0027 also seems to contribute to cellobiose transport, because an *lmo2683-2685/lmo0027* double mutant grew significantly more slowly on cellobiose than a mutant which was deleted only for *lmo2683-2685*. In addition, cellobiose is also taken up by a non-PTS transporter, possibly one of the GlcU proteins, because a *ptsI* mutant was still able to slowly ferment cellobiose (E. Milohanic and J. Deutscher, unpublished results).

L. monocytogenes possesses five other entire PTS of the lactose/cellobiose class. One of them probably transports the β-1,6-glucoside gentiobiose (6-O-β-D-glucopyranosyl-D-glucose), which is efficiently metabolized by the wild-type strain EGD-e, but only slowly by a *ptsI* mutant derived from it (E. Milohanic and J. Deutscher, unpublished results). The utilization of gentiobiose by the *ptsI* mutant suggests that the disaccharide is also slowly taken up by a non-PTS transporter. In the first catabolic step, cellobiose and gentiobiose taken up and phosphorylated by a PTS are usually hydrolyzed by a 6-P-β-glucosidase to glucose-6-P and glucose. The resulting glucose is subsequently also phosphorylated to glucose-6-P by a glucokinase or hexokinase. There is one presumed operon present in *L. monocytogenes* that contains the genes for all these proteins: *lmo2765* (EIIA), *lmo2763* (EIIC) and *lmo2762* (EIIB) encode a PTS of the lactose/cellobiose class. These genes are part of a presumed operon containing also the genes for a 6-P-β-glucosidase- (*lmo2761*) and a glucokinase-like (*lmo2764*) enzyme. It is therefore likely that this PTS is the transporter for gentiobiose; alternatively it might be another cellobiose uptake and phosphorylation system.

In addition to Lmo2761, *L. monocytogenes* contains ten potential 6-P-β-glucosidases exhibiting significant similarity to BglH of *B. subtilis* and two proteins, Lmo0521 and Lmo0536, strongly resembling the 6-P-β-glucosidase LicH of this organism. The genes encoding three of these 6-P-β-glucosidase-like proteins, Lmo0300, Lmo0917 and Lmo0372 are associated with a lactose/cellobiose class PTS (Lmo0298/0299/3001, Lmo0914-0916 and Lmo0373/Lmo0374; the latter lacks an EIIA component). These PTS might transport one or several of the β-glucosides arbutin, salicin, amygdalin and esculin. These heterosides are all fermented by the *L. monocytogenes* wild-type strain, but not by the *ptsI* mutant (E. Milohanic and J. Deutscher, unpublished results), indicating that they are exclusively taken up via a PTS. In *Streptococcus pneumoniae*, cellobiose, gentiobiose, arbutin, amygdalin and esculin are all transported by a single PTS of the lactose/cellobiose class [47].

The presumed operon of one of the remaining PTS of the lactose/cellobiose class, *lmo0874-0876*, contains a gene encoding a protein resembling glucosamine-6-P deaminase/isomerase, NagB. It is therefore tempting to assume that this PTS transports glucosamine, which is used as carbon source by *L. monocytogenes* [48]. However, this operon contains two additional genes encoding an oxidoreductase- (Lmo0878) and a sugar-phosphate epimerase-like protein (Lmo0879). It is therefore possible that 2-amino-2-deoxy-D-galactitol (also called galactosaminitol) might be an additional substrate for this PTS. After its transport and phosphorylation this compound might be oxidized to 2-amino-2-deoxy-D-galactose-6-P by the oxidoreductase Lmo0878, subsequently converted into glucosamine-6-P by the sugar-phosphate epimerase-like protein Lmo0879 and finally converted by the NagB-like enzyme into fructose-6-P or glucose-6-P.

Finally, the presumed operon encoding the last entire PTS of the lactose/cellobiose class (Lmo2783, EIIC; Lmo2782, EIIB; Lmo2780, EIIA) contains a gene encoding a presumed membrane-anchored β-glucosidase (Lmo2781) attached to the outer surface of the membrane. This β-glucosidase might catalyze the breakdown of larger polysaccharides, such as chitobiose, into mono- or oligo-saccharides, which are subsequently transported by the PTS into the cells. This PTS is not present in *L. innocua*.

The function of most incomplete PTS of the lactose/cellobiose class remains unknown. Some of them might carry out regulatory functions instead of being involved in carbohydrate uptake.

Members of the Glucose/glucoside Class PTS

L. monocytogenes also possesses five complete and two incomplete PTS belonging to the glucose/glucoside class. However, none of them plays a major role in the transport of glucose, which is primarily taken up by a mannose class PTS, as was mentioned before. Several members of the glucose/glucoside class PTS have been shown to be involved in the uptake of cellobiose or other β-glucosides, including aryl-β-glucosides, such as salicin and arbutin. One is the BvrB (Lmo2787) protein, an EII composed of B, C and A domains in the indicated order (EIIBCA). Expression of the gene encoding this protein is controlled by an antiterminator, BvrA, and it is induced by the presence of cellobiose and salicin, but not by arbutin. In agreement with this result, inactivation of the *bvr* operon prevented the repressive effect of cellobiose and salicin, but not of arbutin, on the expression of the virulence gene *plcB* [49]. These results suggest that BvrB is a transporter specific for cellobiose and salicin, whereas arbutin seems to be taken up by another PTS. However, it should be noted that in a later study expression of the *lmo2787* gene was not induced by the presence of cellobiose but was similar to that measured in cells grown either on glucose, mannose or glycerol [19]. The *bvr* operon is present only in about half of the *L. monocytogens* strains (including EGD-e) and is absent from all other sequenced listerial species.

In the preceding chapter we already discussed a potential contribution of Lmo0027, an EIIBCA protein, to cellobiose transport. It is encoded by an isolated gene with no catabolic genes being associated. Cellobiose-6-P formed during the uptake via Lmo0027 might therefore be hydrolyzed by one of the few 6-P-β-glucosidases present in *L. monocytogenes* and the resulting glucose and glucose-6-P catabolized via the Embden Meyerhof pathway.

Other PTS probably involved in β-glycoside transport are Lmo0738 and Lmo2772, both with the domain order EIIBCA. In addition, both contain in their presumed operon a gene encoding a 6-P-β-glucosidase-like protein, Lmo0739 and Lmo2771, respectively. Interestingly, *lmo2772* and *lmo2771* were found to be strongly (about a 100-fold) overexpressed in an *hprK* mutant when grown in glucose-containing BHI medium compared to the wild-type strain [50]. The two genes were only slightly (6-fold) upregulated in a *ccpA* mutant, suggesting that P-Ser-HPr plays a specific role in the regulation of their expression. While no other gene is associated with *lmo2772*, the presumed operon of *lmo0738* contains in addition a gene encoding a protein resembling ribulose-5-P 3-epimerase (Lmo0735) and another gene encoding a ribose-5-phosphate isomerase-like protein (Lmo0736). The presence of the genes encoding these two enzymes suggests that this PTS transports a ribose containing glycoside. Hydrolysis of the glycoside after PTS-mediated transport and hydrolysis by the 6-P-β-glucosidase-like protein might produce ribose-5-P, which could be isomerized to ribulose-5-P by Lmo0736 and subsequently converted into xylulose-5-P by the enzyme ribulose-5-P 3-epimerase Lmo0735.

The α-disaccharide trehalose seems to be exclusively taken up by *L. monocytogenes* via a PTS transporter, which is most likely encoded by the *lmo1255* gene. Indeed, inactivation of *ptsI* leads to a complete loss of trehalose fermentation when carrying out a Galerie API 50 CH carbohydrate fermentation assay (Biomérieux, Lyon, France) (E. Milohanic and J. Deutscher, unpublished results). The transporter is composed of only two domains (EIIBC) and an EIIA is lacking. The system might be complemented with the glucose/glucoside class EIIA encoded by the mono-cistronic *lmo1017* or an enzyme IIA component of another glucose/glucoside class PTS. The transporter gene is followed by a gene encoding the enzyme α,α-(1,1) phosphotrehalase, TreA, which has been shown to hydrolyze trehalose-6-P into glucose and glucose-6-P [51]. Inactivation of the *treA* gene leads to the accumulation of intracellular trehalose-6-P, which was found to increase the stress resistance of the bacterium.

The last PTS of the glucose/glucoside family present in *L. monocytogenes* EGD-e is the EIICBA protein encoded by *lmo1035*. This gene forms a presumed operon with genes encoding a fucose isomerase-like protein (Lmo1031), two genes encoding the N- and C-terminal parts of a transketolase (Lmo1032 and Lmo1033) and a protein resembling glycerol kinase (Lmo1034), which, however, lacks the activation loop with the phosphorylatable His typical for GlpKs from firmicutes (see chapter "Carbohydrate uptake via facilitated diffusion"). Lmo1034 is therefore unlikely to phosphorylate glycerol, but rather modifies another carbohydrate. Finally, *lmo1036* encodes a protein exhibiting similarity to glycosidases. The PTS protein encoded by *lmo1035* might therefore catalyze the uptake of a fucoside in a transport step resembling that reported for fucosyl-α-1,3-N-acetylglucosamine, which is taken up by *L. casei* via a mannose/glucose class PTS without phosphorylation [52]. Fucose is a 6-deoxy-galactose and therefore lacks the usually phosphorylated hydroxyl group at the C-6 position. In fact, fucosyl-α-1,3-N-acetylglucosamine is so far the only carbohydrate reported to be taken up by a PTS without phosphorylation. The intracellular fucoside transported by the PTS might subsequently be hydrolyzed by the presumed glycosidase and the resulting L-fucose might be converted into L-fuculose by the enzyme fucose isomerase Lmo1031 (Figure 2). The kinase encoded by *lmo1034* exhibits similarity to fuculose kinases and might therefore transform fuculose into fuculose-1-P, which might subsequently be cleaved by an aldolase into dihydroxyacetone phosphate and L-lactaldehyde. There is no fuculose-1-P aldolase-encoding gene present in the *lmo1031-1036* operon. It is therefore likely that the fuculose-1-P aldolase-like protein encoded by the operon also containing the gene for a presumed fucose-specific PTS permease (Lmo2093) catalyzes the catabolic step (Figure 2). Nothing can be concluded about the identity of the sugar moiety or the aglycon connected to fucose via an α- or β-glycosidic bond.

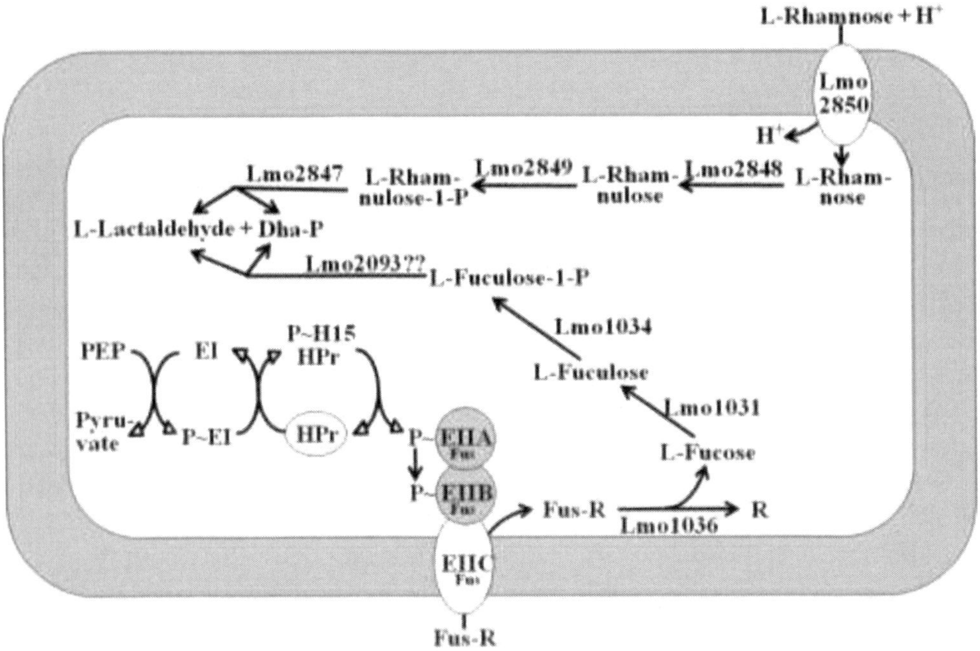

Figure 2. Schematic presentation of the transport and catabolism of D-rhamnose by *L. monocytogenes* and of a presumed fucoside-transporting PTS and the catabolism of the fucoside, which is presumed to be similar to that of D-rhamnose. After its uptake via a H^+ symporter, D-rhamnose is first isomerized to D-rhamnulose by the rhamnose isomerase Lmo2848, subsequently phosphorylated to D-rhamnulose-1-P by the enzyme rhamnulokinase (Lmo2849), and in the last step cleaved to Dha-P and D-lactaldehyde by the enzyme D-rhamnulose-1-P aldolase (Lmo2847) [117]. D-lactaldehyde is reduced to 1,2-propanediol probably by the enzyme L-lactaldehyde dehydrogenase (Lmo1171), which explains why growth on D-rhamnose leads to the induction of the propanediol operon.
Similar to the *L. casei* fucoside transporter [52], the *L. monocytogenes* PTS Lmo1035 is also assumed to take up a fucoside without phosphorylation. The glycosidase Lmo1036 is thought to cleave the fucoside into L-fucose and an unknown second moiety. The catabolism of L-fucose follows the same enzymatic steps operative for D-rhamnose (isomerisation to fuculose catalyzed by Lmo1031 and phosphorylation to L-fuculose-1-P catalyzed by Lmo1034). *L. monocytogenes* contains another *pts* operon assumed to encode the enzymes for the transport and catabolism of fucose, including a fuculose-1-P aldolase (Lmo2093). This enzyme might also participate in the catabolism of fucosides taken up via Lmo1035.

Members of the Fructose/Mannitol Class PTS

One of the members of the fructose/mannitol class PTS is the fructose-specific EIIABC transport protein, which is encoded by the *fruA* gene (*lmo2335*). FruA proteins usually do not phosphorylate the 6-position, but rather the 1-position of fructose. Accordingly, the operon also contains a *fruK* gene which encodes a 1-P-fructokinase, which specifically phosphorylates the 6-position of fructose-1-P leading to the formation of fructose-1,6-bisphosphate, which is metabolized via glycolysis. Indeed, PEP-dependent phosphorylation of fructose at the 1-position was observed when membrane fragments derived from fructose-grown *L. monocyogenes* cells were complemented with EI and HPr [17]. 1-P-fructokinase (FruK) activity was also demonstrated in crude extracts prepared from fructose-grown *L. monocytogenes* cells. Although neither FruK nor FruA were identified, it is likely that they are encoded by the two genes *lmo2335* and *lmo2334*. Fructose is also transported with low affinity by certain PTS of the mannose class. In addition, we observed that a *ptsI* mutant is still able to ferment fructose (E. Milohanic and J. Deutscher, unpublished results), indicating that there is also a non-PTS transporter specific for fructose present in *L. monocytogenes.*

L. monocytogenes contains seven more PTS belonging to the fructose class. Most of them have so far not been studied and their substrate specificity can therefore only be guessed. For example, the genes *lmo0426-0428* encode a fructose class PTS with three distinct proteins, EIIA, EIIB and EIIC, respectively. The presumed operon contains one additional gene, *lmo0429*, which codes for a protein resembling α-mannosidases. It is therefore likely that this PTS transports an α-mannoside, that is phosphorylated by the PTS and subsequently cleaved into mannose-6-P and the aglycon or the second carbohydrate moiety. The four genes encoding the three PTS components and the glycosyl hydrolase were deleted in *L. monocytogenes* strain F2365 [53]. The system seems to play a role in the general stress response, because expression of the operon was induced by high pressure and inhibited by elevated salt concentrations. In addition, in the presence of 5% NaCl the mutant grew significantly faster than the wild-type strain.

The *lmo0631-0633* genes encoding another fructose/mannitol class PTS are followed by a gene encoding a tagatose-6-P kinase, suggesting that this PTS might transport tagatose. However, *L. monocytogenes* is not able to ferment tagatose (E. Milohanic and J. Deutscher, unpublished results). Another ketohexose might therefore be taken up by this PTS. Alternatively, this PTS might be induced only under specific conditions. The genes encoding the EIIA (Lmo2797) and EIICB (Lmo2799) components of another fructose/mannitol class PTS are associated with three genes encoding enzymes resembling an N-acetyl-D-mannosamine-6-P epimerase (Lmo2801), an oxidoreductase (Lmo2800) and a phosphatase (Lmo2798). This PTS might therefore transport N-acetyl-D-mannosamine, which after its transport and phosphorylation by the PTS would be converted into N-acetyl-D-glucosamine-6-P by the enzyme N-acetyl-D-mannosamine-6-P epimerase. Nevertheless, the roles of the oxidoreductase and phosphatase remain obscure.

Two PTS of the fructose/mannitol class contain in their presumed operon one (*lmo0359*) or two (*lmo2133* and *lmo2134*) genes encoding fructose-1,6-bisphosphate aldolase-like proteins. The aldolases might not necessarily recognize fructose-1,6-bisphosphate but the bisphosphate of another ketohexose or ketopentose. As a consequence, no predictions about the substrate specificity can be made. The gene encoding another fructose/mannitol class

PTS, *lmo2733*, is associated with two genes encoding a presumed sugar α-hydrolase-like protein (Lmo2734) and a sucrose phosphorylase (Lmo2735). The trisaccharide erlose, which is also called glucosylsucrose and which is composed of a glucose molecule followed by a sucrose moiety [glucose-α(1–4)-glucose-α(1–2)-fructose], might be taken up by this PTS. The triholoside erlose is found in honey and in royal jelly. After its transport and phosphorylation by the PTS, erlose-6''-P might be cleaved into glucose-6-P and sucrose by Lmo2734. Sucrose might subsequently be phosphorolyzed by Lmo2735 into fructose and α-D-glucose-1-P.

In contrast to most firmicutes, *L. monocytogenes* does not contain a mannitol-specific PTS, which is usually composed of an EIICB and an EIIA component [54]. During the transport step, the PTS converts mannitol into mannitol-1-P, which is oxidized to fructose-6-P by the enzyme mannitol-1-P dehydrogenase (MtlD). No protein exhibiting significant similarity to *B. subtilis* MtlD could be detected in *L. monocytogenes* strain EGD-e. This finding is in agreement with the observation that *L. monocytogenes* strain EGD-e is not able to ferment mannitol (E. Milohanic and J. Deutscher, unpublished results).

Members of the Galactitol Class PTS

L. monocytogenes possesses three members of the galactitol class PTS. One of them, *lmo2667-2665*, has been shown to be required for the uptake of the pentitol D-arabitol. The D-arabitol PTS (PTS^{Atl}) is composed of the three proteins $EIIA^{Atl}$ (Lmo2667), $EIIB^{Atl}$ (Lmo2666) and $EIIC^{Atl}$ (Lmo2665). Insertion of a transposon into *lmo2665* prevented growth of *L. monocytogenes* on D-arabitol [55]. In addition, deletion of the *ptsI* gene prevents fermentation of D-arabitol, confirming that the pentitol is exclusively transported by a PTS. The *atl* operon also contains the genes for two dehydrogenases, Lmo2663 and Lmo2664, one of which might convert D-arabitol-1-P into xylulose-5-P and the other D-arabitol-5-P into ribulose-5-P. A D-arabitol-5-P dehydrogenase with a dual substrate specificity for D-arabitol-1-P and D-arabitol-5-P has been isolated from *Enterococcus avium* [56]. The D-arabitol-specific PTS of this organism was suggested to phosphorylate the pentitol either at the 1- or the 5-position. The protein encoded by *lmo2662* has been annotated as ribose-5-P 3-epimerase, but we found that it rather exhibits ribose-5-P isomerase activity (E. Milohanic, T. Kentache and J. Deutscher, unpublished results), thus allowing the transformation of ribulose-5-P into ribose-5-P. Finally, the transketolase-like enzyme Lmo2660 can convert ribose-5-P and xylulose-5-P into glyceraldehyde-3-P and sedoheptulose-7-P. D-arabitol therefore seems to be mainly metabolized via the pentose phosphate pathway.

The second galactitol class PTS encoded by the genes *lmo0503*, *0507* and *0508* strongly resembles the D-arabitol-specific PTS. The operon also encodes enzymes of the pentose phosphate pathway, but in place of a transketolase it contains the gene for a transaldolase-like protein (*lmo0500*). In addition, we observed that growth on D-arabitol led to the induction of not only the *lmo2667-2659* operon but also of *lmo0498-0508* (E. Milohanic, T. Kentache and J. Deutscher, unpublished results). This operon might therefore also contribute to D-arabitol uptake and catabolism. However, its physiological role remains elusive, especially seen that disruption of the $EIIC^{Atl}$ component Lmo2665 caused a complete loss of D-arabitol utilization [55].

The third galactitol class PTS is also composed of three proteins (Lmo2098, 2097 and 2096) corresponding to the EIIA, EIIB and EIIC components, respectively. The genes downstream from the PTS genes encode an enzyme resembling 1-P-fructokinase and an aldolase, that resembles L-fuculose-1-P aldolase. It is therefore tempting to assume that similar to a *L. casei* α-fucoside PTS [52] this PTS catalyzes the uptake of the 6-deoxy sugar fucose or fuculose without phosphorylation. Intracellular fucose/fuculose could then be metabolized as outlined in Figure 2 by phosphorylating it to L-fuculose-1-P (Lmo2095), which might be cleaved by the aldolase Lmo2094 into Dha-P and L-lactaldehyde. This PTS might therefore be responsible for the observed fucose utilization during intracellular growth of *L. monocytogenes*n [57]. However, it should be noted that the presumed *fuc* operon lacks a gene encoding fucose isomerase. This activity might be provided by the protein Lmo1031 (Figure 2)

Members of the Ascorbate Class PTS

L. monocytogenes possesses two PTS belonging to the ascorbate class. One is composed of the Lmo1973, Lmo1972 and Lmo1971 proteins, which correspond to the EIIA, EIIB and EIIC components, respectively. The genes encoding these PTS proteins are organized in an operon together with three additional genes *lmo1970-1968* (Q.M. Ma Pham, E. Milohanic and J. Deutscher, unpublished results). They encode enzymes resembling phosphotriesterases, 2-keto-3-deoxygluconate-6-P aldolases and creatinine amidohydrolases, respectively. We purified Lmo1969 and confirmed its 2-keto-3-deoxygluconate-6-P aldolase activity. Like Eda, the 2-keto-3-deoxygluconate-6-P aldolase of *E. coli*, it therefore cleaves 2-keto-3-deoxygluconate-6-P into glyceraldehyde-3-P and pyruvate (Q.M. Ma Pham, E. Milohanic and J. Deutscher, unpublished results). It is therefore likely that this PTS transports and phosphorylates 2-keto-3-deoxygluconate or a derivative of it, which might resemble 2-amino-2-deoxy-D-gluconic acid, that is transported via a PTS by *Salmonella enterica* [58]. The phosphorylated substrate might subsequently be converted into 2-keto-3-deoxygluconate-6-P by the phosphotriesterase- and creatinine amidohydrolases-like enzymes. Alternatively, these two enzymes might modify 2-keto-3-deoxygluconate-6-P to convert it into a yet unknown metabolite. Interestingly, expression of the genes of this operon was 4- to 8-fold upregulated during intracellular growth of *L. monocytogenes* [59].

The components of the second listerial PTS of the ascorbate class, Lmo2651-2649, strongly resemble the corresponding components of the presumed 2-keto-3-deoxygluconate-related PTS with sequence identities ranging from 35% to 40%. In addition, this operon also contains the genes encoding phosphotriesterase- and creatinine amidohydrolase-like enzymes (*lmo2647* and *lmo2648*, respectively). The gene *lmo2646* also seems to encode an aldolase, the amino acid sequence of which is, however, completely different from that of Lmo1969. It rather resembles an aldolase that cleaves the unphosphorylated five-carbon compound 4-hydroxy-2-oxoglutarate into pyruvate and glyoxylate. However, it can presently not be excluded that this enzyme also recognizes 2-keto-3-deoxygluconate-6-P as substrate.

Member of the Glucitol Class PTS

The *L. monocytogenes* genes *lmo0544-0542* encode the components of a typical glucitol PTS. The gene order is the same as that reported for the glucitol operon of *Lactobacillus casei* (*gutFRMCBA*) [60], with the exception that in *L. monocytogenes* the transcription regulator gene *gutR* is not included in the operon, but is located upstream from it. It nevertheless contains *gutM*, which encodes a protein necessary for the expression of the *gut* operon. The first gene, *gutF*, encodes a glucitol-6-P dehydrogenase. Members of the glucitol class PTS have a unique organisation with two short membrane proteins composed of about 160 (GutC) and 300 (GutB) amino acids. In fact, the membrane-spanning part of GutB comprises only about 120 amino acids, because its N-terminal part corresponds to the soluble EIIB domain with the phosphorylatable Cys located around position 70. The last gene of the operon, *gutA*, encodes the EIIA component. Surprisingly, although all PTS proteins and the glucitol-6-P dehydrogenase seem to contain no obvious defect or truncation and the EIIA and EIIB components possess the conserved phosphorylation sites when compared to the corresponding glucitol-specific *L. casei* PTS components, *L. monocytogenes* is nevertheless not able to grow on glucitol as the sole carbon source or to ferment the hexitol. The failure to utilize glucitol might therefore be caused by defects in the regulatory proteins controlling the expression of the *gut* operon. Alternatively, the operon might only be induced when *L. monocytogenes* is exposed to specific environmental conditions.

Carbohydrate Uptake via non-PTS Transporters

Carbohydrate Uptake via Facilitated Diffusion

Glycerol and probably also dihydroxyacetone enter into *L. monocytogenes* via energy-independent facilitated diffusion catalyzed by the pore-forming membrane protein GlpF [61]. The *lmo1539* gene encodes a protein exhibiting 61% sequence identity to *Bacillus subtilis* GlpF. After entering the cell, glycerol can either be phosphorylated to glycerol-3-P by glycerol kinase (GlpK) or oxidized to dihydroxyacetone (Figure 3). Indeed, the gene donwstream from *lmo1539* encodes a protein strongly resembling GlpK of firmicutes. Lmo1538 contains the typical phosphorylatable histidine in position 231, which is surrounded by three aromatic amino acids and which becomes phosphorylated by P~His-HPr at its N-ε2 position for activation of the GlpK function [62, 63]. It is therefore most likely that the two genes *lmo1539* and *lmo1538* form the *glpFK* operon, although the encoded proteins have not yet been characterized. In other firmicutes it has been shown that the presence of glucose or other carbohydrates efficiently transported via the PTS leads to dephosphorylation of P~GlpK (competition for the common phosphoryl donor P~His-HPr) and consequently to inducer exclusion [63]. Owing to the low activity of unphosphorylated GlpK glycerol enters the cells only very slowly. The *L. monocytogenes glpFK* operon lacks the gene for a glycerol-3-P dehydrogenase or glycerol-3-P oxidase, which convert glycerol-3-P into dihydroxyacetone-P (Figure 3). In most firmicutes a *glpD* or *glpO* gene is located downstream from *glpFK*. In *L. monocytogenes* the gene *lmo1293,* which is located far from *glpFK*, probably encodes GlpD, because this protein exhibits 60% sequence identity when compared to GlpD of *B. subtilis*.

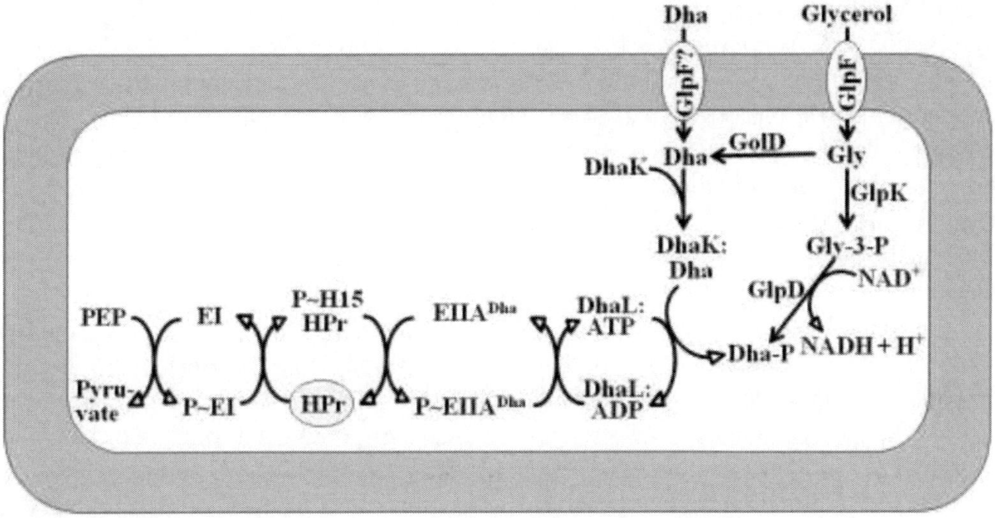

Figure 3. The different pathways of glycerol catabolism in *L. monocytogenes*. Glycerol enters *L. monocytogenes* cells by facilitated diffusion catalyzed by the protein GlpF. Intracellular glycerol is either phosphorylated to glycerol-3-P in an ATP-requiring reaction catalyzed by the enzyme GlpK and subsequently oxidized to Dha-P in an NAD-requiring reaction catalyzed by the enzyme glycerol-3-P dehydrogenase (GlpD); or glycerol is first oxidized to Dha by the enzyme glycerol dehydrogenase (GolD) in an NAD-requiring reaction and subsequently phosphorylated in a PEP-dependent reaction catalyzed by the PTS proteins EI, HPr and $EIIA^{Dha}$ as well as the enzymes DhaL and DhaK. EI and HPr catalyze the phosphorylation of $EIIA^{Dha}$, and $P{\sim}EIIA^{Dha}$ transfers the phosphoryl group to an ADP molecule bound to DhaL. In the last step, $EIIA^{Dha}$:ATP phosphorylates Dha bound to the protein DhaK. It is likely that Dha also enters *L. monocytogenes* cells via facilitated diffusion catalyzed by GlpF or the presumed 1,2-propanediol transporter Lmo1167.

The first step of the second pathway of glycerol assimilation, i.e., its oxidation to dihydroxyacetone, is catalyzed by the enzyme glycerol dehydrogenase (Figure 3). *L. monocytogenes* possesses a new type of glycerol dehydrogenase, GolD (Lmo0344), which belongs to the large family of NAD^+-dependent short chain alcohol dehydro-genases /reductases [64] and is therefore completely different from GldA, the glycerol dehydrogenase of *Escherichia coli*. The resulting dihydroxyacetone is phosphorylated to dihydroxyacetone-3-P by a complex protein system requiring phosphoenolpyruvate (PEP) as phosphoryl donor. It involves the PTS components enzyme I (EI), HPr and the mannose class $EIIA^{Dha}$-2, as well as the proteins DhaK-2 and DhaL-2 [64] (Figure 3). DhaL-2 carries a tightly bound ADP, which is phosphorylated by $P{\sim}EIIA^{Dha}$-2 thus providing ATP, which is used to phosphorylate dihydroxyacetone bound to DhaK-2 [65]. In addition to the genes encoding the proteins necessary for dihydroxyacetone phosphorylation the *gol* operon also contains several genes encoding triose-P isomerase as well as enzymes of the pentose phosphate pathway, suggesting that dihydroxyacetone formed by this system is primarily metabolized via the pentose phosphate pathway. This system was studied in *L. innocua*, but a nearly identical system exists in *L. monocytogenes*. The *L. monocytogenes gol* operon is induced during growth on glycerol, especially during late log phase [11]. This operon is also induced during growth on soil extracts [66] but not during growth within host cells [59]. In contrast, the *glpFK* operon is induced during growth in host cells and its inactivation leads to a significant loss of virulence, confirming the importance of glycerol utilization for intracellular proliferation of *L. monocytogenes* [57].

The tri-cistronic *L. monocytogenes* operon *lmo2695-2697* encodes only the proteins required for dihydroxyacetone phosphorylation, i.e., EIIA^{Dha}-1, DhaK-1 and DhaL-1. It thus resembles the *dha* operons found not only in numerous bacteria but also in certain archaea [67]. Together with EI and HPr, these three proteins probably catalyze the PEP-dependent phosphorylatation of dihydroxyacetone after its entry into *L. monocytogenes* probably via GlpF-catalyzed facilitated diffusion (Figure 3). This operon is also upregulated during intracellular growth [59].

The gene *lmo1167* was erroneously annotated as *glpF*. Indeed, it shows significant sequence identity to GlpF from *L. monocytogenes* and other organisms (up to 57%). However, it is not associated with *glpK*, *glpD* or *glpO*, but is part of a presumed huge operon (*lmo1151-1170*) encoding proteins strongly resembling enzymes catalyzing the metabolism of propanediol and ethanolamine in *Salmonella enterica* serovar typhimurium [68]. It is therefore likely that Lmo1167 does not catalyze the facilitated diffusion of glycerol, but of the closely related 1,2-propanediol and/or ethanolamine. The latter compound contributes to the proliferation of *S. enterica* during intracellular gowth [69] and might have a similar effect on intracellular proliferation of *L. monocytogenes* [70]. The corresponding operon of *L. innocua* (*lin1115-1134*) was indeed reported to be induced by the presence of 1,2-propanediol in the growth medium [71]. As will be explained in the following paragraph “ion-driven carbohydrate transporters”, the metabolism of rhamnose leads to the formation of intracellular 1,2-propanediol and growth on rhamnose therefore also induces the *lin1115-1134* operon. 1,2-Propanediol is catabolized to either *n*-propanol or propionate. In *S. enterica*, propionyl-CoA, an intermediate of the propionate degradation pathway, can be converted into 2-methyl-citric acid by fusing it to oxaloacetate. This reaction is catalyzed by the enzyme 2-methyl-citrate synthase [71]. However, this pathway does not seem to be operative in listeriae because they lack a homologue of 2-methyl-citrate synthase.

Ion-driven Carbohydrate Transporters

L-Rhamnose, also called 6-deoxy-L-mannose, is taken up by the major facilitator superfamily protein Lmo2850 via a proton symport mechanism (Figure 2). As a consequence, deletion of *ptsI* does not affect L-rhamnose fermentation (E. Milohanic and J. Deutscher, unpublished results). Growth of *L. monocytogenes* in L-rhamnose-containing medium leads to the induction of the *lmo2850-2846* operon [72]. Expression of this operon is probably controlled by the AraC-type regulator Lmo2851. In addition, deletion of *lmo0753*, which encodes a Crp/Fnr-like transcription activator, prevented L-rhamnose utilization by strain EGD-e, but had no effect in strain 10403S [73]. Lmo2850 exhibits around 70% sequence identity when compared to the transport protein encoded by the rhamnose operon of several enterococci (*Enterococcus malodoratus*), lactobacilli (*Lactobacillus salivarius*, *Lactobacillus rhamnosus*) and pediococci (*Pediococcus acidilactici*), but does not exhibit similarity to the *E. coli* L-rhamnose transporter RhaT. The other enzymes encoded by the *lmo2850-2846* operon catalyze the degradation of L-rhamnose to dihydroxyacetone phosphate and L-lactaldehyde (Figure 2). They include L-rhamnose-1-epimerase (Lmo2846); L-rhamnose isomerase (Lmo2848), which converts L-rhamnose into L-rhamnulose; L-rhamnulose kinase

(Lmo2849), which phosphorylates L-rhamnulose at its 1-position; and L-rhamnulose-1-P aldolase (Lmo2847), which splits L-rhamnulose-1-P into dihydroxyacetone phosphate and L-lactaldehyde. The catabolic pathway therefore strongly resembles the catabolism of fucose, which has been described in the paragraph "Members of the glucose/glucoside class PTS" (see also Figure 2). In contrast to the transport protein, the catabolic enzymes exhibit between 45 and 56% sequence identity not only to the corresponding enzymes of lactobacillales, but also to the *E. coli* homologues [74]. As already mentioned in the preceding chapter, in an additional probably NADPH-dependent reaction catalyzed by the enzyme PduQ (L-lactaldehyde dehydrogenase, Lmo1171) L-lactaldehyde is reduced to 1,2-propanediol, which is subsequently degraded to *n*-propanol or propionic acid.

As mentioned before, several carbohydrates transported by a PTS, such as glucose, mannose, fructose, cellobiose and gentiobiose, were still fermented by a *L. monocytogenes ptsI* mutant, indicating that these sugars are also taken up via non-PTS transporters. Likely candidates for such non-PTS transporters are three proteins, which resemble GlcU of *B. subtilis*, *Staphylococcus xylosus* [75] and *Lactococcus lactis* [25]. The *L. lactis* transporter was shown to be driven by proton motive force. The three GlcU-like transporters of *L. monocytogenes* seem to have broad substrate specificity and low affinity for their substrates thus catalyzing only slow uptake of mannose, glucose [14], fructose, cellobiose and gentiobiose in a *ptsI* deletion mutant.

L. monocytogenes takes up glucose-6-P and other hexose phosphates via the UhpT protein (uptake hexose phosphate, also called Hpt), another H^+ symporter first identified in *E. coli* [76]. This transporter seems to be very important for proliferation in host cells [77]. This is also supported by the finding that expression of the *uhpT* gene is controlled by PrfA, the transcription activator of the *L. monocytogenes* virulence genes [4]. Expression of the virulence genes is upregulated during intracellular growth and accordingly *uhpT* expression was also elevated during intracellular proliferation. *L. monocytogenes* wild-type cells injected into the cytosol of Caco-2 cells are able to replicate, whereas a *uhpT* mutant had lost this capacity [78]. Mutants carrying a *uhpT* deletion are therefore impaired in virulence [79]. UhpT also catalyzes the transport of the antibiotic fosfomycin. Because *uhpT* is not expressed during extracellular growth, *L. monocytogenes* is resistant to this antibiotic. However, bacteria growing inside host cells or producing a constitutively active PrfA* are highly sensitive to fosfomycin [80].

Carbohydrate-specific ABC Transporters

L. monocytogenes possesses eight ATP binding cassette (ABC) transporters for carbohydrates. They are usually composed of a sugar binding protein anchored at the outside of the cytoplasmic membrane and two membrane-spanning permeases. In *L. monocytogenes*, most operons encoding carbohydrate-specific ABC transporters lack the gene for the ATP-hydrolyzing protein. Instead they use Lmo0278, which similar to MsmX in *B. subtilis* functions with most listerial sugar-transporting ABC systems. Lmo0278 probably interacts with the cytoplasmic loops of the membrane-spanning components of seven listerial ABC systems transporting carbohydrates [81]. Only the ABC transporter encoded by Lmo0808-0810 possesses its own ATP-hydrolyzing protein Lmo0807.

Maltose and maltodextrose have been shown to be taken up by *L. monocytogenes* via the ABC transport system composed of Lmo2123, Lmo2124 and Lmo2125, which exhibit strong similarity to the *L. casei* membrane-spanning MalG and MalF proteins and the maltose-binding protein MalE, respectively [82, 83]. The operon lacks a gene for the ATP hydrolyzing enzyme MalK and the above mentioned Lmo0278 was shown to function as ATPase during maltose and maltodextrose uptake. The maltose operon contains three additional genes encoding enzymes resembling maltogenic amylase (Lmo2126), the maltodextrose utilization protein MalA (Lmo2122), and maltose phosphorylase (Lmo2121). The results published in [84] suggested that the ABC transporter would be the only uptake system for maltose present in *L. monocytogenes*. Surprisingly, when carrying out a fermentation test with the *L. monocytogenes* wild type strain EGD-e and a *ptsI* deletion mutant derived from it (the latter lacks EI) we observed that the *ptsI* mutant had lost the capacity to ferment maltose (E. Milohanic and J. Deutscher, unpublished results). However, we could not detect a *L. monocytogenes* gene encoding a protein resembling either *B. subtilis* 6-P-α-glucosidase [85], which hydrolyzes maltose-6'-P formed during PTS uptake, or *E. faecalis* maltose-6'-P phosphatase, which dephosphorylates maltose-6'-P [86] thus allowing the subsequent phosphorolysis of maltose to glucose and glucose-1-P. One possible explanation for the loss of maltose fermentation in the *ptsI* mutant might be that the absence of EI indirectly affects maltose uptake. The presence of glucose was found to inhibit maltose uptake by *L. monocytogenes*, probably by an inducer exclusion mechanism. Similar as in *L. casei* [83, 87] the inhibitory effect of glucose is apparently mediated via P-Ser-HPr, because in a *L. monocytogenes hprK* mutant maltose uptake was no longer inhibited by the presence of glucose [84]. A mutant lacking EI most likely contains elevated amounts of P-Ser-HPr because it cannot form P~His-HPr, which probably represents the major fraction of the different HPr forms in the maltose-utilizing wild-type strain. In the *ptsI* mutant, part of the P~His-HPr fraction is converted into P-Ser-HPr, which might be sufficient to inhibit maltose uptake via the ABC transport system.

The other seven ABC transport systems of *L. monocytogenes* have not been studied and their substrate specificity is unknown. Nevertheless, based on the activity of the catabolic enzymes usually also encoded by the ABC transporter operons substrate specificities can be predicted. Most of them seem to transport disaccharides or longer oligosaccharides. For example, the ABC transporter encoded by *lmo2839-2837* might transport maltitol, a polyol produced by plants and composed of a glucose and a glucitol moiety connected via an α-1,4-glycosidic linkage (4-O-α-D-glucopyranosyl-D-glucitol). The presumed maltitol operon contains also the genes encoding proteins resembling maltose phosphorylase (Lmo2833), which might transform intracellular maltitol into glucose-1-P and glucitol. The resulting glucose-1-P is subsequently converted into glucose-6-P by one of the phosphoglucomutases present in *L. monocytogenes*. Glucitol might be oxidized to fructose by the oxidoreductase Lmo2834 or the alcohol dehydrogenase Lmo2836. A kinase of the glycerate kinase family, which might phosphorylate fructose to fructose-6-P, is encoded by Lmo2832. Two proteins presumed to exhibit oxidoreductase and cellobiose phosphorylase activity are also present upstream from the genes *lmo1730-1732*, which encode another sugar ABC transport system. It is tempting to assume that Lmo1730-1732 might transport cellobiitol, which similar to maltitol is also composed of glucose and glucitol, with the only difference that maltitol contains an α- and cellobiitol a β-glucosidic linkage (4-O-β-D-glucopyranosyl-D-glucitol).

Cellobiitol is probably degraded via a pathway similar to that proposed for maltitol including its phosphorolysis to glucose-1-P and glucitol and the subsequent oxidation of glucitol to fructose.

The operon *lmo0766-0769* encodes an ABC transporter and an enzyme resembling α-1,6-mannanases, suggesting that this system, which lacks a sugar binding subunit, transports oligo-mannosides derived from mannanes present in plants. Lmo0179-0181 is likely to transport the degradation products of branched-chain oligosaccharides, because the presumed operon also encodes enzymes resembling α-1,4- and α-1,6- xylosidases and glucosidases. Degradation products of a branched-chain polysaccharide might also be transported by the Lmo2009-2007 ABC transporter, because the presumed operon contains a gene encoding a protein resembling the membrane-embedded enzyme catalyzing the degradation of the complex polysaccharide rhamnogalacturonan. Lmo0859-0861 might transport an oligosaccharide containing a mannose moiety connected via its α-1-position to the 6-position of the second glycose moiety or an aglycone. This assumption is based on the observation that the presumed operon contains also genes encoding an α-1,6-glycosidase- and a phospho-mannomutase-like protein. The substrate is probably not methyl-α-D-mannopyranoside, because this compound is likely to be transported by a PTS. Fermentation of methyl-α-D-mannopyranoside was abolished in a *ptsI* deletion mutant (E. Milohanic and J. Deutscher, unpublished results). No reliable predictions can be made for the substrate of the ABC transporter encoded by Lmo0809-0811. This system possesses its own ATP hydrolyzing protein and the operon also encodes enzymes resembling carbonic anhydrase, fructokinase and oxidoreductase. The presumed fructokinase might not phosphorylate fructose, but possibly one of the other three ketohexoses (sorbose, tagatose, psicose) or a ketopentose.

Carbon Source Mediated Repression of Virulence Genes

Like in most other firmicutes, the efficient utilization of a carbon source by *L. monocytogenes* leads to carbon catabolite repression, i.e., the genes encoding the proteins required for the uptake and metabolism of less efficient carbon sources are not expressed when glucose, fructose, mannose or other rapidly metabolized carbohydrates are present. Several different mechanisms are used by bacteria to exert carbon catabolite repression [21]. The main mechanism in firmicutes is mediated by a complex formed between the catabolite control protein A (CcpA) and seryl-phosphorylated HPr (P-Ser-HPr) [88]. The signal for the formation of P-Ser-HPr is provided by metabolites, which accumulate (FBP, ATP) or vanish (P_i) during the utilization of an efficient carbon source [89]. Metabolites such as FBP and ATP stimulate and inorganic phosphate (P_i) inhibits the activity of the bifunctional HPr kinase/phosphorylase (HprK/P) [90]. This enzyme catalyzes the phosphorylation of HPr and dephosphorylation of P-Ser-HPr in response to changes of the intracellular concentration of the above-mentioned metabolites [89]. Dephosphorylation of P-Ser-HPr follows a unique phosphorolysis mechanism [91]. When glucose or other efficiently metabolized carbohydrates are present, the elevated amounts of FBP and ATP and the low P_i concentration will activate the kinase function and inhibit the phosphorylase activity of HprK/P and HPr will become phosphorylated in an ATP-requiring reaction at Ser-46 [92]. P-Ser-HPr interacts with CcpA [93] and the complex binds to specific operator sites on the DNA called catabolite response elements (*cre*) [94, 95], which are usually located in the non-coding part of catabolic genes and operons and frequently overlap the promoter. Binding of the CcpA/P-Ser-HPr complex to the *cre* sites [96] prevents the expression of numerous catabolic genes. *L. monocytogenes* contains all the necessary components for CcpA-mediated carbon catabolite repression (HPr, CcpA, HprK/P and *cre* sites), and this mechanism has indeed been shown to be operative in the pathogen [10, 97].

P-Ser-HPr is also involved in two other, CcpA-independent mechanisms leading to the repression of catabolic genes. In the first CcpA-independent carbon catabolite repression mechanism the expression of numerous operons encoding PTS proteins is controlled by PTS regulation domain (PRD)-containing transcription activators or antiterminators. These regulators usually need to be phosphorylated by PEP, EI and HPr in order to be active. Phosphorylation of HPr at Ser-46 drastically slows the PTS-mediated phosphoryl group transfer [98]. Formation of P-Ser-HPr therefore also slows phosphorylation of PRD-containing transcription activators and antiterminators which consequently are barely active. In mutants deleted for *ccpA* several catabolic operons were still repressed by rapidly metabolizable carbon sources. In *B. subtilis* the CcpA-independent carbon catabolite repression disappeared when in addition either the phosphorylatable Ser-46 in HPr was replaced with Ala or the *hprK* gene was deleted [99], or when a mutant form of an antiterminator was used that did not require activation via phosphorylation by PEP, EI and HPr [100]. The LicT protein controls the expression of the *bglPH* operon of *B. subtilis*. Deletion of CcpA prevented glucose repression of *bglPH* transcription initiation, but in the presence of glucose mRNA synthesis still stopped at the terminator preceding *bglPH*, thus leading to the synthesis of only a small RNA fragment. Only when in the *ccpA* mutant Ser-46 of HPr was replaced with Ala, the long *bglPH* transcript was synthesized even when glucose was present [99]. *L. monocytogenes* possesses an antiterminator-controlled operon (*lmo2772-2771*) resembling that of *bglPH*, which is preceded by a gene (*lmo2773*) encoding a LicT-like antiterminator. It is therefore likely that the regulation of *lmo2772-lmo2771* expression resembles that of *B. subtilis bglPH*. In addition, there is a second *L. monocytogenes* antiterminator (Lmo2436), which also strongly resembles LicT, for which, however, the target gene or operon is not known. Finally, genetic studies with *manR* mutants suggested that the transcription activator of the *L. monocytogenes* glucose/mannose operon, ManR, needs to be phosphorylated by EI and HPr in order to be active [101]. In addition, expression of the *man* operon was strongly diminished in a *ptsI* deletion mutant unable to form P~His-HPr [14].

The second CcpA-independent carbon catabolite repression mechanism affects gene expression only indirectly by inhibiting carbohydrate uptake. Because the transported carbohydrate itself or one of its catabolites acts as inducer of the corresponding operon, this regulatory phenomenon was called inducer exclusion. In enterobacteria inducer exclusion seems to be the major mechanism of carbon catabolite repression [21]. Unphosphorylated $EIIA^{Glc}$ interacts with numerous transport proteins and inhibits their activity. In firmicutes, inducer exclusion seems to be mediated by P-Ser-HPr. The presence of glucose entirely inhibits maltose uptake in *L. casei* [87] and maltose and ribose uptake in *Lactococcus lactis* [102]. However, mutants unable to form P-Ser-HPr take up maltose even in the presence of glucose. In contrast, a *L. casei* mutant converting most of its HPr into P-Ser-HPr was not able to utilize maltose, even when glucose was absent [83]. Maltose utilization by *L. monocytogenes* also seems to be regulated via inducer exclusion [84]. Deletion of *ccpA* did not prevent the repressive effect of glucose on the expression of the *malEFG* operon. However, when *hprK* was disrupted, glucose no longer repressed *malEFG* expression and no longer inhibited maltose uptake. Because the *hprK* mutant is not able to form P-Ser-HPr both, CcpA-mediated carbon catabolite repression and inducer exclusion, are no longer functional. Finally, glucose kinase was also reported to be involved in carbon catabolite repression of certain organisms [103, 104]. However, the underlying mechanism is still poorly understood.

Several virulence genes of *L. monocytogenes* are also repressed by the presence of rapidly metabolizable PTS carbon sources, such as cellobiose, glucose, mannose or fructose, with cellobiose exhibiting the strongest repressing effect [8, 9]. The non-PTS sugars glucose-6-P and glycerol [11] do not repress the expression of virulence genes. Carbon source-mediated virulence gene expression was found to affect the activity of the Crp-like virulence gene activator PrfA and not the expression level of the *prfA* gene, thus suggesting a direct interaction of the repressor/activator compound with PrfA. PrfA binds to conserved 14 bp sequences of dyad symmetry preceding the promoter sequences in its target genes and operons [6, 105, 106]. Certain mutations were found to enhance the transcription activity of PrfA [107, 108], whereas other naturally occurring mutations inhibited it [109]. Interestingly, one of the stimulating mutations (Gly145Ser) was in the same region of the protein as one of the activating mutations (Gly141Ser) identified for Crp [107]. Because several distinct mutations led to the activation of PrfA, it was assumed that the PrfA effector compound would stimulate PrfA activity. The stimulating mutations are thought to mimic the structural changes which occur when the effector molecule binds to PrfA. However, it cannot be excluded that the stimulating mutations might prevent the binding of an inhibiting effector compound. A PrfA activating factor (Paf) present in *L. monocytogenes* crude extracts was found to enhance binding of PrfA to its DNA targets [110]. However, Paf was later identified as RNA polymerase. Because PrfA is essential for virulence gene expression [6], a relief from repression of the *L. monocytogenes* virulence genes can only be observed in strains in which the transcription activator PrfA is functional [14]. Inhibition of virulence gene expression by carbon sources is not mediated via the main carbon catabolite repression mechanism because it is CcpA independent [10] and P-Ser-HPr also does not seem to play a direct role [50, 111]. Certain results suggested that two semi-independent mechanisms of carbon source-mediated virulence gene repression might exist in *L. monocytogenes* [112]. One mechanism was thought to specifically control cellobiose-mediated virulence gene repression, whereas the second mechanism was assumed to be operative for all other repressing carbohydrates. The cellobiose-specific inhibition of the hemolysin-encoding *hly* gene was strongly reduced when the *csrA* gene (*csr* = cellobiose-specific regulation) became inactivated. In fact, it was later found that the PRD-containing transcription activator CsrA controls the expression of the genes encoding the major cellobiose transporter (*lmo2683-2685*) and similar to the *lmo2683-2685* deletion strain [19] the *csrA* mutant probably utilizes cellobiose less efficiently, which might explain the observed partial relief from virulence gene repression. A second mutation (*gcr*, for **g**eneral **c**atabolite **r**egulation) prevented repression of virulence gene expression by several carbon sources, including glucose. Nevertheless, in a *csrA*, *gcr* double mutant α-glucosidase activity was still repressed by glucose and other efficiently utilized carbon sources, confirming that carbon catabolite repression and virulence gene repression are mediated by two distinct mechanisms. The double mutant also exhibited complete relief from cellobiose repression. Unfortunately, the gene altered by the *gcr* mutation could not be identified.

A transposon insertion affecting the PRD-containing antiterminator BvrA, which controls the expression of the *bvrABC* operon, and BvrB, an EIIBCA PTS component of the glucose/glucoside family, prevented repression of the virulence gene *plcB* by cellobiose and the β-glucoside salicin [49]. The closely related β-heteroside arbutin still repressed virulence gene expression because it is probably transported by a different PTS. The reason why the transposon insertion causes a relief from virulence gene expression is not understood. The findings that mutations in transcription regulators of PTS genes affected PrfA activity suggested that PTS components might be directly or indirectly involved in virulence gene repression.

Because PTS components are barely phosphorylated at their conserved cysteine or histidine residue when glucose or other efficiently metabolizable carbon sources are utilized [113, 114], it was assumed that either an unphosphorylated PTS component might inhibit the activity of PrfA or a phosphorylated PTS protein might stimulate the transcription activator. The latter possibility became unlikely when it was found that deletion of the *L. monocytogenes* EI-encoding *ptsI* gene led to strong expression of virulence genes when the cells were grown in glucose-containing medium [14] or in the presence of cellobiose or fructose (F. Aké, E. Milohanic and J. Deutscher, unpublished results). In a *L. monocytogenes ptsI* mutant, all PTS proteins are present in dephosphorylated form, similar as in cells growing on an efficiently utilized PTS substrate. Deletion of *ptsI* was therefore rather expected to cause permanent inhibition of virulence gene expression and not a relief from repression. One possible explanation for the high level expression of the virulence genes in the *ptsI* mutant is that unphosphorylated EI might inhibit PrfA activity in a direct or indirect way, which, of course, is prevented when EI is deleted. A more likely explanation for the phenotype of the *ptsI* mutant is that this strain is not able to efficiently transport PTS sugars and as a consequence virulence genes are not repressed. This concept is supported by the finding that a *ptsH* mutant, which is also unable to transport PTS sugars, exhibited elevated expression of the virulence genes similar to the *ptsI* mutant [50]. Nevertheless, HPr or one of its phosphorylated derivatives might play a role in virulence gene repression. When an *hprK* deletion mutant, which is unable to form P-Ser-HPr, was grown in glucose-containing medium expression of the *L. monocytogenes* virulence genes was strongly upregulated. Complementation of the mutant with an *hprK* wild-type allele restored glucose-mediated repression of the virulence genes. This suggests that P-Ser-HPr, the amount of which drastically increases during the uptake and metabolism of efficient carbon sources [114], might function as direct or indirect inhibitor of PrfA activity. Based on expression studies of a PrfA-controlled reporter gene in a PrfA-producing *B. subtilis* strain and in *hprK*, *ptsI* and *ptsH* mutants derived from it, P-Ser-HPr had already previously been proposed to play a role in PrfA regulation [111]. However, when *L. monocytogenes* wild-type cells were grown in the presence of various carbon sources, the amount of P-Ser-HPr formed during utilization of the different carbohydrates did not correlate with the repressive effect on PrfA activity [50]. Specifically, during growth on cellobiose, which has the strongest repressive effect on virulence gene expression, only a small amount of P-Ser-HPr was formed. It was therefore concluded by the authors that P-Ser-HPr does not play a role in the repression of virulence gene expression.

Because none of the general PTS components seemed to be involved in carbon source-mediated repression of *L. monocytogenes* virulence genes, sugar-specific EII's were considered to play a role in the regulation of PrfA activity. A potential interaction of components of the PTS^{Man} with PrfA was suggested by the observation that overexpression of PrfA inhibited glucose uptake and consequently also growth of *L. monocytogenes* in glucose-containing medium [115]. In a later study, the $EIIAB^{Man}$ component of the major mannose and glucose transporter was proposed to carry out this function, because relief from glucose repression was stronger in a *manL* deletion strain (encodes $EIIAB^{Man}$) than in a mutant lacking *manM* (encodes $EIIC^{Man}$), although both strains should show an identical loss of their capacity to transport glucose via the PTS^{Man} [14]. However, the *manL* deletion caused a relief from virulence gene repression only for cells grown on glucose or mannose, but not for cells grown on cellobiose or fructose (F. Aké, E. Milohanic and J. Deutscher, unpublished results), thus excluding the possibility that $EIIAB^{Man}$ might function as a general repressor of PrfA activity.

The possibility was therefore also considered that a metabolite, most likely a glycolytic intermediate, might control the activity of PrfA. The intracellular concentration of certain glycolytic intermediates, such as glucose-6-P and fructose-1,6-bisphosphate, changes drastically when an efficiently metabolizable carbon source is utilized by bacteria [116]. However, no metabolite has so far been identified that would affect the binding of PrfA to its DNA targets in electrophoretic mobility shift assays (F. Aké, E. Milohanic and J. Deutscher, unpublished results); [11]. Second messengers were also suspected to regulate PrfA activity, especially as Crp is regulated by cAMP. The effect of cAMP, cGMP, cyclic-di-AMP and others on the binding of PrfA to its DNA targets was also tested (F. Aké, E. Milohanic and J. Deutscher, unpublished results). Again, none of these compounds had an effect on the interaction of PrfA with its DNA targets during electrophoretic mobility shift assays with PrfA and appropriate DNA fragments.

In conclusion, although first reported about 20 years ago, the mechanism leading to carbon source-mediated repression of most of the *L. monocytogenes* virulence genes is still not understood.

References

[1] N. E. Freitag, G. C. Port and M. D. Miner, *Nat. Rev. Microbiol.* 7, 623 (2009).

[2] P. Glaser, L. Frangeul, C. Buchrieser, C. Rusniok, A. Amend, F. Baquero, P. Berche, H. Bloecker, P. Brandt, T. Chakraborty, A. Charbit, F. Chetouani, E. Couve, A. de Daruvar, P. Dehoux, E. Domann, G. Dominguez-Bernal, E. Duchaud, L. L. Durant, O. Dussurget, K. D. Entian, H. Fsihi, F. Garcia-del Portillo, P. Garrido, L. Gautier, W. Goebel, N. Gomez-Lopez, T. Hain, J. Hauf, D. Jackson, L. M. Jones, U. Kaerst, J. Kreft, M. Kuhn, F. Kunst, G. Kurapkat, E. Madueno, A. Maitournam, J. M. Vicente, E. Ng, H. Nedjari, G. Nordsiek, S. Novella, B. de Pablos, J. C. Perez-Diaz, R. Purcell, B. Remmel, M. Rose, T. Schlueter, N. Simoes, A. Tierrez, J. A. Vazquez-Boland, H. Voss, J. Wehland and P. Cossart, *Science* 294, 849 (2001).

[3] J. A. Vázquez-Boland, M. Kuhn, P. Berche, T. Chakraborty, G. Domínguez-Bernal, W. Goebel, B. González-Zorn, J. Wehland and J. Kreft, *Clin. Microbiol. Rev.* 14, 584 (2001).

[4] O. Disson and M. Lecuit, *Microbes Infect.* in press (2013).

[5] M. Scortti, H. J. Monzó, L. Lacharme-Lora, D. A. Lewis and J. A. Vázquez-Boland, *Microbes Infect.* 9, 1196 (2007).
[6] M. Leimeister-Wächter, C. Haffner, E. Domann, W. Goebel and T. Chakraborty, *Proc. Natl. Acad. Sci. USA* 87, 8336 (1990).
[7] J. Johansson, P. Mandin, A. Renzoni, C. Chiaruttini, M. Springer and P. Cossart, *Cell* 110, 551 (2002).
[8] S. F. Park and R. G. Kroll, *Mol. Microbiol.* 8, 653 (1993).
[9] A. A. Milenbachs, D. P. Brown, M. Moors and P. Youngman, *Mol. Microbiol.* 23, 1075 (1997).
[10] J. Behari and P. Youngman, *J. Bacteriol.* 180, 6316 (1998).
[11] B. Joseph, S. Mertins, R. Stoll, J. Schär, K. R. Umesha, Q. Luo, S. Müller-Altrock and W. Goebel, *J. Bacteriol.* 190, 5412 (2008).
[12] B. Joseph and W. Goebel, *Microbes Infect.* 9, 1188 (2007).
[13] E. Eylert, J. Schär, S. Mertins, R. Stoll, A. Bacher, W. Goebel and W. Eisenreich, *Mol. Microbiol.* 69, 1008 (2008).
[14] F. M. Aké, P. Joyet, J. Deutscher and E. Milohanic, *Mol. Microbiol.* 81, 274 (2011).
[15] A. Toledo-Arana, O. Dussurget, G. Nikitas, N. Sesto, H. Guet-Revillet, D. Balestrino, E. Loh, J. Gripenland, T. Tiensuu, K. Vaitkevicius, M. Barthelemy, M. Vergassola, M. A. Nahori, G. Soubigou, B. Régnault, J. Y. Coppée, L. M., J. Johansson and P. Cossart, *Nature* 459, 950 (2009).
[16] D. P. Christensen, A. K. Benson and R. W. Hutkins, *Appl. Environ. Microbiol.* 64, 3147 (1998).
[17] W. J. Mitchell, J. Reizer, C. Herring, C. Hoischen and M. H. Saier Jr, *J. Bacteriol.* 175, 2758 (1993).
[18] D. P. Christensen, A. K. Benson and R. W. Hutkins, *Appl. Environ. Microbiol.* 65, 2112 (1999).
[19] R. Stoll and W. Goebel, *Microbiology* 156, 1069 (2010).
[20] P. W. Postma, J. W. Lengeler and G. R. Jacobson, *Microbiol. Rev.* 57, 543 (1993).
[21] J. Deutscher, C. Franke and P. W. Postma, *Microbiol. Mol. Biol. Rev.* 70, 939 (2006).
[22] M. H. Saier Jr, *Mol. Microbiol.* 35, 699 (2000).
[23] R. D. Barabote and M. H. Saier, Jr., *Microbiol. Mol. Biol. Rev.* 69, 608 (2005).
[24] M. J. Yebra, V. Monedero, M. Zúñiga, J. Deutscher and G. Pérez-Martínez, *Microbiology* 152, 95 (2006).
[25] R. Castro, A. R. Neves, L. L. Fonseca, W. A. Pool, J. Kok, O. P. Kuipers and H. Santos, *Mol. Microbiol.* 71, 795 (2009).
[26] B. Stolz, M. Huber, Z. Marković-Housley and B. Erni, *J. Biol. Chem.* 268, 27094 (1993).
[27] P. Sliz, R. Engelmann, W. Hengstenberg and E. F. Pai, *Structure* 5, 775 (1997).
[28] L. Volpon, C. R. Young, A. Matte and K. Gehring, *Protein Sci.* 15, 2435 (2006).
[29] J. Márquez, S. Reinelt, B. Koch, R. Engelmann, W. Hengstenberg and K. Scheffzek, *J. Biol. Chem.* 281, 32508 (2006).
[30] A. E. Oberholzer, M. Bumann, P. Schneider, C. Bächler, C. Siebold, U. Baumann and B. Erni, *J. Mol. Biol.* 346, 521 (2005).
[31] H. R. Kalbitzer, W. Hengstenberg, P. Rösch, P. Muss, P. Bernsmann, R. Engelmann, M. Dörschug and J. Deutscher, *Biochemistry* 21, 2879 (1982).

[32] O. Herzberg, P. Reddy, S. Sutrina, M. H. Saier, Jr., J. Reizer and G. Kapadia, *Proc. Natl. Acad. Sci. USA* 89, 2499 (1992).

[33] G. T. Robillard and J. S. Lolkema, *Biochim. Biophys. Acta* 947, 493 (1988).

[34] G. Cornilescu, B. R. Lee, C. C. Cornilescu, G. Wang, A. Peterkofsky and G. M. Clore, *J. Biol. Chem.* 277, 42289 (2002).

[35] A. Peterkofsky, G. Wang, D. S. Garrett, B. R. Lee, Y. J. Seok and G. M. Clore, *J. Mol. Microbiol. Biotechnol.* 3, 347 (2001).

[36] J. Lei, L. F. Li and X. D. Su, *J. Mol. Biol.* 386, 465 (2009).

[37] Y. Cao, X. Jin, E. J. Levin, H. Huang, Y. Zong, M. Quick, J. Weng, Y. Pan, J. Love, M. Punta, B. Rost, W. A. Hendrickson, J. A. Javitch, K. R. Rajashankar and M. Zhou, *Nature* 473, 50 (2011).

[38] K. Dalet, Y. Cenatiempo, P. Cossart and Y. Héchard, *Microbiology* 147, 3263 (2001).

[39] S. Arous, K. Dalet and Y. Héchard, *FEMS Microbiol. Lett.* 238, 37 (2004).

[40] D. B. Diep, M. Skaugen, Z. Salehian, H. Holo and I. F. Nes, *Proc. Natl. Acad. Sci. USA* 104, 2384 (2007).

[41] M. Ramnath, S. Arous, A. Gravesen, J. W. Hastings and Y. Héchard, *Microbiology* 150, 2663 (2004).

[42] I. Martin-Verstraete, V. Charrier, J. Stülke, A. Galinier, B. Erni, G. Rapoport and J. Deutscher, *Mol. Microbiol.* 28, 293 (1998).

[43] M. Débarbouillé, I. Martin-Verstraete, A. Klier and G. Rapoport, *Proc. Natl. Acad. Sci. USA* 88, 2212 (1991).

[44] D. S. Weiss, J. Batut, K. E. Klose, J. Keener and S. Kustu, *Cell* 67, 155 (1991).

[45] S. Austin and R. Dixon, *EMBO J.* 11, 2219 (1992).

[46] K. Dalet, S. Arous, Y. Cenatiempo and Y. Héchard, *Biochimie* 85, 633 (2003).

[47] A. Bidossi, L. Mulas, F. Decorosi, L. Colomba, S. Ricci, G. Pozzi, J. Deutscher, C. Viti and M. R. Oggioni, *PLoS One* 7, e33320 (2012).

[48] M. Popowska, M. Osińska and M. Rzeczkowska, *Arch. Microbiol.* 194, 255 (2012).

[49] K. Brehm, M. T. Ripio, J. Kreft and J. A. Vazquez-Boland, *J. Bacteriol.* 181, 5024 (1999).

[50] S. Mertins, B. Joseph, M. Goetz, R. Ecke, G. Seidel, M. Sprehe, W. Hillen, W. Goebel and S. Müller-Altrock, *J. Bacteriol.* 189, 473 (2007).

[51] T. C. Ells and L. Truelstrup Hansen, *Appl. Environ. Microbiol.* 77, 6841 (2011).

[52] J. Rodríguez-Díaz, A. Rubio-del-Campo and M. J. Yebra, *Appl. Environ. Microbiol.* 78, 4613 (2012).

[53] Y. Liu, M. Ceruso, Y. Jiang, A. R. Datta, L. Carter, E. Strain, T. Pepe, A. Anastasi and P. Fratamico, *J. Food Sci.* 78, M1392 (2013).

[54] H. Bouraoui, M. Ventroux, M. F. Noirot-Gros, J. Deutscher and P. Joyet, *Mol. Microbiol.* 87, 789 (2013).

[55] H. Saklani-Jusforgues, E. Fontan and P. L. Goossens, *Res. Microbiol.* 152, 175 (2001).

[56] M. Povelainen, E. V. Eneyskaya, A. A. Kulminskaya, D. R. Ivanen, N. Kalkkinen, K. N. Neustroev and A. N. Miasnikov, *Biochem. J.* 371, 191 (2003).

[57] K. Schauer, G. Geginat, C. Liang, W. Goebel, T. Dandekar and T. M. Fuchs, *BMC Genomics* 11, 573. doi: 10.1186/1471 (2010).

[58] K. A. Miller, R. S. Phillips, J. Mrázek and T. R. Hoover, *J. Bacteriol.* 195, 4057 (2013).

[59] B. Joseph, K. Przybilla, C. Stühler, K. Schauer, J. Slaghuis, T. M. Fuchs and W. Goebel, *J. Bacteriol.* 188, 556 (2006).

[60] C. Alcántara, L. A. Sarmiento-Rubiano, V. Monedero, J. Deutscher, G. Pérez-Martínez and M. J. Yebra, *Appl. Environ. Microbiol.* 74, 5731 (2008).
[61] C. Maurel, J. Reizer, J. I. Schroeder, M. J. Chrispeels and M. H. Saier Jr., *J. Biol. Chem.* 269, 11869 (1994).
[62] J. Deutscher, B. Bauer and H. Sauerwald, *J. Bacteriol.* 175, 3730 (1993).
[63] E. Darbon, P. Servant, S. Poncet and J. Deutscher, *Mol. Microbiol.* 43, 1039 (2002).
[64] C. Monniot, A. C. Zébré, F. M. Aké, J. Deutscher and E. Milohanic, *J. Bacteriol.* 194, 4972 (2012).
[65] C. Bächler, K. Flükiger-Brühwiler, P. Schneider, P. Bähler and B. Erni, *J. Biol. Chem.* 280, 18321 (2005).
[66] P. Piveteau, G. Depret, B. Pivato, D. Garmyn and A. Hartmann, *PLoS One* 6, e24881 (2011).
[67] P. A. Kirkland, M. A. Gil, I. M. Karadzic and J. A. Maupin-Furlow, *J. Bacteriol.* 190, 193 (2008).
[68] G. D. Havemann and T. A. Bobik, *J. Bacteriol.* 185, 5086 (2003).
[69] S. Srikumar and T. M. Fuchs, *Appl. Environ. Microbiol.* 77, 281 (2011).
[70] C. Buchrieser, C. Rusniok, F. Kunst, P. Cossart, P. Glaser and a. t. L. Consortium, *FEMS Immunol. Med. Microbiol.* 35, 207 (2003).
[71] J. Xue, C. M. Murrieta, D. C. Rule and K. W. Miller, *Appl. Environ. Microbiol.* 74, 7073 (2008).
[72] L. Fieseler, S. Schmitter, J. Teiserskas and M. J. Loessner, *PLoS One* 7, e43444. doi: 10.1371/journal.pone.0043444. (2012).
[73] W. Z. Salazar JK, McMullen PD, Luo Q, Freitag NE, Tortorello ML, Hu S, Zhang W., *Appl. Environ. Microbiol.* 79, 5584 (2013).
[74] P. Moralejo, S. M. Egan, E. Hidalgo and J. Aguilar, *J. Bacteriol.* 175, 5585 (1993).
[75] H. Fiegler, J. Bassias, I. Jankovic and R. Brückner, *J. Bacteriol.* 181, 4929 (1999).
[76] D. M. Shattuck-Eidens and R. J. Kadner, *J. Bacteriol.* 155, 1062 (1983).
[77] I. Chico-Calero, M. Suárez, B. González-Zorn, M. Scortti, J. Slaghuis, W. Goebel, J. A. Vázquez-Boland and E. L. G. Consortium, *Proc. Natl. Acad. Sci. USA* 99, 431 (2002).
[78] M. Goetz, A. Bubert, G. Wang, I. Chico-Calero, J. A. Vazquez-Boland, M. Beck, J. Slaghuis, A. A. Szalay and W. Goebel, *Proc. Natl. Acad. Sci. USA* 98, 12221 (2001).
[79] C. M. Taylor, M. Beresford, H. A. Epton, D. C. Sigee, G. Shama, P. W. Andrew and I. S. Roberts, *J. Bacteriol.* 184, 621 (2002).
[80] M. Scortti, L. Lacharme-Lora, M. Wagner, I. Chico-Calero, P. Losito and J. Vázquez-Boland, *Nat. Med.* 12, 515 (2006).
[81] S. Chen, M. L. Oldham, A. L. Davidson and J. Chen, *Nature* 499, 364 (2013).
[82] A. Mazé, G. Boël, M. Zúñiga, A. Bourand, V. Loux, M. J. Yebra, V. Monedero, K. Correia, N. Jacques, S. Beaufils, S. Poncet, P. Joyet, E. Milohanic, S. Casarégola, Y. Auffray, G. Pérez-Martínez, J. F. Gibrat, M. Zagorec, C. Francke, A. Hartke and J. Deutscher, *J. Bacteriol.* 192, 2647 (2010).
[83] V. Monedero, M. J. Yebra, S. Poncet and J. Deutscher, *Res. Microbiol.* 159, 94 (2008).
[84] S. Gopal, D. Berg, N. Hagen, E. M. Schriefer, R. Stoll, W. Goebel and J. Kreft, *PLoS One* 5, e10349 (2010).
[85] J. Thompson, A. Pikis, S. B. Ruvinov, B. Henrissat, H. Yamamoto and J. Sekiguchi, *J. Biol. Chem.* 273, 27347 (1998).

[86] A. Mokhtari, V. S. Blancato, G. D. Repizo, C. Henry, A. Pikis, A. Bourand, M. F. Álvarez, S. Immel, A. Mechakra-Maza, A. Hartke, J. Thompson, C. Magni and J. Deutscher, *Mol. Microbiol.* 88, 234 (2013).

[87] R. Viana, V. Monedero, V. Dossonnet, C. Vadeboncoeur, G. Pérez-Martínez and J. Deutscher, *Mol. Microbiol.* 36, 570 (2000).

[88] J. Deutscher, J. Reizer, C. Fischer, A. Galinier, M. H. Saier, Jr. and M. Steinmetz, *J. Bacteriol.* 176, 3336 (1994).

[89] J. Deutscher and R. Engelmann, *FEMS Microbiol. Lett.* 23, 157 (1984).

[90] M. Kravanja, R. Engelmann, V. Dossonnet, M. Blüggel, H. E. Meyer, R. Frank, A. Galinier, J. Deutscher, N. Schnell and W. Hengstenberg, *Mol. Microbiol.* 31, 59 (1999).

[91] I. Mijakovic, S. Poncet, A. Galinier, V. Monedero, S. Fieulaine, J. Janin, S. Nessler, J. A. Marquez, K. Scheffzek, S. Hasenbein, W. Hengstenberg and J. Deutscher, *Proc. Natl. Acad. Sci. USA* 99, 13442 (2002).

[92] J. Deutscher, B. Pevec, K. Beyreuther, H.-H. Kiltz and W. Hengstenberg, *Biochemistry* 25, 6543 (1986).

[93] J. Deutscher, E. Küster, U. Bergstedt, V. Charrier and W. Hillen, *Mol. Microbiol.* 15, 1049 (1995).

[94] Y. Fujita, Y. Miwa, A. Galinier and J. Deutscher, *Mol. Microbiol.* 17, 953 (1995).

[95] B. E. Jones, V. Dossonnet, E. Küster, W. Hillen, J. Deutscher and R. E. Klevit, *J. Biol. Chem.* 272, 26530 (1997).

[96] M. A. Schumacher, G. S. Allen, M. Diel, G. Seidel, W. Hillen and R. G. Brennan, *Cell* 118, 731 (2004).

[97] S. E. Gilbreth, A. K. Benson and R. W. Hutkins, *Curr. Microbiol.* 49, 95 (2004).

[98] J. Deutscher, U. Kessler, C. A. Alpert and W. Hengstenberg, *Biochemistry* 23, 4455 (1984).

[99] S. Krüger, S. Gertz and M. Hecker, *J. Bacteriol.* 178, 2637 (1996).

[100] C. Lindner, M. Hecker, D. Le Coq and J. Deutscher, *J. Bacteriol.* 184, 4819 (2002).

[101] J. Xue and K. W. Miller, *Appl. Environ. Microbiol.* 73, 5648 (2007).

[102] V. Monedero, O. P. Kuipers, E. Jamet and J. Deutscher, *J. Bacteriol.* 183, 3391 (2001).

[103] E. Wagner, S. Marcandier, O. Egeter, J. Deutscher, F. Götz and R. Brückner, *J. Bacteriol.* 177, 6144 (1995).

[104] J. Gubbens, M. Janus, B. I. Florea, H. S. Overkleeft and G. P. van Wezel, *Mol. Microbiol.* 86, 1490 (2012).

[105] N. E. Freitag, P. Youngman and D. A. Portnoy, *J. Bacteriol.* 174, 1293 (1992).

[106] T. Chakraborty, M. Leimeister-Wächter, E. Domann, M. Hartl, W. Goebel, T. Nichterlein and S. Notermans, *J. Bacteriol.* 174, 568 (1992).

[107] Y. Vega, C. Dickneite, M. T. Ripio, R. Böckmann, B. González-Zorn, S. Novella, G. Domínguez-Bernal, W. Goebel and J. A. Vázquez-Boland, *J. Bacteriol.* 180, 6655 (1998).

[108] Y. Vega, M. Rauch, M. J. Banfield, S. Ermolaeva, M. Scortti, W. Goebel and J. A. Vázquez-Boland, *Mol. Microbiol.* 52, 1553 (2004).

[109] P. Velge, M. Herler, J. Johansson, S. M. Roche, S. Témoin, A. A. Fedorov, P. Gracieux, S. C. Almo, W. Goebel and P. Cossart, *Microbiology* 153, 995 (2007).

[110] C. Dickneite, R. Böckmann, A. Spory, W. Goebel and Z. Sokolovic, *Mol. Microbiol.* 27, 915 (1998).

[111] R. Herro, S. Poncet, P. Cossart, C. Buchrieser, E. Gouin, P. Glaser and J. Deutscher, *J. Mol. Microbiol. Biotechnol.* 9, 224 (2005).
[112] A. Milenbachs Lukowiak, K. J. Mueller, N. E. Freitag and P. Youngman, *Microbiology* 150, 321 (2004).
[113] B. M. Hogema, J. C. Arents, R. Bader, K. Eijkemans, H. Yoshida, H. Takahashi, H. Aiba and P. W. Postma, *Mol. Microbiol.* 30, 487 (1998).
[114] V. Monedero, S. Poncet, I. Mijakovic, S. Fieulaine, V. Dossonnet, I. Martin-Verstraete, S. Nessler and J. Deutscher, *EMBO J.* 20, 3928 (2001).
[115] A. K. Marr, B. Joseph, S. Mertins, R. Ecke, S. Müller-Altrock and W. Goebel, *J. Bacteriol.* 188, 3887 (2006).
[116] J. Thompson and D. A. Torchia, *J. Bacteriol.* 158, 791 (1984).
[117] J. Nishitani and G. Wilcox, *Gene* 105, 37 (1991).

In: *Listeria monocytogenes*
Editor: Edmund C. Hambrick
ISBN: 978-1-63117-054-6

Chapter 4

OZONE AND ATMOSPHERIC COLD PLASMA FOR CONTROL OF *LISTERIA MONOCYTOGENES*

Sonal Patil and Paula Bourke*
School of Food Science and Environmental Health,
Dublin Institute of Technology, Dublin, Ireland

ABSTRACT

Listeria monocytogenes is the causative agent of listeriosis which can survive under a wide range of adverse conditions. This gram positive organism is of major concern to the food industry due to the regulatory environment and concerns for consumer food safety. Due to its ubiquitous presence in the environment and psychrotolerant nature, its presence is anticipated even under refrigerated conditions. Adaptation to environmental stresses such as heat, acid, salt, cold provides protection to *Listeria* cells against wide range of other stresses. Biofilm formation by this organism presents a major risk specifically in the food industry, further introducing the possibility of post-process contamination. Thus, this pathogen is of great concern to manufacturers due to its resistance to a number of food preservation practices. In this chapter we describe the survival strategies employed by *L. monocytogenes* and how the novel non-thermal approaches of ozone and atmospheric cold plasma (ACP) may prove useful for control and inactivation of this pathogen. Studies presenting the effects of system and process control parameters associated with *Listeria* inactivation efficacy and the resulting control measures that may be considered for effective processing applications are highlighted. Further exploration of the potential to optimise both ozone and ACP for control of *Listeria* spp is given in the context of the wider research in the area.

* Corresponding Author: Paula Bourke, School of Food Science and Environmental Health, Dublin Institute of Technology, Cathal Brugha Street, Dublin 1, Ireland. E-mail: paula.bourke@dit.ie.

Introduction

Listeria monocytogenes is a gram positive, psychrotrophic organism, that can be found in a wide variety of raw and processed foods. This bacteria is widely distributed in nature and has the ability to survive and grow at temperatures in the range of 1 and 45°C, pH levels 4-9 and low salt concentration up to 10% (Yousef and Carlstrom, 2003).

Foodborne diseases are major public health problem that cause a number of illnesses worldwide, with an estimated 48 million illnesses annually in the US alone (Mead et al., 1999; Scallan et al., 2011). Food products including milk, cheese, meat, fish and vegetables have been linked with listeriosis. Listeriosis caused by *L. monocytognes* is a foodborne disease, reported to affect most likely pregnant women, newborns and immunocompromised people (Schuchat et al., 1991). Between 2009-2011 1,651 cases of listeriosis were reported in the US involving persons in higher risk groups, mainly adults aged ≥65 years and pregnant women (CDC, 2013).

Listeria is a major cause of product recalls worldwide (Gray et al., 2004) and has also been reported as one of the leading causes of death in the US from food borne illness with the highest case fatality rate (Behravesh et al., 2011). In the food industry, a variety of surfaces such as stainless steel, glass, polypropylene and rubber are used and bacteria such as *L. monocytogenes* have the ability to adhere and grow on such surfaces (Lunden et al., 2000). Biofilm formation by bacteria makes them resistant to different antimicrobial agents and processing conditions. *L.monocytogenes* has the ability to survive at low temperature, under acid stress, osmotic stress and adapt to different environmental stresses. These factors allow this bacteria to overcome or counteract the efficacy of food preservation hurdles.

There is a consumer led demand for high quality and safe food products, but with natural flavours and taste. Traditional thermal pasteurisation for preservation ensures microbial safety but it is difficult to maintain fresh-like food quality parameters. Non thermal technologies are gaining interest and wider consideration for food preservation applications towards ensuring microbiological safety without compromising organoleptic and nutritional characteristics of a food product. In this chapter, we describe survival strategies employed by *L. monocytogenes* under different environmental stresses and the potential application of non thermal technologies including ozone and atmospheric cold plasma for their elimination.

Stress Survival of *L. Monocytogenes*

Low Temperature

Microbial growth is dependent on strict temperature control (Zwietering et al., 1992). *L. monocytogenes* has the ability to survive and grow in foods preserved at refrigeration temperatures such as ready-to-eat (RTE) foods, depending on product formulation, storage conditions as well as strain characteristics (Pal et al., 2008). Furthermore, the expression of key virulence genes in *L. monocytogenes* has been shown to be associated with temperature (McGann et al., 2007). Conte et al., (1994) evaluated the role of growth temperature in the virulence of *L. monocytogenes* and concluded that this bacterium produces virulence factors essential to invade intestinal cells even at low temperature. Duodu et al., (2010) investigated influence of storage temperature (0, 4, and 22°C) on gene expression and virulence properties of *L. monocytogenes* strains grown on raw salmon and suggested that abuse temperature conditions can increase the infective potential of low virulence strains in salmon, further indicating the importance of risk assessment of sushi and other dishes of raw fish. The ability of *L. monocytogenes* to grow at low temperatures makes the control of this bacterium a very challenging real or theoretical risk in the food industry and it is thus an important food safety problem. Arguedas-Villa et al., (2010) evaluated the responses of *L. monocytogenes* strains, from human listeriosis cases and those recovered from minced meat products and their related production environments, associated with cold growth phenotypes and cold adaptation gene expression. The study showed a dependence of enhanced and poor cold-growth phenotypes with particular strain capacity to activate important cold stress gene expression responses upon transition of bacteria into low temperature environments. Pre-exposure of *L. monocytogenes* to a sub-lethal level of an environmental stress results in enhanced resistance to lethal levels of the same stress (Lou and Yousef, 1997).

Miladi et al., (2008) examined the survival of *L. monocytogenes* in artificially contaminated cold fresh-salmon under freezing conditions. The inoculated pieces of fish were stored at -20°C for ten months and it was demonstrated that frozen storage of *L. monocytogenes* in salmon did not decrease the potential survival of bacteria over a period of ten months. Miladi et al., (2013) studied responses of three strains of *L. monocytogenes* to refrigeration (4°C) and freezing (-20°C) stress and demonstrated different behaviours of *L. monocytogenes* strains regarding low temperature stress to maintain optimum membrane fluidity. *L. monocytogenes* modifies its plasma membrane structure in response to low temperature to have an optimum membrane fluidity essential for survival (Tasara and Stephen, 2006; Najjar et al., 2007).

The attachment of *L. monocytogenes* to food contact and non-contact surfaces can be affected by environmental parameters such as temperature and nutrient status (Chae and Schraft, 2000; Mai and Conner 2007; Di Bonaventura et al., 2008). Efficient attachment of *L. monocytogenes* to a surface at earlier stages will result in strong adherence over time, prior to cleaning and disinfection procedures (Lunden et al., 2000). The ability of *L. monocytogenes* to survive on the surface of fresh intact or cut strawberries throughout the shelf life of the fruit and on frozen strawberries for periods of 4 weeks has been reported (Flessa et al., 2005). Ben Slama et al., (2012) demonstrated that cold stressed (-20°C) *L. monocytogenes* strains were more adhesive to different types of abiotic material like polyethylene, glass, polyvinyl chloride and staineless steel surfaces than non stressed cells. This ability of *L. monocytogenes* to grow and survive under refrigerated conditions within a biofilm structure further illustrates the challenge for control.

Pricope-Ciolacu et al., (2013) studied the effect of milk components and storage conditions, on the in vitro virulence of *L. monocytogenes*. Compared to raw milk, incubation of *L. monocytogenes* in pasteurized milk at refrigeration temperature (4°C) showed a higher invasion, and intracellular proliferation of four different strains, using human intestinal epithelial Caco-2 cells. The in vitro virulence of *L. monocytogenes* was influenced by the milk environment and different storage conditions and both factors were suggested for consideration in the risk assessment of the contaminated food (Pricope-Ciolacu et al., 2013). Low temperature is usually present in food production and storage environments. Thus adaptation to cold stress is an important strategy of *L. monocytogenes* enabling survival and growth to reach minimum infectious levels on refrigerated foods (Tsara and Stephen, 2006).

The control of cold stress including various cold stress response mechanisms in *L. monocytogenes*, and their implications for food safety has been reviewed by Tsara and Stephen (2006). Schmid et al., (2009) further concluded that cold shock proteins (Csps) are required for efficient adaptation of *L.monocytogenes* against cold and osmotic stress. Adaptation to these stresses have a major implication for microbial control in the food processing environment as combined or sequential exposure to these stresses might induce cross-protective responses.

Acid Stress

Exposure of *L. monocytogenes* to adverse conditions can result in cross-protection against additional stresses such as its entry and colonization in the host (Hill et al., 2002; Garner et al., 2006). The ability of *L. monocytogenes* to respond to low pH conditions plays an integral role in its survival and resistance to acidic foods (Cotter et al., 2000), thus impacting on food processing and preservation protocols. The organism can become highly resistant to even extremely acidic conditions due to stress hardening (Lou and Yousef, 1997).

Mild acid or salt stress exposure permit survival under lethal stress conditions by activating physiological and genetic changes (Cataldo et al., 2006). Acid tolerance response and invasion efficiency of acid adapted bacteria was higher than non adapted cells, even when bacteria were simultaneously pre-exposed to increasing salt stress (Cataldo et al., 2006). The protection of *L. monocytogenes* from the acid barrier of the stomach due to the induction of acid tolerance response (ATR) increases its potential to cause disease by allowing its survival in the GI tract (Stopforth et al., 2005; Garner et al., 2006; Olesen et al., 2009). Mazzotta (2001) reported increased resistance of *L. monocytogenes* to heat following acid adaptation of cells in fruit juice adjusted to pH 3.9. Caggia et al., (2009) reported that orange juice and minimally processed orange juice slices supported the growth of acid adapted *L. monocytogenes* and that the low acidic conditions present before pasteurisation which can induce an acid tolerance response, can then result in increased thermal tolerance of bacteria. Recently, Melo et al., (2013) concluded that *L. monocytogenes* strains respond differently to environmental stress conditions such as acid and salt stress with different sets of proteome assembled to overcome different stresses. Skandamis et al.,(2012) studied adaptive ATR of *L. monocytogenes* strains under planktonic and immobilized growth conditions and suggested that cells grown on food surfaces or in viscous matrices may show higher resistance to lethal acid conditions than those grown in liquid substrates. This underlines the importance of using appropriate challenge conditions to ensure food safety. Ilhak et al., (2011) showed that acid adapted *L. monocytogenes* cells survive at higher salt concentration of 6.7% at 4°C for at least 15 days in white cheese and can also survive in simulated gastric fluid (pH~ 2.3) after 4 h. If acid adaptation occurs prior to contamination of high salt cheeses, the inherent food safety hurdle of high levels in some cheeses may not provide a barrier against acid adapted cells.

Lin et al., (2011) reported that acid adaptation significantly enhanced the viability of *L. monocytogenes* exposed to tested disinfectants. Of further relevance to food production conditions, Chorianopoulos et al., (2011) demonstrated that acid adaptation of *L. monocytogenes* cells enhanced subsequent attachment to stainless steel (incubated at 16°C for up to 15 days under mild pH of 4.5), and improved resistance of sessile cells to extreme acidic treatment at pH 2, concluding that prior to application of acidic decontamination strategies for eradication of *L. monocytogenes* cells attached to food processing equipment, ATR should be considered. Werbrouck et al., (2009) evaluated the influence of food related acid stress on the virulence capacity of *L. monocytogenes* and indicated that low pH and acetic acid might influence the virulence potential of *L. monocytogenes*. The mechanisms employed by *L. monocytogenes* for survival at low pH and regulatory responses such as LisRK 2- component regulatory system, SOS, changes in membrane fluidity, enzymatic systems regulating internal hydrogen ion concentration have been described by Smith et al. (2013). Phan-Thanh et al., (2000) observed that the lowest pH value which *L. monocytogenes* could resist was dependent on the strain and kind of acid used. Ryan et al., (2008) described acid stress responses in *L. monocytogenes*, detailing ATR, cross protection, listerial survival in foods, acid resistance and pathogenesis of *L. monocytogenes* and mechanism of acid resistance. Liu et al., (2005) examined acid, salt and alkali tolerance in *L. monocytogenes* virulent and avirulent strains and highlighted that these treatments that are commonly employed in food product processing may not be sufficient to eliminate *L. monocytogenes* and that strict quality measures are thus required. Recently, Pongkanpai et al., (2013) showed an enhanced acid and heat tolerance following prior acid stress of *L. monocytogenes* in broth but not in a shrimp model. Heat or acid stress on *L. monocytogenes* does induce tolerance to other stresses as well as thermal tolerance (Lin and Chou, 2004; Hassani et al., 2007; Skandamis et al.,2008).

Osmotic Stress

Exposure of bacteria such as *L. monocytogenes* to osmotic stress can occur in the food environment or in the gastrointestinal tract of the host. Osmotic stress adaptation might provide cross protection against other environmental stresses. A number of studies have reported osmotic stress influence on *L.monocytogenes* viability, virulence and invasiveness (Garner et al., 2006a; Olesen et al., 2009; Lorentzen et al., 2011; Bae et al., 2012; Walecka-Zacharska et al., 2013). Bae et al., (2012) analysed the growth and global gene expression profiles of *L. monocytogenes* strain F2365 under salt stress and found decreased transcript levels of phosphoenolpyruvate-dependent sugar phosphotransferase system (PTS) genes and reduced cell growth as a result of salt stress. Exposure of *L. monocytogenes* to osmotic stress at 37°C increased its resistance to subsequent exposure to bile salts (Begley et al. 2002). Bae et al., (2010) reported increased thermal resistance of *L. monocytogenes* due to pre-exposure to 5% NaCl. Cold shock proteins (CspA, CspB and CspD) are reported to be necessary for both cold and osmotic stress tolerance in *L. monocytogenes* (Schmid et al., 2009). Bergholz et al., (2012) assessed genome wide changes during either short term or long term adaptation of *L. monocytogenes* to salt stress at 7°C and 37°C and found significantly increased hydrogen peroxide (H_2O_2) resistance at both temperatures following short-term exposure to salt stress. Recently, Hingston et al., (2013) investigated the effect of presence of salt and fatty food soils on desiccation survival of *L. monocytogenes* on stainless steel coupons. A significantly increased survival of *L. monocytogenes* during desiccation as a result of previous osmoadaptation, exposure to lard (20%-60%) or salt (5%) was reported, with important implications for food processing. Gandhi and Chikindas (2007) reviewed the ability of *L. monocytogenes* to survive in adverse environments, survival mechanisms employed and also discussed development of technologies to prevent and control *Listeria* contamination in food processing facilities. Similarly, Bergholz et al., (2013) observed increased resistance in *L. monocytogenes* to antimicrobial nisin due to exposure to salt stress. The wild type strains showed increased resistance to nisin due to prior exposure to salt stress whereas mutant strains (isogenic deletion mutations in liaR i.e., ΔliaR) were significantly more sensitive to nisin. It was suggested that *L. monocytogenes* resistance to antimicrobials such as nisin can be influenced by environmental stresses similar to those encountered on foods. Bergholz et al., (2013) also indicated that pre-exposure of *L. monocytogenes* to salt stress at 7°C, increased the expression of genes involved in nisin resistance including the response regulator liaR. Zhang et al., (2011) proposed a crucial role of SigB in *Listeria* survival under bile salt environment in the gastrointestinal tract.

Biofilm

Biofilm is defined as a microbial community attached to a substratum or to each other, which are engulfed or embedded in a self produced extracellular polymeric matrix (O'Toole et al., 2000; Donlan and Costerton, 2002). *L. monocytogenes* has the ability to adhere to different surfaces and form biofilms, thus presenting a great risk of contamination in food processing industry. The literature reports the ability of *L. monocytogenes* to attach and form biofilms on surfaces such as polystyrene, stainless steel, polymers, plastic, teflon, rubber (Blackman and Frank, 1996; Borucki et al., 2003; Møretrø and Langsrud, 2004). Cells in a biofilm form are resistant to number of stresses such as desiccation, refrigeration, acidity, disinfectants, antimicrobials and sanitizers as compared to planktonic cells (Kumar and Anand, 1998; Borucki et al., 2003; Møretrø and Langsrud, 2004; Folsom and Frank, 2007; Brooks and Flint, 2008; Giaouris et al., 2013) thus, biofilm formation comprises a significant food safety hazard. Previous studies reported that a variety of FDA approved disinfectants were not very effective for control of *L. monocytogenes* biofilms in the presence of organic matter and low temperature (Holah et al., 2002; Romanova et al., 2002; Heir et al., 2004; Pan et al., 2006). However, recently, Upadhyay et al., (2013) reported the efficacy of antimicrobials to inhibit *L. monocytogenes* biofilm formation and inactivate mature biofilms even at 4°C. Møretrø and Langsrud (2004) reviewed biofilm formation by *L. monocytogenes* and its persistence in food processing environments. There are a number of studies evaluating the efficacy of antimicrobials including essential oils, sanitation methods such as quaternary ammonium compound, peroxyacetic acid, chlorine, neutral electrolyzed water and processing techniques like ultrasound against *L. monocytogenes* biofilms (Romanova et al., 2007; Sandasi et al., 2008; Leonard et al., 2010; Ölmez and Temur 2010; Vaid et al., 2010; Ibusquiza et al., 2011; Arevalos-Sánchez et al., 2012; Chang et al. 2012; Nyila et al., 2012; Arevalos-Sánchez et al., 2013; Upadhyaya et al., 2013). Recently, Kadam et al., (2013) studied the biofilm forming behaviour of 143 *L. monocytogenes* strains of various origin including dairy, meat, industrial environment, human and animal and evaluated the impact of conditions like growth medium and temperature on biofilm formation. The authors concluded temperature was the most important factor impacting biofilm formation; that medium and serotype also significantly affected biofilm level formed but that the impact of strain origin was insignificant. Bonsaglia et al., (2014) analyzed *L. monocytogenes* biofilm formation capacity on different surfaces that are often used in food industry and retail and at different temperatures (4, 20 and30°C) and concluded that the surface composition may be an important factor facilitating rapid biofilm development. The ability of *L. monocytogenes* to attach firmly on different surfaces in food processing environment with subsequent biofilm formation, provides protection against various chemical and physical stresses (Norwood and Gilmour, 2000; Pan et al., 2006). Removal or eradication of biofilms remains a major challenge in the food industry to ensure the safety of food products supplied to customers.

In this chapter, we further discuss the potential of non thermal technologies, namely; ozone and atmospheric cold plasma for removal of *L. monocytogenes.* The collective information has been based on literature reporting application of these technologies for elimination of this gram positive pathogen.

NON-THERMAL TECHNOLOGIES

Ozone

Ozone is a triatomic form of oxygen and is a potent disinfectant for water treatment, food processing and preservation purposes. The approval of ozone as an antimicrobial agent for food treatment, storage and processing by the US Food and Drug Administration (FDA) triggered interest in ozone applications. It is a powerful broad-spectrum antimicrobial agent active against bacteria, fungi, viruses, protozoa, and bacterial and fungal spores (Khadre et al., 2001). Ozone is a bluish gas with a pungent odour and strong oxidizing properties. Generally electrical (corona) discharge method is used for production of ozone at a commercial level. Besides this, there are other methods that can be or are used for generation of ozone are described in the literature (Muthukumarappan et al., 2000, 2009). The oxidizing action of ozone for food application can be affected by some environmental factors, such as pH of medium, temperature, relative humidity and organic compounds in the medium (Kim et al., 1999). The influence of these factors on ozone efficiency is also discussed in the literature (Khadre et al., 2001, Patil and Bourke 2012). Ozone as a potent sanitizer, is environmentally safe and promising alternative to chlorine and hydrogen peroxide leaving no residues on food surfaces. Application of ozone does not require heat and may reduce water consumption, thus saving energy (Khadre et al., 2001). Ozone application at the appropriate levels does not interfere with sensory and nutritional quality of fresh and minimally processed produce (Ölmez 2012). Ozone as an antimicrobial agent has been tested for elimination of *L. monocytogenes* present on different type of food products. Muthukumar and Muthuchamy (2013) assessed the efficacy of gaseous ozone for inactivation of *L. monocytogenes* (10^8CFU/ml) inoculated on fresh chicken samples and recorded that treatment at specific dose of 33 mg/min for 9 min could be effective for inactivating 2 x 10^6 CFU/g of *L. monocytogenes* on chicken samples. Yuk et al., (2007) tested the efficacy of ozonated water alone at different concentrations (1, 3 and 5 ppm) for control of *L. monocytogenes* on enoki mushroom over a range of exposure times (0.5, 1, 3 or 5min), and recorded only 0.5 log count reductions. However, the combined effect of 3 ppm ozone with 1 % organic acids enhanced inactivation efficacy achieving 1.32 log count reductions of *L. monocytogenes* on enoki mushroom. Recently, inactivation of challenge popularions of *L. monocytogenes* in broccoli florets using combined non thermal treatments was evaluated (Severino et al., 2013). Different essential oils were incorporated in native and modified chitosan coating formulations to test the antilisterial effect. In addition, the combined effect of coating formulation with ozonated water, UV-C and γ-ray treatments during storage of inoculated broccoli florets at 4°C for 13 days was evaluated. At day 1 and 3, high antilisterial effects as a result of combination of coating with ozonated water was recorded, whereas the best effect was obtained in combination with γ-ray treatments (Severino et al., 2013). Nicholas et al., (2013) investigated the effect of gaseous ozone on *L. monocytogenes* surface attached on common food contact surfaces such as stainless steel, polypropylene and polished granite and exposed to ozone concentrations of 2, 5, 10 ppm (stainless steel surfaces only) and 45 ppm (all surfaces) for 1h. Ozonation at 10 ppm gave < 1 log reductions whereas at 45 ppm, 3.41 log reduction on stainless steel surface was reported. In addition, *L. monocytogenes* biofilms with initial cell density of 2 X 10^8 CFU/cm^2 were exposed to ozone at a concentration of 45 ppm for 1h. Compared to stainless steel surface attached bacteria, biofilm organisms provided a greater challenge to the treatment. Changes in cell surface structure before and after ozone exposure were analysed using scanning electron microscopy (SEM) and atomic force microscopy (AFM) and different membrane and cell surface modifications were demonstrated. Other than ozone, open air factor (OAF) effects on surface attached cells and biofilm of *L. monocytogenes* were also investigated. OAF was observed to be significantly better than ozone at reducing the number of biofilm microorganisms. Kang et al., (2013) suggested the combination of ozonated water and metal ions may be useful as an antimicrobial agent, reporting 2.53 and 4.71 log reductions of *L. monocytogenes* in a metal solution containing 0.2 ppm amd 0.4 ppm ozone for 30 min. Choi et al., (2012) concluded the efficacy of gaseous ozone for inactivation of food borne pathogens including *L. monocytogenes* in apple juice was dependent on the solids content of the juice sample. Patil et al., (2010) assessed the ozone inactivation of *L. monocytogenes* and *L. innocua* in orange juice. Gaseous ozone treatment of acid stressed cells at 0.098 mg/min/ml for different time periods (0-8) was examined. The bacterial population reduction of 5 log cycles within a time range varied between 5 and 9 min was reported depending on strain type and acid stress conditions. Based on these findings, the potential of ozone for control of *Listeria* populations in fruit juices was highlighted, but that adaptive stress responses should still be considered for optimal process design. Fan et al., (2007) reported the efficacy of gaseous ozone at a concentration of 50 and 100 nl/l during short exposure times at 5 and 20°C to control *L. innocua* populations on solid media. Robbins et al., (2005) evaluated the efficacy of ozone, chlorine and hydrogen peroxide against two test strains (Scott A and 10403S) of *L. monocytogenes* planktonic cells and biofilms. Complete destruction of planktonic cells of Scott A strain (8.29 log reduction) and strain 10403S (8.16 log reduction) was recorded after exposure to 0.25 ppm and 1ppm of ozone, respectively. However, to destroy biofilm cells of Scott A strain, a 16 fold increase in sanitizer concentration was required. Attached cells of strain 10403S were eliminated at ozone concentration of 4 ppm. Wade et al., (2003) studied the efficacy of aqueous ozone treatment against *Listeria monocytogenes* on alfalfa seeds and sprouts, with significant reductions observed following continuous sparging of sprouts with ozonated water for 5 or 10 min, by comparison with soaking in water. However, during storage at 4°C up to 11 days, a significant deterioration in the sensory quality was also noted.

The literature on the efficiency of ozone for inactivation of *Listeria* in food products varies (Rodgers et al., 2004; Vaz-Velho et al., 2006; Ölmez and Akbas 2009).

Atmospheric Cold Plasma

Generally, plasma is referred as a fourth state of matter, sharing the properties of both gases and liquids. Plasma can be classified as thermal or non thermal plasma depending on the basis of conditions in which they are generated. Plasma behaviour is decided by the plasma chemistry, which is a very complex process comprising of chemical species, charged particles, radicals, atoms, positive ions, negative ions and electrons. Plasma can be generated in air but also in a range of other gasses and importantly for the food industry, in different gas mixtures. The complexity of the plasma species generated may be a function of the working gas type used. More than 50 chemical reactions with at least 10 species were reported to be involved in the mechanism of ozone generation in oxygen plasmas (Lee et al., 2004). On the other hand, more than 75 species and almost 500 species are believed to be involved in air chemistry (Gordillo-Vázquez 2008). Plasma discharge results in generation of a number of known anti-microbial agents including reactive oxygen species (ROS), reactive nitrogen species (RNS), ultraviolet (UV) radiation, energetic ions, and charged particles. The generation and type of reactive species depend on the type of working gas utilised for plasma discharge which can have a significant effect on microbial inactivation efficacy (Lerouge et al., 2000; Purevdorj et al., 2003). Gas mixtures commonly employed for plasma generation include helium, argon, sulphur hexafluoride, nitrogen/CO_2 mixture and air. Furthermore, the power input and mode of plasma exposure are major factors deciding the generation and extent of species formation. There are numerous ways to obtain non thermal plasma, however, plasma generation at atmospheric pressure is advantageous, making the decontamination process less expensive and easier (Kim et al., 2011). As plasma generation at atmospheric pressure does not have specific requirements, it is of great interest to the food industries (Misra et al., 2011). Atmospheric cold plasma (ACP) is typically obtained by a number of ways such as corona discharge, micro hollow cathode discharge, atmospheric pressure plasma jet, gliding arc discharge, one atmospheric uniform glow discharge, dielectric barrier discharge, and plasma needle, all having important technological applications (Nehra et al., 2008). ACP has a non-uniform distribution of energy, i.e., nonequilibrium, among the constituent species (Niemira 2012). The most commonly used forms of ACP are dielectric barrier discharge (DBD) and plasma jets.

ACP has been proved effective for microbial inactivation (Deng et al., 2007; Joshi et al., 2011; Ziuzina et al., 2013). A wide range of applications for DBD-ACP have been described for inactivation of bacteria, spores, viruses, bacteriophage and fungi (Laroussi 2002; Venezia et al., 2008; Hati et al., 2012). Recently, the potential of atmospheric pressure plasma (APP) jets to enhance the safety and extend the shelf life of cooked egg white and yolk has been reported (Lee et al., 2012). Effect of APP on inactivation of *L.* monocytogenes as well as the quality and genotoxicity of cooked egg white and yolk was evaluated and the treatment was conducted using different gases. APP treatment in He gas resulted in 5 decimal reduction in the number of *L. monocytogenes* in cooked egg white and application of APP in He + O_2, N_2, N_2 + O_2 further decreased the number, to undetectable levels. All APP treatments were reported to be effective for cooked egg yolk achieving undetectable levels of inoculated *L.* monocytogenes. The effectiveness of APP for inactivation of *L. monocytogenes* inoculated on sliced ham and cheese has also been reported by Song et al., (2009). The influence of different process parameters including input power and plasma exposure time were evaluated, where increased microbial reduction was observed with increasing input power and exposure time. However, significantly different inactivation effects were noted based on the type of food, indicating the APP inactivation effects on *L. monocytogenes* are strongly dependent on type of food. Furthermore, authors suggested more research is still needed to assess nutritional and chemical changes in food products treated with APP. Yun et al., (2010) investigated the effect of APP for inactivation of *L. monocytogenes* inoculated onto disposable food containers including disposable plastic trays, aluminium foil and paper cups. Effect of APP system and process parameters such as input power (75, 100, 125, and 150 W) and exposure time (60, 90, and 120 s) on *L. monocytogenes* inactivation was determined, at the gap distance of 6 mm, between the powered electrode and treatment surface. Again, the reduction was reported to be associated with increased input power and exposure time with efficacy also highly dependent on surface characteristics of storage material. Microbial inactivation by plasma treatment can be dependent on different factors including plasma type, microorganism, culture age, exposure power, temperature, pH, exposure time, media composition and material type. Critzer et al., (2007) evaluated the potential of atmospheric plasma for inactivation of foodborne pathogens including *E. coli* O157:H7, Salmonella and *L. monocytogenes* on apples, cantaloupes and lettuce and reported >3 and >5 log reductions of *L. monocytogenes* populations after 3 and 5 min of plasma exposure. Authors suggested the capability of atmospheric plasma as a nonthermal processing technology to be used for reducing microbial populations on produce surfaces. Rød et al., (2012) indicated the potential of cold atmospheric plasma for surface decontamination of pre-packed RTE food products in regard to *L. monocytogenes,* but further indicated the possibility of oxidation issues in some products due to treatment. The application of atmospheric pressure plasma jet for inactivation of *L. innocua* and *E. coli* inoculated on polysaccharide surfaces was studied by Fröhling et al., (2012). APP treatment for 4 min with 20 W operating power achieved > 6 log inactivation with similar inactivation achieved in only 1.5 min with the higher 40 W operating power, however, for food surface decontamination use of 20 W was suggested. Kim et al., (2011) investigated APP treatment for inactivation of challenge pathogens; *L. monocytogenes*, *E. coli* and Salmonella Typhimurium on bacon. The effect of input power (75, 100, 125 W), treatment time (60, 90 s) and different gases (helium, mixture of helium and oxygen) on microbial reduction was evaluated. Effective microbial reduction was observed after APP treatment in helium/oxygen mixture at 125 W for 90 s achieving 5.79 log CFU/g reduction of *L. monocytogenes*. APP was suggested as a promising non-thermal cold pasteurization method to improve the safety of processed meat products, but that there was still a need to further investigate the quality changes of treated foods. Lee et al., (2011) investigated different gas mixtures for plasma generation and reported $N_2 + O_2$ gas mixture was the best for *L. monocytogenes* inactivation. Noriega et al., (2011) found that higher voltage, excitation frequency and the presence of oxygen in the gas composition for generation of plasma, resulted in higher inactivation rates of *L. innocua* on chicken muscle and skin. Furthermore, these authors indicated the surface topography had a significant role for inactivation, where 10 s of plasma treatment gave > 3 log reductions of *L. innocua* on membrane filters, but longer treatment time was required to achieve similar levels of reduction on chicken skin and muscle.

The application of APP for decontamination of processing equipment in the meat industry was also reported. APP treatment of 340 s achieved 5 log reduction of *L. innocua* inoculated on the surface of a rotating knife (Leipold et al., 2010). Further interactions with processing technology have been investigated where APP inactivation efficacy for bacteria including *Listeria* spp. inside a sealed container has been reported (Chiper et al., 2011; Leipold et al., 2011; Rød et al., 2012). A 6 log reduction of *L. innocua* inside a low density polyethylene bag following 15 min of APP exposure was achieved (Leipold et al., 2011). Recently, we investigated in-package inactivation of *L. monocytogenes* in a large volume of buffered liquid (unpublished data). The influence of different key system and process parameters including treatment time, post-treatment storage time, gas mixtures and mode of exposure (direct or indirect) on bacterial inactivation was evaluated. Direct atmospheric cold plasma (ACP) treatment at $56kV_{RMS}$ for 1 min achieved 8.3 log reduction of *L. monocytogenes* cells following 24 h of post-treatment storage time. Further, to determine cell surface changes following ACP, SEM was performed. SEM showed noticeable alteration in the cell surface with roughness and wrinkles appearing on the surface which was more signifycant with longer storage time. The ACP yielded useful reductions in buffered liquid volumes of 10ml, which is significant for industrial processing applications. The increasing voltage level recorded a strong effect on inactivation. Greater reduction was observed for *L. monocytogenes* using higher voltage of $70kV_{RMS}$. Similarly, a working gas mixture with higher oxygen content was noted to be very effective. ACP treatment for 30 s in gas mixture with high oxygen content, rendered *L. monocytogenes* undetectable with up to 8 log reductions achieved, irrespective of the mode of exposure. This illustrates the great potential of ACP for bacterial inactivation on different surfaces, including in-package decontamination of food products, with minimal post-process contamination risk.

CONCLUSION

L. monocytogenes is known to cause listeriosis and survive under a number of environmental stresses. Its ubiquitous presence in nature, wide variety of foods in addition to the strong survival strategies facilitates overriding the majority of subsequent hurdles that can be presented. Adaptation to low temperature, low pH, salt concentrations, starvation provides them cross protection or increased tolerance against stress such as heat. This organism of public health concern, poses a major risk as a food safety challenge. Additionally, ability of this pathogen to attach onto different variety of food contact and non-contact surfaces influenced by different environmental parameters, results in biofilm formation in a food processing facility. Biofilm formation increases the risk of food contamination, thus, it is of great concern to the food industry. Resistance of biofilms towards antimicrobials and disinfectants highlight the immense challenge of biofilms eradication. Ozone and ACP are technologies that have been reported as effective for reduction or elimination of this resistant pathogen. Optimisation of these technologies in terms of process and system associated parameters, would significantly improve the inactivation efficiency. Application of these technologies for eliminating Listeria spp. is advantageous, making the product safe for consumption and ensuring the food safety. However, tandem investigations to determine any effects on nutritional, quality and organoleptic characteristics of treated food products are always necessary for food applications. Moreover, the ACP mechanism of action is a complex process, that is not fully understood. Hence, further elucidating the mechanism of action of ACP in the context of the controllable system and process parameters, should allow process design targeted to the risks identified e.g., cell types, cell components, thus further enhancing both applicability and efficacy for strategically adaptive microorganisms such as Listeria spp.

REFERENCES

Arevalos-Sanchez, M., Regalado, C., Martin, S.E., Dominguez-Dominguez, J. and Garcia-Almendarez, B.E. (2012) Effect of neutral electrolyzed water and nisin on *Listeria monocytogenes* biofilms, and on listeriolysin O activity. *Food Control* 24, 116-122.

Arevalos-Sánchez, M., Regalado, C., Martin, S.E., Meas-Vong, Y., Cadena-Moreno, E. and García-Almendárez, B.E. (2013) Effect of neutral electrolyzed water on lux-tagged *Listeria monocytogenes* EGDe biofilms adhered to stainless steel and visualization with destructive and non-destructive microscopy techniques. *Food Control* 34, 472-477.

Arguedas-Villa, C., Stephan, R. and Tasara, T. (2010) Evaluation of cold growth and related gene transcription responses associated with *Listeria monocytogenes* strains of different origins. *Food Microbiology* 27, 653-660.

Bae, D., Liu, C., Zhang, T., Jones, M., Peterson, S.N. and Wang, C.L. (2012) Global Gene Expression of *Listeria monocytogenes* to Salt Stress. *Journal of Food Protection* 75, 906-912.

Bae, Y.M. and Lee, S.Y. (2010) Effect of pre-exposure to sodium chloride on the resistance of pathogens to thermal and acid stresses. *Journal of Food Safety* 30, 1016-1025.

Begley, M., Gahan, C.G. and Hill, C. (2002). Bile stress response in *Listeria monocytogenes* LO28: adaptation, cross-protection, and identification of genetic loci involved in bile resistance. *Applied Environmental Microbiology*, 68, 6005– 6012.

Behravesh, C.B., Jones, T.F., Vugia, D.J., Long, C., Marcus, R., Smith, K., Thomas, S., Zansky, S., Fullerton, K.E., Henao, O.L., Scallan, E. and FoodNet Working, G. (2011) Deaths Associated With Bacterial Pathogens Transmitted Commonly Through Food: Foodborne Diseases Active Surveillance Network (FoodNet), 1996-2005. *Journal of Infectious Diseases* 204, 263-267.

Blackman, I. C. and Frank, J. F. (1996). Growth of *Listeria monocytogenes* as a biofilm on various food-processing surfaces. *Journal of Food Protection*, 59(8), 827–831.

Ben Slama, R., Bekir, K., Miladi, H., Noumi, A. and Bakhrouf, A. (2012) Adhesive ability and biofilm metabolic activity of *Listeria monocytogenes* strains before and after cold stress. *African Journal of Biotechnology*, 11 (61), 12475-12482.

Bergholz, T.M., Bowen, B., Wiedmann, M. and Boor, K.J. (2012) *Listeria monocytogenes* shows temperature-dependent and -independent responses to salt stress, including responses that induce cross-protection against other stresses. *Applied and Environmental Microbiology* 78, 2602-2612.

Bergholz, T.M., Tang, S.L., Wiedmann, M. and Boor, K.J. (2013) Nisin resistance of *Listeria monocytogenes* is increased by exposure to salt stress and is mediated via liar. *Applied and Environmental Microbiology* 79, 5682-5688.

Bonsaglia, E.C.R., Silva, N.C.C., Fernades Júnior, A., Araújo Júnior, J.P., Tsunemi, M.H. and Rall, V.L.M. (2014) Production of biofilm by *Listeria monocytogenes* in different materials and temperatures. *Food Control* 35, 386-391.

Borucki, M.K., Peppin, J.D., White, D., Loge, F. and Call, D.R. (2003). Variation in biofilm formation among strains of *Listeria monocytogenes*. *Applied and Environmental Microbiology*, 69, 7336-7342.

Brooks, J. D. and Flint, S. H. (2008). Biofilms in the food industry: Problems and potential solutions. *International Journal of Food Science and Technology*, 43(12), 2163–2176.

Caggia, C., Ombretta Scifò, G., Restuccia, C., and Randazzo, C. L. (2009). Growth of acidadapted *Listeria monocytogenes* in orange juice and in minimally processed orange slices. *Food Control*, 20(1), 59-66.

Cataldo, G., Conte, M.P., Chiarini, F., Seganti, L., Ammendolia, M.G., Superti, F. and Longhi, C. (2006). Acid adaptation and survival of *Listeria monocytogenes* in Italian-style soft cheeses. *Journal of Applied Microbiology* 103:185–193.

CDC. (2013). Vital signs: *Listeria* illnness, deaths and outbreaks - United States, 2009-2011. *Morbidity Mortality Weekly Report* 66(22):448-452. http://www.cdc.gov/mmwr/preview /mmwrhtml/mm6222a4.htm?s_cid=mm6222a4_w

Chae, M.S. and Schraft, H. (2000). Comparative evaluation of adhesion and biofilm formation of different *Listeria monocytogenes* strains. *International Journal of Food Microbiololgy* 62, 103–111.

Chang, Y.H., Gu, W.M. and McLandsborough, L. (2012) Low concentration of ethylenediaminetetraacetic acid (EDTA) affects biofilm formation of *Listeria monocytogenes* by inhibiting its initial adherence. *Food Microbiology* 29, 10-17.

Chiper, A.S., Chen, W., Mejlholm, O., Dalgaard, P. and Stamate, E. (2011) Atmospheric pressure plasma produced inside a closed package by dielectric barrier discharge in Ar/CO2 for bacterial inactivation of biological samples. *Plasma Sources Science Technology* 20, doi:10.1088/0963-0252/20/2/025008

Choi, M.R., Liu, Q., Lee, S.Y., Jin, J.H., Ryu, S. and Kang, D.H. (2012). Inactivation of *Escherichia coli* O157:H7, *Salmonella typhimurium* and *Listeria monocytogenes* in apple juice with gaseous ozone. *Food Microbiology*, 32(1), 191-195.

Chorianopoulos, N., Giaouris, E., Grigoraki, I., Skandamis, P. and Nychas, G.J. (2011) Effect of acid tolerance response (ATR) on attachment of *Listeria monocytogenes* Scott A to stainless steel under extended exposure to acid or/and salt stress and resistance of sessile cells to subsequent strong acid challenge. *International Journal of Food Microbiology* 145, 400-406.

Conte, M.P., Longhi, C., Petrone, G., Polidoro, M., Valenti, P. and Seganti, L. (1994) *Listeria monocytogenes* infection of Caco-2 cells: role of growth temperature. *Research in Microbiology* 145, 677-682.

Cotter, P.D., Gahan, C.G.M. and Hill, C. (2000) Analysis of the role of the Listeria monocytogenes F0F1-ATPase operon in the acid tolerance response. *International Journal of Food Microbiology* 60, 137-146.

Critzer, F.J., Kelly-Wintenberg, K., South, S.L. and Golden, D.A. (2007). Atmospheric plasma inactivation of foodborne pathogens on fresh produce surfaces. *Journal of Food Protection*, 70(10), 2290-2296.

Deng, S.B., Ruan, R., Mok, C.K., Huang, G.W., Lin, X.Y. and Chen, P. (2007) Inactivation of Escherichia coli on almonds using nonthermal plasma. *Journal of Food Science* 72, M62-M66.

Di Bonaventura, G., Piccolomini, R., Paludi, D., D'Orio, V., Vergara, A., Conter, M. and Ianieri, A. (2008) Influence of temperature on biofilm formation by *Listeria monocytogenes* on various food-contact surfaces: relationship with motility and cell surface hydrophobicity. *Journal of Applied Microbiology* 104, 1552-1561.

Donlan, R. M. and Costerton, J. W. (2002). Biofilms: Survival mechanisms of clinically relevant microorganisms. *Clinical Microbiology Reviews*, 15(2), 167–193.

Duodu, S., Holst-Jensen, A., Skjerdal, T., Cappelier, J.M., Pilet, M.F. and Loncarevic, S. (2010) Influence of storage temperature on gene expression and virulence potential of *Listeria monocytogenes* strains grown in a salmon matrix. *Food Microbiology* 27, 795-801.

Fan, L., Song, J., McRae, K. B., Walker, B. A. and Sharpe, D. (2007). Gaseous ozone treatment inactivates *Listeria innocua* in vitro. *Journal of Applied Microbiology*, 103(6), 2657-2663.

Flessa, S., Lusk, D.M. and Harris, L.J. (2005). Survival of *Listeria monocytogenes* on fresh and frozen strawberries. *International Journal of Food Microbiology*, 101, 255– 262.

Folsom, J.P. and Frank, J.F. (2007). Proteomic analysis of a hypochlorous acid-tolerant *Listeria monocytogenes* cultural variant exhibiting enhanced biofilm production. *Journal of Food Protection*, 5, 1129-1136.

Fröhling, A., Baier, M., Ehlbeck, J., Knorr, D. and Schluter, O. (2012) Atmospheric pressure plasma treatment of Listeria innocua and Escherichia coli at polysaccharide surfaces: Inactivation kinetics and flow cytometric characterization. *Innovative Food Science & Emerging Technologies* 13, 142-150.

Gandhi, M. and Chikindas, M.L. (2007) Listeria: A foodborne pathogen that knows how to survive. *International Journal of Food Microbiology* 113, 1-15.

Garner, M.R., James, K.E., Callahan, M.C., Wiedmann, M. and Boor, K.J. (2006a) Exposure to salt and organic acids increases the ability of *Listeria monocytogenes* to invade Caco-2 cells but decreases its ability to survive gastric stress. *Applied and Environmental Microbiology* 72, 5384-5395.

Garner, M.R., Njaa, B.L., Wiedmann, M., Boor, K.J. (2006). Sigma B contributes to *Listeria monocytogenes* gastrointestinal infection but not to systemic spread in the guinea pig infection model. *Infection and Immunity*, 74,876–886.

Giaouris, E., Heir, E., Hébraud, M., Chorianopoulos, N., Langsrud, S., Møretrø, T., Habimana, O., Desvaux, M., Renier, S. and Nychas, G.-J. (2013). Attachment and biofilm formation by foodborne bacteria in meat processing environments: Causes, implications, role of bacterial interactions and control by alternative novel methods. *Meat Science*. http://dx.doi.org/10.1016/j.meatsci.2013.05.023

Gordillo-Vázquez, F.J. (2008). Air plasma kinetics under the influence of sprites. *Journal of Physics D: Applied Physics*, 41(23), p. 234016. doi:10.1088/0022-3727/41/23/234016

Gray, M.J., Zadoks, R.N., Fortes, E.D., Dogan, B., Cai, S., Chen, Y., Scott, V.N., Gombas, D.E., Boor, K.J. and Wiedmann, M. (2004). *Listeria monocytogenes* isolates from foods and humans form distinct but overlapping populations. *Applied Environmental Microbiology*, 70 (10), 5833–5841.

Hassani, M., Manas, P., Pagan, R. and Condon, S. (2007) Effect of a previous heat shock on the thermal resistance of *Listeria monocytogenes* and *Pseudomonas aeruginosa* at different pHs. *International Journal of Food Microbiology* 116, 228-238.

Hati, S., Mandal, S., Vij, S., Minz, P.S., Basu, S., Khetra, Y., Yadav, D. and Dahiya, M. (2012). Nonthermal plasma technology and its potential applications against foodborne microorganisms. *Journal of Food Processing and Preservation*, 36(6), 518-524.

Heir, E., Lindstedt, B.A., Rotterud, O.J., Vardund, T., Kapperud, G. and Nesbakken, T., (2004). Molecular epidemiology and disinfectant susceptibility of *Listeria monocytogenes* from meat processing plants and human infections. *International Journal of Food Microbiology*, 96, 85-96.

Hill, C., Cotter, P.D., Sleator, R.D. and Gahan, CGM. (2002) Bacterial stress response in *Listeria monocytogenes*: jumping the hurdles imposed by minimal processing. *International Dairy Journal*, 12, 273–283.

Hingston, P.A., Stea, E.C., Knochel, S. and Hansen, T. (2013) Role of initial contamination levels, biofilm maturity and presence of salt and fat on desiccation survival of *Listeria monocytogenes* on stainless steel surfaces. *Food Microbiology* 36, 46-56.

Holah, J.T., Taylor, J.H., Dawson, D.J. and Hall, K.E. (2002). Biocide use in the food industry and the disinfectant resistance of persistent strains of *Listeria monocytogenes* and *Escherichia coli*. *Journal of Applied Microbiology*, 92, 111-120.

Ibusquiza, P.S., Herrera, J.J.R. and Cabo, M.L. (2011) Resistance to benzalkonium chloride, peracetic acid and nisin during formation of mature biofilms by *Listeria monocytogenes*. *Food Microbiology* 28, 418-425.

Ilhak, O.I., Oksuztepe, G., Calicioglu, M. and Patir, B. (2011) Effect of acid adaptation and different salt concentrations on survival of *Listeria monocytogenes* in turkish white cheese. *Journal of Food Quality* 34, 379-385.

Joshi, S. G., Cooper, M., Yost, A., Paff, M., Ercan, U. K., Fridman, G., Fridman, G., Fridman, A. and Brooks, A. D. (2011). Nonthermal dielectric-barrier discharge plasma induced inactivation involves oxidative DNA damage and membrane lipid peroxidation in *Escherichia coli*. *Antimicrobial Agents and Chemotherapy*, 55, 1053 – 1062.

Kadam, S.R., den Besten, H.M.W., van der Veen, S., Zwietering, M.H., Moezelaar, R. and Abee, T. (2013) Diversity assessment of *Listeria monocytogenes* biofilm formation: Impact of growth condition, serotype and strain origin. *International Journal of Food Microbiology* 165, 259-264.

Kang, S.N., Kim, K.J., Park, J.H. and Lee, O.H. (2013). Effect of a combination of low level ozone and metal ions on reducing *Escherichia coli* O157:H7 and *Listeria monocytogenes*. *Molecules*, 18(4), 4018-4025.

Khadre, M. A., Yousef A. E. and. Kim, J. G. (2001). Microbiological aspects of ozone applications in food: a review. *JFS: Concise Reviews in Food Science*, 66(9), 1242-1252.

Kim, J.G., Yousef, A.E. and Dave, S. (1999). Application of ozone for enhancing the microbiological safety and quality of foods: A review. *Journal of Food Protection*, 62(9), 1071–1087.

Kim, B., Yun, H., Jung, S., Jung, Y., Jung, H., Choe, W. and Jo, C. (2011) Effect of atmospheric pressure plasma on inactivation of pathogens inoculated onto bacon using two different gas compositions. *Food Microbiology* 28, 9-13.

Kumar, C. G. and Anand, S. K. (1998). Significance of microbial biofilms in food industry: A review. *International Journal of Food Microbiology*, 42(1–2), 9–27.

Laroussi, M. (2002). Nonthermal decontamination of biological media by atmospheric-pressure plasmas: review, analysis and prospects. *IEEE T Plasma Science*, 30 (4), 1409-1415.

Lee, H.M., Chang, M.B. and Wei, T.C. (2004). *Kinetic modeling of ozone generation via dielectric* barrier discharges. *Ozone: Science and Engineering*, 26(6), 551-562.

Lee, H.J., Jung, H., Choe, W., Ham, J.S., Lee, J.H. and Jo, C. (2011) Inactivation of *Listeria monocytogenes* on agar and processed meat surfaces by atmospheric pressure plasma jets. *Food Microbiology* 28, 1468-1471.

Lee, H.J., Song, H.P., Jung, H., Choe, W., Ham, J.S., Lee, J.H. and Jo, C. (2012) Effect of atmospheric pressure plasma jet on inactivation of *Listeria monocytogenes*, quality, and genotoxicity of cooked egg white and yolk. *Korean Journal for Food Science of Animal Resources* 32, 561-570.

Leipold, F., Kusano, Y., Hansen, F. and Jacobsen, T. (2010) Decontamination of a rotating cutting tool during operation by means of atmospheric pressure plasmas. *Food Control* 21, 1194-1198.

Leipold, F., Schultz-Jensen, N., Kusano, Y., Bindslev, H. and Jacobsen, T. (2011) Decontamination of objects in a sealed container by means of atmospheric pressure plasmas. *Food Control* 22, 1296-1301.

Leonard, C.M., Virijevic, S., Regnier, T. and Combrinck, S. (2010) Bioactivity of selected essential oils and some components on *Listeria monocytogenes* biofilms. *South African Journal of Botany* 76, 676-680.

Lerouge, S., Wertheimer, M.R., Marchand, R., Tabrizian, M. and Yahia, L.H.(2000). Effect of gas composition on spore mortality and etching during low-pressure plasma sterilization. *Journal of Biomedical Material Research*, 51, 128–135.

Lin, M.H., Lee, S.L. and Chou, C.C. (2011) Acid adaptation affects the viability of *Listeria monocytogenes* BCRC 14846 and Salmonella Typhimurium BCRC 10747 exposed to disinfectants at 25 degrees C and 40 degrees C. *Foodborne Pathogens and Disease* 8, 1077-1081.

Lin, Y.D. and Chou, C.C. (2004) Effect of heat shock on thermal tolerance and susceptibility of *Listeria monocytogenes* to other environmental stresses. *Food Microbiology* 21, 605-610.

Liu, D., Lawrence, M.L., Ainsworth, A.J. and Austin, F.W. (2005) Comparative assessment of acid, alkali and salt tolerance in *Listeria monocytogenes* virulent and avirulent strains. *FEMS Microbiology Letters* 243, 373-378.

Lorentzen, G., Mennen, S., Olsen, R. and Skjerdal, T. (2011). Invasiveness of *Listeria monocytogenes* strains of Caco-2 cells in response to a period of extreme salt stress reflecting salt-curing and rehydration of cod (*Gadus morhua* L.). *Food Control*, 22 (7), 1040-1045.

Lou, Y.Q. and Yousef, A.E. (1997) Adaptation to sublethal environmental stresses protects *Listeria monocytogenes* against lethal preservation factors. *Applied and Environmental Microbiology* 63, 1252-1255.

Lunden, J., Miettinen, M., Autio, T., Korkeala, H., 2000. Persistent *Listeria monocytogenes* strains show enhanced adherence to food contact surface after short contact times. *Journal of Food Protection*, 63, 1204–1207.

Mai, T.L. and Conner, D.E. (2007) Effect of temperature and growth media on the attachment of *Listeria monocytogenes* to stainless steel. *International Journal of Food Microbiology* 120, 282-286.

Mazzotta, A. S. (2001). Thermal inactivation of stationary-phase and acid-adapted *Escherichia coli* O157:H7, Salmonella, and *Listeria monocytogenes* in fruit juices. *Journal of Food Protection*, 64, 315-320.

McGann, P., Wiedmann, M. and Boor, K.J. (2007). The alternative sigma factor σB and the virulence gene regulator PrfA both regulate transcription of *Listeria monocytogenes* internalins. *Applied and Environmental Microbiology*, 73, 2919-2930.

Mead, P.S., Slutsker, L., Dietz, V., McCaig, L.F., Bresee, J.S., Shapiro, C., Griffin, P.M. and Tauxe, R.V. (1999) Food-related illness and death in the United States. *Emerging Infectious Diseases* 5, 607-625.

Melo, J., Andrew, P.W. and Faleiro, M.L. (2013) Different assembly of acid and salt tolerance response in two dairy *Listeria monocytogenes* wild strains. *Archives of Microbiology* 195, 339-348.

Miladi, H., Bakhrouf, A. and Ammar, E. (2013). Cellular lipid fatty acid profiles of reference and food isolates *Listeria monocytogenes* as a response to refrigeration and freezing stress. Journal of Food Biochemistry, 37, 136-143.

Miladi, H., Chaieb, K., Bakhrouf, A.,Elmnasser, N. and Ammar, E. 2008. Freezing effects on survival of *Listeria monocytogenes* in artificially contaminated cold fresh-salmon. *Annals of Microbiology*, 58, 471–476.

Misra, N.N., Tiwari, B.K., Raghavarao, KSMS., Cullen, P.J. (2011). Nonthermal plasma inactivation of food-borne pathogens. *Food Engineering Reviews*, 3(3-4), 159-170.

Møretrø, T. and Langsrud, S. (2004). *Listeria monocytogenes*: Biofilm formation and persistence in food-processing environments. *Biofilms*, 1, 107–121.

Muthukumarappan, K., Halaweish F. and Naidu. A.S. (2000). Ozone. pp. 783-800. *In*: Naidu, A.S. (Ed.). Natural Food Anti-Microbial Systems. CRC Press, Boca Raton, FL.

Muthukumarappan, M., O'Donnell, C.P. and Cullen, P.J. (2009). Ozone treatment of food materials. pp. 263-280. *In*: Jun, S. and J.M. Irudayaraj (eds.). Food Processing Operations Modeling Design and Analysis. CRC Press Taylor and Francis Group, Boca Raton, Florida.

Muthukumar, A. and Muthuchamy, M. (2013). Optimization of ozone in gaseous phase to inactivate *Listeria monocytogenes* on raw chicken samples. *Food Research International,* 54(1), 1128-1130.

Najjar, B.M., Chikindas, M. and Montville, T.J. (2007). Changes in *Listeria monocytogenes* membrane fluidity in response to temperature stress. *Applied Environmental Microbiology*, 73, 6429–6435.

Nehra, V., Kumar, A. and Dwivedi, H.K. (2008). Atmospheric Non-Thermal Plasma Sources. *International Journal of Engineering*, 2(1), 53-68.

Nicholas, R., Dunton, P., Tatham, A. and Fielding, L. (2013). The effect of ozone and open air factor on surface-attached and biofilm environmental *Listeria monocytogenes*. *Journal of Applied Microbiology*, 115(2), 555-564.

Niemira, B.A. (2012). Cold Plasma Decontamination of Foods. *Annual Review of Food Science and Technology*, 3, 125-142.

Noriega, E., Shama, G., Laca, A., Díaz, M. and Kong, M.G. (2011) Cold atmospheric gas plasma disinfection of chicken meat and chicken skin contaminated with *Listeria innocua*. *Food Microbiology* 28, 1293-1300.

Norwood, D.E. and Gilmour, A. (2000) The growth and resistance to sodium hypochlorite of *Listeria monocytogenes* in a steady-state multispecies biofilm. *Journal of Applied Microbiology* 88, 512-520.

Nyila, M.A., Leonard, C.M., Hussein, A.A. and La, N. (2012) Activity of South African medicinal plants against *Listeria monocytogenes* biofilms, and isolation of active compounds from Acacia karroo. *South African Journal of Botany* 78, 220-227.

Olesen, I, Vogensen, F.K. and Jespersen, L. (2009). Gene transcription and virulence potential of *Listeria monocytogenes* strains after exposure to acidic and NaCl stress. *Foodborne Pathogen Diseases*, 6, 669–680.

Ölmez, H. (2012). Ozone. Produce contamination. In Decontamination of Fresh and Minimally Processed Produce ed. Gomez-Lopez, V. M. pp. 177-191. USA: Wiley-Blackwell Publishing.

Ölmez, H. and Akbas, M. Y. (2009). Optimization of ozone treatment of fresh-cut green leaf lettuce. *Journal of Food Engineering*, 90(4), 487-494.

Ölmez, H. and Temur, S.D. (2010) Effects of different sanitizing treatments on biofilms and attachment of Escherichia coli and *Listeria monocytogenes* on green leaf lettuce. *Lwt-Food Science and Technology* 43, 964-970.

O'Toole G., Kaplan, H.B. and Kolter R. (2000). Biofilm formation as a microbial development. *Annual Reviews of Microbiology*, 54,49-79.

Pan, Y.K., Breidt Jr., Kathariou, S. (2006). Resistance of *Listeria monocytogenes* bio-films to sanitizing agent in a simulated food-processing environment. *Applied and Environmental Microbiology*, 72, 7711-7717.

Pal, A., Labuza,T.P. and Diez-Gonzalez,F. (2008). Evaluating the growth of *Listeria monocytogenes* in refrigerated ready-to-eat frankfurters: influence of strain, temperature, packaging, lactate and diacetate, and background microflora. *Journal of Food Protection*, 71, 1806-1816.

Patil, S. and Bourke, P. (2012). Ozone processing of fluid foods. pp: 225-261. *In*: P. J. Cullen, B. K. Tiwari and V. P. Valdramidis (eds.). Novel thermal and non-thermal technologies for fluid foods. Elsevier Inc.

Patil, S., Valdramidis, V.P., Cullen, P.J., Frias, J.M. and Bourke, P. (2010) Ozone inactivation of acid stressed Listeria monocytogenes and Listeria innocua in orange juice using a bubble column. *Food Control* 21, 1723-1730.

Phan-Thanh, L., Mahouin, F., & Alige, S. (2000). Acid responses of *Listeria monocytogenes*. *International Journal of Food Microbiology*, 55(1–3), 121–126.

Pongkanpai, V., Makakarnchanakul, W. and Garnjanagoonchorn, W. (2013) Acid and heat tolerance of acid-stressed *Listeria monocytogenes* Inoculated in broth and shrimp model. *Journal of Pure and Applied Microbiology* 7, 837-843.

Pricope-Ciolacu, L., Nicolau, A.I., Wagner, M. and Rychli, K. (2013) The effect of milk components and storage conditions on the virulence of *Listeria monocytogenes* as determined by a Caco-2 cell assay. *International Journal of Food Microbiology* 166, 59-64.

Purevdorj, D., Igura, N., Ariyada, O. and Hayakawa, I. (2003). Effect of feed gas composition of gas discharge plasmas on Bacillus pumilus spore mortality. *Letters in Applied Microbiology*, 37(1), 31-34.

Robbins, J.B., Fisher, C.W., Moltz, A.G. and Martin, S.E. (2005). Elimination of *Listeria monocytogenes* biofilms by ozone, chlorine and hydrogen peroxide. *Journal of Food Protection*, 68(3): 494-498.

Rodgers, S. L., Cash, J. N., Sıddìq, M. and Ryser, E. T. (2004). A comparison of different chemical sanitizers for inactivating *Escherichia coli* O157:H7 and *Listeria monocytogenes* in solution and on apples, lettuce, strawberries & cantaloupe. *Journal of Food Protection*, 67, 721-731.

Romanova, N., Farvin, S. and Griffiths, M.W. (2002). Sensitivity of *Listeria monocytogenes* to sanitizers used in the meat processing industry. *Applied and Environmental Microbiology*, 68, 6405-6409.

Romanova, N.A., Gawande, P.V., Brovko, L.Y. and Griffiths, M.W. (2007) Rapid methods to assess sanitizing efficacy of benzalkonium chloride to *Listeria monocytogenes* biofilms. *Journal of Microbiological Methods* 71, 231-237.

Rød, S.K., Hansen, F., Leipold, F. and Knøchel, S. (2012) Cold atmospheric pressure plasma treatment of ready-to-eat meat: Inactivation of Listeria innocua and changes in product quality. *Food Microbiology* 30, 233-238.

Ryan, S., Hill, C. and Gahan, C.G.M. (2008). Acid stress responses in *Listeria monocytogenes*. *Advances in Applied Microbiology*, 65, 67–91.

Sandasi, M., Leonard, C.M. and Viljoen, A.M. (2008) The effect of five common essential oil components on *Listeria monocytogenes* biofilms. *Food Control* 19, 1070-1075.

Scallan, E., Hoekstra, R., Angulo, F., Tauxe, R.V., Widdowson, M. A., Roy, S. L., Jones, J.L. and Griffin, P.M. (2011). Foodborne illness acquired in the United States: major pathogens. *Emerging Infectious Diseases*, 17, 7–15.

Schmid, B., Klumpp, J., Raimann, E., Loessner, M.J., Stephan, R. and Tasara, T. (2009) Role of cold shock proteins in growth of *Listeria monocytogenes* under cold and osmotic stress conditions. *Applied and Environmental Microbiology* 75, 1621-1627.

Schuchat, A., B. Swaminathan, and C. V. Broome. (1991). Epidemiology of human listeriosis. *Clinical Microbiology Reviews*, 4, 169–183.

Severino, R., Dang Vu, K., Donsì, F., Salmieri, S., Ferrari, G. and Lacroix, M. (2013). Antimicrobial effects of different combined non-thermal treatments against *Listeria monocytogenes* in broccoli florets. *Journal of food Engineering*, 124, 1-10.

Skandamis, P.N., Gounadaki, A.S., Geornaras, I. and Sofos, J.N. (2012) Adaptive acid tolerance response of *Listeria monocytogenes* strains under planktonic and immobilized growth conditions. *International Journal of Food Microbiology* 159, 160-166.

Skandamis, P.N., Yoon, Y., Stopforth, J.D., Kendall, P.A. and Sofos, J.N. (2008) Heat and acid tolerance of *Listeria monocytogenes* after exposure to single and multiple sublethal stresses. *Food Microbiology* 25, 294-303.

Smith, J.L., Liu, Y.H. and Paoli, G.C. (2013) How does *Listeria monocytogenes* combat acid conditions? *Canadian Journal of Microbiology* 59, 141-152.

Song, H.P., Kim, B., Choe, J.H., Jung, S., Moon, S.Y., Choe, W. and Jo, C. (2009). Evaluation of atmospheric pressure plasma to improve the safety of sliced cheese and ham inoculated by 3-strain cocktail *Listeria monocytogenes*. *Food Microbiology*, 26, 432–436.

Stopforth, J.D., Yoon, Y., Barmpalia, I.M., Samelis, J., Skandamis, P.N. and Sofos, J.N. (2005). Reduction of *Listeria monocytogenes* populations during exposure to a simulated gastric fluid following storage of inoculated frankfurters formulated and treated with preservatives. *International Journal of Food Microbiology*, 99, 309–319.

Tasara, T. and Stephen, R. (2006). Cold stress tolerance of *Listeria monocytogenes*: A review of molecular adaptive mechanisms and food safety implications. *Journal of Food Protection*, 69, 1473–1484.

Upadhyay, A., Upadhyaya, I., Kollanoor-Johny, A. and Venkitanarayanan, K. (2013) Antibiofilm effect of plant derived antimicrobials on *Listeria monocytogenes*. *Food Microbiology* 36, 79-89.

Vaid, R., Linton, R.H. and Morgan, M.T. (2010) Comparison of inactivation of *Listeria monocytogenes* within a biofilm matrix using chlorine dioxide gas, aqueous chlorine dioxide and sodium hypochlorite treatments. *Food Microbiology* 27, 979-984.

Vaz-Velho, M., Silva, M., Pessoa, J. and Gibbs, P. (2006). Inactivation by ozone of *Listeria innocua* on salmon-trout during cold-smoke processing. *Food Control*, 17 (8), 609-616.

Venezia, R.A., Orrico, M., Houston, E., Yin, S.M., Naumova, Y.Y. (2008) Lethal activity of nonthermal plasma sterilization against microorganisms. *Infection Control Hospital Epidemiology*, 29(5), 430-436.

Wade, W.N., Scouten, A.J., McWatters, K.H., Wick, R.L., Demirci, A., Fett, W.F. and Beuchat, L.R. (2003). Efficacy of ozone in killing *Listeria monocytogenes* on alfalfa seeds and sprouts and effects on sensory quality of sprouts. *Journal of Food protection*, 66(1): 44-51.

Walecka-Zacharska, E., Kosek-Paszkowska, K., Bania, J., Karpiskova, R. and Stefaniak, T. (2013) Salt stress-induced invasiveness of major *Listeria monocytogenes* serotypes. *Letters in Applied Microbiology* 56, 216-221.

Werbrouck, H., Vermeulen, A., Van Coillie, E., Messens, W., Herman, L., Devlieghere, F. and Uyttendaele, M. (2009). Influence of acid stress on survival, expression of virulence genes and invasion capacity into Caco-2 cells of *Listeria monocytogenes* strains of different origins. *International Journal of Food Microbiology*, 134, 140–146.

Yousef, A.E., Carlstrom, C., (2003). *Listeria monocytogenes*. In: Food Microbiology: a Laboratory Manual. John Wiley & Sons Inc, Hoboken, NJ, pp. 138-139.

Yuk, H.-G., Yoo, M.-Y., Yoon, J.-W., Marshall, D.L. and Oh, D.-H. (2007) Effect of combined ozone and organic acid treatment for control of Escherichia coli O157:H7 and *Listeria monocytogenes* on enoki mushroom. *Food Control* 18, 548-553.

Yun, H., Kim, B., Jung, S., Kruk, Z.A., Kim, D.B., Choe, W. and Jo, C. (2010) Inactivation of *Listeria monocytogenes* inoculated on disposable plastic tray, aluminum foil, and paper cup by atmospheric pressure plasma. *Food Control* 21, 1182-1186.

Zhang, Q., Feng, Y., Deng, L., Feng, F., Wang, L. Z. Q. and Luo, Q. (2011). SigB plays a major role in *Listeria monocytogenes* stress to bile stress. *International Journal of Food Microbiology*, 145(1): 238-243.

Ziuzina, D., Patil, S., Cullen, P.J., Keener, K.M. and P. Bourke. (2013). Atmospheric cold plasma inactivation of *Escherichia coli* in liquid media inside a sealed package. *Journal of applied microbiology*, 114, 778-787.

Zwietering, M.H., Wijtzes, T., de Witt, J.C. and van't Riet, K. (1992). A decision support system for prediction of the microbial spoilage in foods. *Journal of Food Protection*, 55, 973-979.

In: *Listeria monocytogenes*
Editor: Edmund C. Hambrick
ISBN: 978-1-63117-054-6

Chapter 5

SEEING THE LIGHT: EXPLORING THE POTENTIAL OF VISIBLE LIGHT AS A MEANS OF CONTROLLING *LISTERIA MONOCYTOGENES* IN THE FOOD CHAIN

Kerrie NicAogáin[1], Beth O'Donoghue[1] and Conor P. O'Byrne[1*]
Bacterial Stress Response Group, Microbiology, School of Natural Sciences,
National University of Ireland Galway, Galway, Ireland

ABSTRACT

The ability of *Listeria monocytogenes* to withstand processes employed by the food industry to inhibit or limit bacterial growth has become a major health and economic problem worldwide. The bacterium's survival is due, in part, to the action of the alternative sigma factor σ^B. This alternative transcription factor is responsible for regulating defence mechanisms in response to stresses such as extremes of temperature, osmotic, acid and ethanol stress, as well as contributing to virulence. Recently, light therapies have been proposed as additional means for the control of contamination, with applications focused on the food and medical device industries. Several variations of light treatment are currently being investigated, among them UV light therapy and photodynamic inactivation therapies. While these therapies are not limited solely to the management of *L. monocytogenes*, the discovery of the blue light photoreceptor Lmo0799 upstream of the *Listerial* σ^B activation cascade has resulted in increased interest in the effects of light on *L. monocytogenes*. This chapter reviews the mechanisms by which light is known to stimulate the σ^B activation cascade, and light treatments currently being investigated as suitable regimes for management of contamination. A detailed understanding of the mechanisms by which light influences virulence and the stress response in *L. monocytogenes*, as well as the mechanisms by which light causes cellular damage, is key for the successful development of new light-based control measures.

Keywords: Listeria, sigmaB, photoinactivation, visible light, Lmo0799, YtvA, porphyrins

[*] Author for correspondence. Email: conor.obyrne@nuigalway.ie; Tel: 00353 91 493957; Fax: 00353 91 494598.

Introduction

Listeria monocytogenes is found widely in the environment and it is well adapted to overcome the harsh preservation conditions found in some read-to eat foods, including high salt concentrations, acidic pH, and low temperatures. Its ability to grow and survive in these conditions is partly due to a protective stress response that is under the control of the alternative sigma factor, sigma B (σ^B) (Hecker et al., 2007; O'Byrne & Karatzas, 2008). This stress pathway is activated in the presence of harsh conditions and offers the cell protection against the stress so that it can survive. Recent studies have shown that σ^B is also activated in response to visible light (Ondrusch & Kreft, 2011; Avila-Pérez et al., 2009). The development of novel methods for controlling the growth of this bacterium in the food chain is an urgent priority for the ready-to-eat food industry. In this article we evaluate the potential of light as a means to control the growth and survival of *L. monocytogenes*. We review different methods of bacterial photoinactivation and discuss the possibility that this method may be useful within the food processing environment. However this approach might be limited by the fact that σ^B activation by light can trigger increased expression of some virulence genes, raising the possibility that light-treated cells might be more virulent than cells not exposed in this way. Therefore the future development of light-based food safety controls will have to balance the potential advantages and risks for each particular food application. This limitation notwithstanding, light offers a potentially cheap and effective way to limit bacterial growth and survival in foods and food processing environments.

Photodynamic Inactivation

In recent years, the investigation of new sterilisation methods for the decontamination of bacteria from food related environments has become a large area of research. One of these new methods is photodynamic inactivation (also called photoinactivation or photosensitization). This is an antimicrobial method that uses a combination of three different agents to inactivate microorganisms; a photoactive compound, light and oxygen (Dai et al., 2009). Photodynamic inactivation could help to combat the problem of bacterial contamination within food related environments such as on surfaces or food packaging. There have also been cases where it has been successfully used for decontaminating the surface of certain fruits such as raspberries and strawberries (Bialka & Demirci, 2008).

Mechanisms of Photoinactivation

Photodynamic inactivation or photosensitization is the inactivation of microorganisms using a photosensitizer in combination with light and oxygen. A photosensitizer is a photoactive compound which can be an exogenous substance taken up by the bacteria (Romanova et al., 2003) or a compound produced endogenously by the organism (Buchovec et al., 2010). Some examples of exogenous compounds which can be used as a photosensitizer include methylene blue, Na-chlorophyllin and hypericin (Lin et al., 2012; Luksiene et al., 2010; Kairyte et al., 2012). Endogenous photoactive compounds include compounds such as porphyrins. It is well known that some bacteria can produce these within the cell through the heme biosynthetic pathway. They are produced from 5-aminolevulinic acid (ALA) which is the precursor from which heme is produced. ALA can be produced in bacteria via two separate pathways. In one pathway succinyl-CoA and glycine are converted into ALA by ALA synthase. In the second pathway glutamate is converted to glutamyl – tRNA which is further converted to glutamate 1-semialdehyde (GSA) by the enzyme glutamyl-tRNA reductase. This enzyme is encoded by the gene *hemA*. ALA is finally synthesised from the conversion of GSA by the enzyme GSA amino-transferase (Panek & Brian, 2002). From here ALA is converted via a number of different intermediates to form heme within the cell. These intermediates include uroporphyrin III, coproporphyrinogen III and protoporphyrin IX and it is a mixture of these intermediates that make up endogenous porphyrins, which can be targeted for photodynamic inactivation within the bacterial cell (Hamblin et al., 2005; Lee & Shalita, 1978).

It is generally accepted that Gram positive microorganisms are less resistant to photosensitization by exogenous compounds than Gram negative organisms (Ashkenazi et al., 2003), probably due to the structure of their cell envelope (Nitzan et al., 2004). Gram positive bacteria have a layer of peptidoglycan and lipoteichoic acid surrounding their cell membrane, which allows for easier absorption of compounds by the bacteria, as it is a porous layer and does not act as a good protective barrier for the bacteria. In Gram negative bacteria the combination of the two membranes and the periplasm layer act as a tough barrier and therefore this restricts the entry of exogenous molecules into the cell (Demidova & Hamblin, 2005). For this reason Gram positive bacteria are better able to take up exogenous photosensitizers and therefore this makes them more susceptible to photoinactivation. However, it is possible that if the production of endogenous porphyrins can be stimulated, then this may allow for more effective photoinactivation of Gram negative bacteria as this method would not involve the absorption of an external substance.

In the photosensitization process, light illuminates and excites the photosensitizer and this causes the photosensitizer to react with molecular oxygen to form radical oxygen species (ROS). The light sources that are used in these experiments typically have irradiance values in the range 10 - 200 mW/cm^2. To put these values into perspective on a sunny day in Ireland in June the average total solar irradiance is approximately 29.7 mW/cm^2 (Data from NUI Galway weather observatory, 2013). Once the photosensitizer has interacted with oxygen, the ROS that is formed can then interact with and cause lethal damage to internal structures within the bacteria. Some of the structures which can be affected by this disruption in the cell include the cell membrane, DNA and different enzymes therefore leading to cell inactivation (Arakane et al., 1996). For this process either UV light or visible light can be effectively used. Visible light ranges from 400nm to 700nm. When visible light is used as a component of photosensitization, it causes the excitation of the photoactive component, whether it is endogenous or exogenous, and causes the generation of reactive species. This occurs by excitation of the photosensitiser leading to the formation of an excited triplet state, which further transfers electrons or hydrogen atoms to form a reactive radical state of the photosensitizer. The radical state can then further react with oxygen to form reactive oxygen species which in turn lead to damage of the cell and can possibly be lethal to the bacterial cell (Luksiene & Zukauskas, 2009).

The UV spectrum covers wavelengths shorter than this from about 100nm to 400nm. UV light is split up into three different types based on wavelength. UVA is defined as light with wavelengths in the range 320nm to 400nm, UVB from 290nm to 320nm and UVC from 100nm to 290nm. The type of damage UV light causes to the cell is dependent on the wavelength of the light and can be either direct or indirect. Direct cell damage refers to direct DNA damage which can subsequently lead to cell mutation or cell death, whereas indirect damage refers to the generation of radical oxygen species which produce secondary effects on the cell. UVC can cause direct damage to the cell whereas UVB can cause both direct and indirect damage. UVA mainly causes indirect damage through the generation of radical oxygen species (Miller et al., 1999) and therefore light with a wavelength in the UVA range is usually used when UV light is required in the photoinactivation procedure.

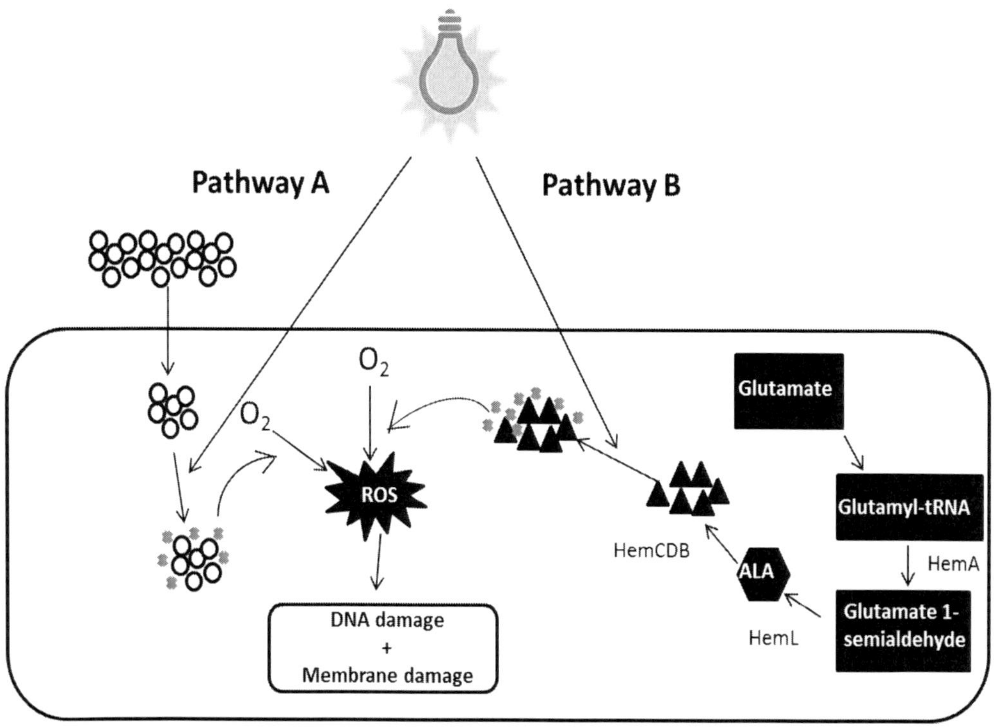

Figure 1. Pathway A shows the uptake of an exogenous photosensitizer into the bacterial cell. When this photosensitizer is activated by light it promotes electron transfer reactions that allow the formation of Reactive Oxygen Species (ROS). The ROS then cause damage to the cell which can lead to cell inactivation. Alternatively Pathway B shows the formation of endogenous porphyrins via the Heme biosynthetic pathway. Enzymes from the heme bisosynthetic pathway that are known to be encoded in the *L. monocytogenes* genome are indicated. Heme is produced from a precursor called 5-aminolevulinic acid (ALA) which is a naturally occuring substance that is produced by bacteria plants and mamals. Different endogenous porphyrins are produces as intermediates of this pathway, including uroporphyrin III, coproporphyrinogen III and protoporphyrin IX. Once the porphyrins are formed they are activated in the same mechanism as an exogenous photosensitizer and therefore can also lead to cell inactivation by stimulating ROS production.

PHOTOINACTIVATION OF *LISTERIA MONOCYTOGENES*

It has been shown in a number of different studies that photodynamic treatment can cause inactivation of the foodborne pathogen *Listeria monocytogenes*. Since this bacterium can survive within food-processing environments, on food packaging and within food, it is important to investigate different approaches that might be used to overcome this problem.

PHOTOINACTIVATION BY EXOGENOUS PHOTOSENSITIZERS

Photoinactivation of *Listeria monocytogenes* can be carried out either using exogenous photosensitizers or endogenous porphyrins. In a series of studies carried out by a group in Lithuania, the effectiveness of different photosensitizers for the inactivation of *L. monocytogenes* was investigated. Na-chlorophyllin was one of these exogenous photoactive compounds. It is an anti-mutagenic and anti-carcinogenic water soluble food additive (E141) which is made up of a mixture of sodium copper salts which are derived from the green pigment chlorophyll. Since it is used as a food additive and has photoactive properties, this makes it a potentially good photosensitizer to combat *L. monocytogenes* in the food chain and within food environments. It has been shown that a concentration as low as 7.5×10^{-7}M of Na-chlorophyllin is sufficient to inactivate *Listeria monocytogenes* by up to 7 log cycles when illuminated with visible light (405nm) with an irradiance of 12mW/cm^2 for up to 30 minutes after the cells are incubated in Na- chlorophyllin for 2 minutes prior to illumination. This was compared with no decrease in bacterial numbers when the culture was exposed to Na-chlorophyllin only without illumination. However they did not compare these results to a culture only illuminated by visible light without the presence of Na-chlorophyllin (Luksiene & Paskeviciute, 2011).

In the same set of studies, the inactivation of *L. monocytogenes* on food packaging was also tested and compared to other methods used for decontamination. It was found that the photosensitization of food packaging led to a reduction of *L. monocytogenes* to an almost undetectable level using 7.5×10^{-7} M Na-chlorophyllin. In comparison it was found that when the packaging was either washed with sterile water or 200ppm Na-hypochlorite the results only showed a 20% and 40% reduction respectively. This result showed that Na-chlorophyllin photoinactivation of vegetative *L. monocytogenes* cells proved to be a very successful method of sterilisation of food packaging. This method of decontamination was also carried out on biofilms of *L. monocytogenes* on food packaging. However, it was shown that higher concentrations of Na-Chlorophyllin (1.5×10^{-4}M) were needed, as biofilms were more resistant to photoinactivation than vegetative cells. It is hypothesised that this is due to the complex structure of *L. monocytogenes* biofilms which are enclosed in a polymeric matrix produced by the cells. It is thought that this matrix may be involved in blocking some of the light from reaching the bacteria (Luksiene et al., 2010; Luksiene & Paskeviciute, 2011).

Other exogenous photosensitizers have also been used for the photoinactivation of *L. monocytogenes* such as methylene blue and hypercin. One study which used methylene blue with visible light showed that when *L. monocytogenes* was exposed to either of these components separately, there was no effect on the viability of the cells. However when the cells were treated with methylene blue (0.5 μg/ml) and exposed to visible light with a power density of 200 mW/cm^2 , there was up to a 7 log reduction after 10 min of exposure. This reduction depended on the concentration of methylene blue that was used. Lower concentrations such as 0.05μg/ml gave up to a 3 log reduction while concentrations higher than 0.2 μg/ml gave up to 7 log reduction in viability of the cells. After incubation with 0.5 μg/ml methylene blue and exposure to light for 10 min, the surface of the cell became distorted when compared to untreated cells. This type of photosensitization was found to cause DNA damage and leads to the leakage of cell components out through the cell membrane (Lin et al., 2012).

ENDOGENOUS PORPHYRINS ARE PRODUCED BY *LISTERIA MONOCYTOGENES*

As well as the possibility of inactivating *L. monocytogenes* using an exogenous photosensitizer, it has been shown that this bacterium produces intracellular porphyrins which may also allow the possibility of photosensitizing the bacterium to cause its deactivation (Buchovec et al., 2010). In the visible light spectrum, light with a wavelength between 400nm and 500nm is most effective at inactivating *L. monocytogenes,* with a 5.1 log reduction seen after cultures are subjected to a dose of 604.8 Jcm^{-1}. This result was recorded in the absence of any exogenous photosensitizer being present (Endarko et al., 2012). This has been further narrowed down by the same group who showed that when a culture is exposed to a dose of 123.3 Jcm^{-1} for each 10nm bandwidth between 400nm and 450nm, the most successful bandwidth is between 400nm and 410nm. They concluded that light with a wavelength of 405nm was the most successful. While this is a significant result, it is important to note that light at this wavelength may also extend into the UV range of the spectrum and therefore UV damage may contribute to the inactivation effect seen. It is also interesting to note that in this same study, when the oxygen scavenger dimethylthiourea (DMTU) was added to a *L. monocytogenes* culture which was illuminated with a dose of 185 Jcm^{-1} there was no loss in viability detected. This result suggests that the mechanism of killing by light at 405 nm is likely to be via the production of reactive oxygen species rather than by direct light damage. To further compare the findings of this previous study, another study was conducted to compare the effects of light at 405nm on *L. monocytogenes* along with 3 other bacteria, *Escherichia coli, Shigella sonnei* and *Salmonella enterica.* It was found that when each of these were streaked on TSA and illuminated with a light dose of 108 Jcm^{-1}, *L. monocytogenes* exhibited the greatest susceptibility to photoinactivation (Murdoch et al., 2012). This observation suggested that Gram negative species may overall exhibit more resistance to photoinactivaton than the Gram positive bacteria. In some cases it is possible to stimulate the production of endogeneous porphyrins by incubating bacterial cells with 5-aminolevulinic acid (ALA). ALA is a naturally occurring substance produced by mammals and plants (Marc et al., 2009). In mammals it is produced as part of the heme biosynthetic pathway and in plants it is produced during the production of chlorophyll. In bacteria ALA is a precursor in the heme biosynthetic pathway which generates porphyrins as intermediates of the pathway as seen in Figure 1 (Nitzan et al., 2004). *L. monocytogenes* contains an operon HemACDBL which encodes five of the genes needed for the biosynthesis of heme (Rea et al., 2005) and it has also been observed in a previous study, that *L. monocytogenes* is capable of producing endogenous porphyrins (Buchovec et al., 2010). Therefore if *L. monocytogenes* is incubated with ALA (7.5mM), it can help to stimulate porphyrin production within the cells. In one study *L. monocytogenes* cells were incubated with ALA for 2h in the dark and then exposed to a visible light dose of $24J/cm^2$. The cells were killed by up to 4 log compared to no killing when the cells with incubated with either ALA or light alone. These findings suggest the basis for a process for the photoinactivation of *L. monocytogenes* within food and food processing plants (Buchovec et al., 2010). Routine illumination of work surfaces, equipment, packaging, and even food surfaces might in future be used, in conjunction with other disinfection regimes, as a means of controlling the growth and survival of this pathogen (and others) in the food chain.

LIGHT EFFECTS WITHIN *L. MONOCYTOGENES*

L. monocytogenes largely owes its ability to withstand and survive food related stresses to the alternative sigma factor sigma B (Hecker et al., 2007). σ^B coordinates the transcriptional response that provides protection against a range of stresses including osmotic stresses, ethanol stress, sub- optimal temperatures and cell wall stresses (Ferreira et al., 2001; Ferreira et al., 2003). Light, as may be inferred from recent photoinactivation studies, is a stress and triggers the activation of the σ^B stress response in *L. monocytogenes* (Murdoch et al., 2012; Ondrush & Kreft, 2011).

In *B. subtilis* the activation of σ^B has been closely studied and is often used as an efficacious model for the *L. monocytogenes* σ^B activation pathway (Hecker et al., 2007; Glaser et al., 2001; Chaturognkal & Boor, 2006). The *L. monoctogenes* system differs from that of *B. subtilis,* in that σ^B in *Listeria* is also responsible for regulation of the activity of a number of virulence genes (Chaturongakul & Boor, 2004; Kaziermczak et al., 2003; Kaziermczak et al., 2006; Nadon et al., 2001; Garner et al., 2006; O'Byrne & Karatzas, 2008). The presence of light has been shown to activate both the σ^B-mediated stress response and virulence in *L. monocytogenes* (Ondrush & Kreft, 2011; Tiensuu et al., 2013). Light therapies are currently being investigated as suitable processes for limiting *Listeria* survival. While light inactivation therapies may show promise, studies into light activation of the σ^B –mediated stress response and virulence activities indicate that virulence priming would be an important consideration in the development of a light-based therapy.

σ^B ACTIVATION PATHWAY

σ^B is present but inactive under unstressed conditions in the cell (Ferreira et al., 2004). It and the genes involved in the σ^B activation pathway are encoded by an eight-gene *sigB* operon: *rsbR, rsbS, rsbT, rsbU, rsbV, rsbW, sigB, rsbX,* with an internal σ^B-dependent promoter preceding the *rsbV* gene (Ferreira et al., 2004). The σ^B activation pathway has two aspects: signalling, and sensing. Sensing occurs via the putative multi-protein stressosome complex, confirmed in the *B. subtilis* σ^B activation pathway (Chen et al., 2003; Delumeau et al., 2006; Martinez et al., 2010).

In the absence of light or other stresses, σ^B is maintained in an inactive state, bound by the anti-sigma factor RsbW (Benson & Haldenwang, 1993). Once a stress signal has been received, conformational changes occur within the stressosome which lead to the dissociation of RsbT proteins from the complex (Marles Wright et al., 2008). The release of RsbT initiates a signalling cascade that results in the release of σ^B. When RsbT is released from the stressosome it activates the phosphatase ability of RsbU (Yang et al., 1996). RsbT interacts with RsbU at the RsbU N-terminal region, and it is thought that this induces conformational changes which expose phosphatase activity at the C-terminal end (Yang et al., 1996; Delumeau et al., 2004). RsbU then removes the phosphate group from RsbV-P (Dufour & Haldenwang, 1994; Yang et al., 1996). RsbU is suggested to act on RsbV in an RsbU-RsbT complex; however it is still not known whether RsbT confers a conformational activation change that is retained for some time after RsbT dissociation (Delumeau et al., 2004). RsbW sequesters σ^B under normal conditions but has a higher binding affinity for the unphosphorylated form of the anti-antisigma factor RsbV (Benson & Haldenwang, 1993; Dufour & Haldenwang, 1994). Once RsbV becomes available, RsbW binds it and σ^B is free to interact with RNA polymerase and the promoters of specific genes, including many that encode proteins involved in stress protection and in virulence (Benson & Haldenwang, 1993; Dufour & Haldenwang, 1994; Hecker et al., 2007). Thus, the phosphorylation state of RsbV is the determining factor in σ^B activation (Dufour & Haldenwang 1993). The reincorporation of RsbT into the RsbR-RsbS complex can only occur after the removal of the phosphate group from RsbS by the RsbX phosphatase (Yang et al., 1996; Smirnova et al., 1998; Chen et al., 2004).

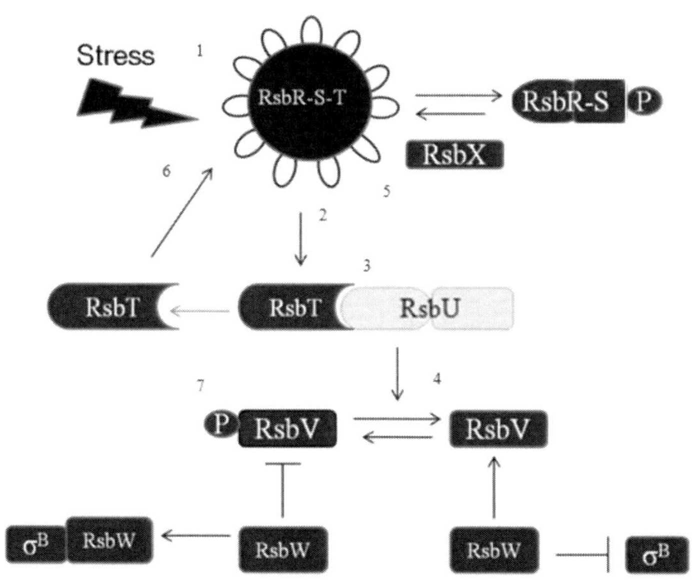

Figure 2. σ^B Activation Cascade. (1) Stress signalling to the (putative) stressosome results in conformational changes at the centre of the complex. (2) RsbT phosphorylates RsbS, dissociates from the complex and activates the RsbU phosphatase (3). RsbU dephosphorylates RsbV-P and (4) RsbW kinase preferentially binds RsbV, leaving σ^B free. (5) RsbX removes phosphate groups from RsbS (and RsbR). (6) The phosphatase activity of RsbX enables RsbT to recomplex with the stressosome. (7) In the unstressed cell the balance of RsbV proteins become phosphorylated and RsbW is then free once again to interact with σ^B.

UPSTREAM OF THE SIGB ACTIVATION PATHWAY: THE STRESSOSOME

Currently the only known model for the activation of σ^B in *L. monoctogenes* is via the (putative) stressosome. This differs from the *B. subtilis* system which has a separate pathway for SigB activation under conditions of energy stress (Vijay et al., 2000; Ferreira et al., 2004; Chaturongakul & Boor, 2006). It is thought that the stressosome is a 1.5 MDa structure composed of multiple RsbS and RsbR proteins at the centre, from which the RsbR N-terminal sensory 'turrets' protrude (Chen et al., 2003; Marles-Wright & Lewis, 2010; Marles-Wright et al., 2008; Ondrusch & Kreft, 2011). Interestingly, the structure of the N-terminal regions of RsbR has been found to be homologous to the same region in RsbU (Hardwick et al., 2007). RsbT molecules interact with the stressosome in the spaces between RsbR N-terminal protrusions and are thought to exist in a formation whereby an RsbT-ATP cap is held just above RsbS Ser ^{59} in the C-terminal domain (Marles-Wright et al., 2008). Neither the RsbS nor the RsbR proteins are capable of sequestering RsbT alone, and can only do so in complex with each other (Chen et al., 2003; Kim et al., 2004). Once a stress is sensed by the RsbR N-terminal regions, it is thought that signalling to the centre of the stressosome causes RsbT to phosphorylate both RsbS and RsbR and the phosphorylation of RsbS triggers the release of RsbT from the complex (Chen et al., 2003; Kim et al., 2004; Marles-Wright & Lewis, 2010; Marles-Wright et al., 2008).

Five separate RsbR-like sensory proteins have been identified in *L. monocytogenes*: RsbR (Lmo0889), Lmo0799, Lmo1642, Lmo0161 and Lmo1842 (Ondrusch & Kreft, 2011; Heavin & O'Byrne, 2011). These RsbR paralogues consist of highly variable N-terminal regions (Murray et al., 2005) connected to C-terminal STAS domains, which are highly conserved in sulphate transporters and anti-sigma factor antagonists (Aravind et al., 2000). It has been observed that the replacement of the native *rsbR* in *B. subtilis* with its *L. monocytogenes* paralogue enabled stressosome mediated stress responses to both energy and environmental stresses (Martinez et al., 2010). The exact function of all but one of these proteins is unknown; Lmo0799 has been confirmed as a blue light sensor as has its homologue YtvA in *B. subtilis*, (Losi et al., 2002; Ondrusch & Kreft, 2011). Interestingly, while Lmo0799 only senses blue light, red light has also been shown to increase σ^B activity levels in *L. monocytogenes,* although to a lesser degree (Ondrusch & Kreft, 2011).

Red light is known to trigger σ^B activity in *B. subtilis* via the RsbP/Q signalling pathway (Avila-Perez et al., 2010). However, this pathway does not exist in *L. monocytogenes* and the mechanism of red-light induced σ^B activation remains unknown. Indeed, it is not known how most stresses are sensed by the stressosome, with some suggestions that each RsbR paralogue sense a different stress, but Shin et al. (2010) propose that the stressosome identifies stresses such as cold, salt and acid by monitoring envelope function, which may be affected by visible light (Kim et al., 2004).

THE LIGHT SENSOR: LMO0799 STRUCTURE AND FUNCTION

Since *L. monocytogenes* is found widely in the environment (Hecker et al., 2007) it is not surprising that *L. monocytogenes* possesses specific factors enabling it to sense and react appropriately to the presence of light. Lmo0799 is unique among the known *L. monocytogenes* RsbR-like proteins in that it carries a LOV domain at its N- terminal region (Ondrusch & Kreft, 2011). LOV (Light, Oxygen, Voltage) domain proteins are blue light-stimulated protein sensors which require a flavin cofactor (Huala, et al., 1997; Salomon et al., 2002; Moglich & Moffat, 2007). LOV proteins are a subgroup of the Per-Arnt-Sim domain superfamily and are found across the bacterial, eukaryotic and archaeal kingdoms (Taylor & Zhulin, 1999; Herrou & Crosson, 2011; Huala, 1997; Losi et al., 2002; Crosson et al., 2003). The first of these sensors to be discovered in bacteria was YtvA, the Lmo0799 homolog in *B. subtilis* (Losi et al., 2002). Analyses have since shown that Lmo0799 shares 50/67% identity/similarity with YtvA (Heavin & O' Byrne, 2011). Lmo0799 is a LOV- STAS protein, a configuration which accounts for approximately 10% of all LOV proteins found in bacteria (Herrou and Crosson, 2011). Based on observations with YtvA, it is thought that Lm0799 is required to form dimers at the N-terminal region in order to function (Moglich & Moffat, 2007; Buttani et al.,2007; Avila-Perez et al., 2009).

The protein binds its flavin cofactor in a pocket formed by a conserved sequence at the LOV domain active site. Upon illumination with blue light, a covalent adduct is formed between the flavin cofactor, and a conserved cysteine residue at position 56 (Swartz et al., 2001; Salomon et al., 2000; Crosson et al., 2003; Ondrush & Kreft, 2011). The resulting conformational changes enable signalling from the STAS output domain to the stressosome core (Crosso*n* et al., 2003; Moglich & Moffat, 2007). Under dark conditions, the protein returns to its original conformation (Salomon et al., 2000). Although flavin adenine dinucleotide (FAD) is available in the cell, Lmo0799 uses flavin mononucleotide (FMN) almost exclusively as its binding chromphore (Chan et al., 2012). It is suggested that the absence of two conserved arginine residues at the LOV domain binding pocket on Lmo0799 provides the protein with far less stability than that of the highly similar YtvA protein (Chan et al., 2012). The replacement of these residues is assumed to account for the reported discharge of significant amounts of FMN during storage, upon photoexcitation and at higher temperatures (>26°C) (Chan et al., 2012; Losi et al., 2003). Chan et al. (2012) suggest that the replacement of these arginine residues prevents the formation of hydrogen bonds which lend stability to the structural conformation at the protein core. The increased flexibility of the protein makes it a poor light sensor, and Chan et al. (2012) posit that this indicates the protein has other functions in addition to light sensing, possibly as a cold sensor.

LIGHT INDUCED STRESS AND VIRULENCE IN *L. MONOCYTOGENES* ARE LINKED BY σ^B

Stress protection and virulence activation are tightly linked in *L. monocytogenes.* While the virulence gene regulator PrfA is the key regulator of genes involved in survival and growth in blood, σ^B controls the expression of proteins required to adapt the bacteria to the intestinal environment (Ferreira et al., 2004; Garner et al., 2006; Toledo-Arana et al., 2009;

O'Byrne & Karatzas, 2008). Kazmierczak and colleagues reported that the internalin genes *inlA, inlB* and the bile salt hydrolase gene *bsh* are all under the control of σ^B (although both *inlA* and *inlB* are also regulated in a σ^B independent manner) (Kazmierczak et al., 2003). *Bsh* expression, which is essential for *Listeria* pathogenesis in the gastrointestinal environment, was induced in the presence of blue light and 0.3M NaCl (Dussurget et al., 2002; Toledo-Arana et al., 2009; Ondrusch & Kreft, 2011). This expression was reduced significantly in an *lmo0799* mutant exposed to similar conditions. *L. monocytogenes* infectivity in human enterocyte-like Caco-2 cells has been shown to increase two- fold upon prior exposure to blue light and 0.3 M salt, compared to the infectivity of bacteria exposed only to salt. This observation can be directly linked to the light response, as these conditions did not affect infectivity levels of an *lmo0799* deletion mutant (Ondrusch & Kreft, 2011). Further study has also found that the presence of blue light specifically increases *L. monocytogenes* EGD-e *inlA* and *inlB* transcription (3 and 4.5-fold, respectively). Exposure of the cells to blue light and a second stress, in the form of salt, raised transcription levels higher again (Ondrusch & Kreft, 2011).

PrfA has three known promoters, a promoter upstream of *plcA*, *prfA*P1 and the σ^B-regulated promoter *prfA*P2 (Freitag et al., 1993; Nadon et al., 2002; Leimeister-Wachter et al., 1992). σ^B does not appear to be absolutely required for infection and dissemination of the bacteria; however it is required for wild-type levels of expression of PrfA (Freitag et al., 1993; Nadon et al., 2002). It seems that the gastrointestinal step in *L. monocytogenes* infection may be the point at which the bacteria switch from σ^B -mediated virulence to a PrfA – regulated mode of virulence and it is suggested that P2*prfA* transcription is an important factor driving the switch (Nadon et al., 2002; Garner et al., 2006; Kazmierczak et al., 2006).

The swimming motility of wild- type *L. monocytogenes* normally observed at 27°C diminishes considerably when exposed to blue light and this loss of motility is not observed with *lmo0799* deletion mutants (Ondrusch & Kreft, 2011; Tiensuu et al., 2013). This correlates with other work which found that the gene encoding the MogR transcriptional repressor of flagellin genes is partially under the control of σ^B. The gene has two identified promoters, of which the P1 promoter is induced by σ^B (Toledo-Arana et al., 2009). Repression of motility is associated with virulence as it also occurs at 37°C, in conjunction with up-regulation of virulence genes.

Lmo0799 contributes to the stress response to produce highly ROS resistant bacteria and it has been observed that the presence of H_2O_2 increases the expression of *lmo0799*, in a σ^B - independent manner. Tiensuu et al., (2013) have shown that bacteria grown during light exposure in a light- dark cycle are 4 times more resistant to H_2O_2, indicating that the amount of ROS required to cause significant damage to bacterial cells is increased after exposure to light. From their observations, it is suggested that Lmo0799 functions in the initial stages of the ROS response, and is replaced by other stronger response processes at later stages (Tiensuu et al., 2013). Due to the strong connection between light sensing and virulence priming in *Listeria monocytogenes,* there is a danger that light inactivation therapies may generate a number of strongly resistant bacteria that are primed for infection (Gomelsky & Hoff, 2011).

Conclusions and Future Perspectives

Initial assays into the use of photoinactivation treatments suggest such procedures may be employed to control the presence of *L. monocytogenes* in food and other food-processing environment. Photoinactivation therapies require light, oxygen and either exogenous or endogenous photosensitizers (photoactive compounds). The generation of ROS is thought to be responsible for cell damage, and additional DNA damage may result from the use of UV (100-399 nm), rather than visible (400-700 nm) light sources. Studies have established several putative suitable exogenous photosensitizers; of particular interest is the water- soluble, anti-mutagenic, food additive Na- chlorophyllin (Lin et al., 2012; Luksiene et al., 2010; Kairyte et al., 2012). In addition, *L. monocytogenes* appears to possess endogenous photosensitzers in the form of porphyrins (Buchovec et al., 2010). Exploitation of endogenous porphyrins may be preferable to the addition of compounds in order to inactivate *Listeria* in foodstuffs. The use of photoinactivation for control of *L. monocytogenes* shows great promise, however there is a concern that such therapies could stimulate the σ^B directed stress- response of *L. monocytogenes* and generate highly resistant and virulent bacteria. Light is recognised as a stress by *L. monocytogenes* and triggers activation of the strong σ^B stress response via the blue light receptor protein Lmo0799 and through other unknown stressosome– related mechanisms (Ondrusch & Kreft, 2011). σ^B is responsible for activating virulence genes involved in the initial stages of infection virulence (Ferreira et al., 2004; Garner et al., 2006; Toledo-Arana et al., 2009; O'Byrne & Karatzas, 2008). Ideally, photoinactivation procedures should employ a light intensity which either kills the total population of *L. monocytogenes* present or inactivates the bacteria without significantly stimulating σ^B activity. It is therefore necessary that investigations into the virulence priming effects of photoinactivation treatments are carried out in parallel with research into the development of such therapies. It is hoped that these investigations would facilitate the development of new control strategies to help eliminate the problem of food-borne listeriosis.

References

Aravind, L., & Koonin, E. V. (1994). The STAS domain — a link between anion antagonists. *Current Biology*, 53–55.

Arakane, K. et al., 1996. Singlet oxygen (1 delta g) generation from coproporphyrin in *Propionibacterium acnes* on irradiation. *Biochemical and biophysical research communications*, 223(3), 578–82.

Ashkenazi, H., Nitzan, Y. & Ga, D.(2003). Photodynamic Effects of Antioxidant Substituted Porphyrin Photosensitizers on Gram-positive and -negative Bacteria, *Photochemistry and Photobiology* 77(2), 186–191.

Avila-Pérez, M., Van der Steen, J. B., Kort, R., & Hellingwerf, K. J. (2010). Red light activates the sigmaB-mediated general stress response of *Bacillus subtilis* via the energy branch of the upstream signaling cascade. *Journal of Bacteriology*, *192*(3), 755–62.

Avila-Pérez, M., Vreede, J., Tang, Y., Bende, O., Losi, A., Gärtner, W., & Hellingwerf, K. (2009). In vivo mutational analysis of YtvA from *Bacillus subtilis:* mechanism of light activation of the general stress response. *The Journal of Biological Chemistry, 284*(37), 24958–64.

Benson, A. K., & Haldenwangt, W. G. (1993). *Bacillus subtilis* SigB is regulated by a binding protein (RsbW) that blocks its association with core RNA polymerase. *Proc.Natl. Acad. Sci USA 90*, 2330–2334.

Bialka, K.L. & Demirci, A., (2008). Efficacy of pulsed UV-light for the decontamination of *Escherichia coli* O157:H7 and *Salmonella* spp. on raspberries and strawberries. *Journal of Food Science*, 73(5), M201–7.

Brigulla, M., Hoffmann, T., Krisp, A., Voelker, A., Bremer, E., & Voelker, U. (2003). Chill induction of the SigB-dependent general stress response in *Bacillus subtilis* and its contribution to low-temperature adaptation. *Journal of Bacteriology, 185*, 4305–4315.

Buchovec, I., Paskeviciute, E. & Luksiene, Z. (2010). Photodynamic Inactivation of Food Pathogen *Listeria monocytogenes. Food technology and Biotechnology* , 48(2), 207–213.

Buttani, V., Losi, A., Polverini, E., & Gärtner, W. (2006). Blue news: NTP binding properties of the blue-light sensitive YtvA protein from *Bacillus subtilis. FEBS letters, 580*(16), 3818–22.

Chan, R. H., Lewis, J. W., & Bogomolni, R. A. (2012). Photocycle of the LOV-STAS protein from the pathogen *Listeria monocytogenes. Photochemistry and Photobiology, 89*(2), 361–9.

Chaturongakul, S., & Boor, K. J. (2004). RsbT and RsbV Contribute to SigB -Dependent Survival under Environmental, Energy, and Intracellular Stress Conditions in *Listeria monocytogenes. Applied and Evironmental Microbiology, 70*(9), 5349–5356.

Chaturongakul, S., & Boor, K. J. (2006). SigmaB activation under environmental and energy stress conditions in *Listeria monocytogenes. Applied and Environmental Microbiology, 72*(8), 5197–203.

Chen, C., Yudkin, M. D., & Delumeau, O. (2004). Phosphorylation and RsbX-Dependent Dephosphorylation of RsbR in the RsbR-RsbS Complex of *Bacillus subtilis. Journal of Bacteriology, 186*(20), 6830-6836.

Chen, C.-C., Lewis, R. J., Harris, R., Yudkin, M. D., & Delumeau, O. (2003). A supramolecular complex in the environmental stress signalling pathway of *Bacillus subtilis. Molecular Microbiology, 49*(6), 1657–1669.

Crosson, S., Rajagopal, S., & Moffat, K. (2003). The LOV Domain Family : Photoresponsive Signaling Modules Coupled to Diverse Output Domains. *Biochemistry*, 42(1) , 2–10.

Dai, T. et al., (2009). Photodynamic therapy for *Acinetobacter baumannii* burn infections in mice. *Antimicrobial Agents and Chemotherapy*, 53(9), 3929–34.

Delumeau, O., Chen, C.-C., Murray, J. W., Yudkin, M. D., & Lewis, R. J. (2006). High-molecular-weight complexes of RsbR and paralogues in the environmental signaling pathway of *Bacillus subtilis. Journal of Bacteriology, 188*(22), 7885–92.

Delumeau, O., Dutta, S., Brigulla, M., Kuhnke, G., Hardwick, S. W., Völker, U., Yudkin, M. D., et al., (2004). Functional and structural characterization of RsbU, a stress signaling protein phosphatase 2C. *The Journal of Biological Chemistry, 279*(39), 40927–37.

Demidova, T.N. & Hamblin, M.R. (2005). Effect of Cell-Photosensitizer Binding and Cell Density on Microbial Photoinactivation. *Antimicrobial agents and Chemotherapy* , 49(6), 2329–2335.

Dufour, A., & Haldenwang, W. G. (1994). Interactions between a *Bacillus subtilis* Anti-Sigma Factor (RsbW) and Its Antagonist (RsbV). *Journal of Bacteriology*, *176*(7), 1813-1820.

Dussurget, O., Cabanes, D., Dehoux, P., Lecuit, M., Buchrieser, C., Glaser, P., & Cossart, P. (2002). *Listeria monocytogenes* bile salt hydrolase is a PrfA-regulated virulence factor involved in the intestinal and hepatic phases of listeriosis. *Molecular microbiology*, *45*(4), 1095–106.

Endarko, E. et al., (2012). High-intensity 405 nm light inactivation of *Listeria monocytogenes. Photochemistry and Photobiology*, 88(5), 1280–6.

Ferreira, A., O' Byrne, C. P. and Boor, K. J. (2001). Role of B in Heat, Ethanol, Acid, and Oxidative Stress Resistance and during Carbon Starvation in *Listeria monocytogenes. Applied and Environmental Microbiology*, *67*(10), 4454–4457.

Ferreira, A., Gray, M., Wiedmann, M., & Boor, K. J. (2004). Comparative Genomic Analysis of the *sigB* Operon in *Listeria monocytogenes* and in Other Gram-Positive Bacteria. *Current Microbiology*, *48*(1), 39–46.

Ferreira, A., Sue, D., Byrne, C. P. O., & Boor, K. J. (2003). Role of *Listeria monocytogenes* sigma(B) in Survival of Lethal Acidic Conditions and in the Acquired Acid Tolerance Response. *Applied and Environmental Microbiology, 69*(5), 2692–2698.

Freitag, N. E., Rong, L., & Portnoy, D. a. (1993). Regulation of the *prfA* transcriptional activator of *Listeria monocytogenes*: multiple promoter elements contribute to intracellular growth and cell-to-cell spread. *Infection and Immunity*, *61*(6), 2537–44.

Garner, M. R., Njaa, B. L., Wiedmann, M., & Boor, K. J. (2006). Sigma B Contributes to *Listeria monocytogenes* Gastrointestinal Infection but Not to Systemic Spread in the Guinea Pig Infection Model. *Infection and Immunity*, 74(2), 876-886.

Glaser, P., Frangeul, L., Buchrieser, C., Rusniok, C., Amend, a, Baquero, F., Berche, P., et al., (2001). Comparative genomics of *Listeria* species. *Science (New York, N.Y.)*, *294*(5543), 849–52.

Gomelsky, M., & Hoff, W. D. (2011). Light helps bacteria make important lifestyle decisions. *Trends in Microbiology*, *19*(9), 441–448.

Hamblin, M.R. et al., (2005). *Helicobacter pylori* Accumulates Photoactive Porphyrins and Is Killed by Visible Light. *Antimicrobial agents and Chemotherapy, 49(7),* 2822-2827.

Hardwick, S. W., Pané-Farré, J., Delumeau, O., Marles-Wright, J., Murray, J. W., Hecker, M., & Lewis, R. J. (2007). Structural and functional characterization of partner switching regulating the environmental stress response in *Bacillus subtilis. The Journal of biological chemistry*, *282*(15), 11562–72.

Heavin, S., & O'Byrne, C. P. (2011). Post-Genomic Insights Into the Regulation of Transcription in the Facultative Intracellular Pathogen *Listeria monocytogenes*. In H. Wong (Ed.), *Stress Response of Foodborne Microorganisms*. Nova Science Publishers.

Hecker, M., Pané-Farré, J., & Völker, U. (2007). SigB-dependent general stress response in *Bacillus subtilis* and related gram-positive bacteria. *Annual Review of Microbiology*, *61*, 215–36.

Herrou, J., & Crosson, S. (2011). Function, structure and mechanism of bacterial photosensory LOV proteins. *Nature reviews. Microbiology*, *9*(10), 713–23.

Huala, E. (1997). *Arabidopsis* NPH1: A Protein Kinase with a Putative Redox-Sensing Domain. *Science, 278*(5346), 2120–2123.

Kairyte, K. et al., (2012). Effective inactivation of food pathogens *Listeria monocytogenes* and *Salmonella enterica* by combined treatment of hypericin-based photosensitization and high power pulsed light. *Journal of Applied Microbiology, 112,*1144–1151.

Kamp, H. D., & Higgins, D. E. (2011). A protein thermometer controls temperature-dependent transcription of flagellar motility genes in *Listeria monocytogenes. PLoS Pathogens, 7*(8), e1002153.

Kazmierczak, M. J., Mithoe, S. C., Boor, K. J., & Wiedmann, M. (2003). *Listeria monocytogenes* σ^B Regulates Stress Response and Virulence Functions. *Journal of Bacteriology, 185*(19).

Kazmierczak, M. J., Wiedmann, M., & Boor, K. J. (2006). Contributions of *Listeria monocytogenes* sigmaB and PrfA to expression of virulence and stress response genes during extra- and intracellular growth. *Microbiology (Reading, England), 152*(6), 1827–38.

Kim, T.-J., Gaidenko, T. a, & Price, C. W. (2004). A multicomponent protein complex mediates environmental stress signaling in *Bacillus subtilis. Journal of Molecular Biology, 341*(1), 135–50.

Lee, W.L. & Shalita, A.R. (1978). Comparative Studies of Porphyrin Production in *Propionibacterium acnes* and *Propionibacterium granulosum. Journal of Bacteriology, 133(2),* 811-815.

Leimeister-Wächter, M., Domann, E., & Chakraborty, T. (1992). The expression of virulence genes in *Listeria monocytogenes* is thermoregulated. *Journal of Bacteriology, 174*(3), 947–52.

Lin, S. et al., (2012). Photodynamic inactivation of methylene blue and tungsten-halogen lamp light against food pathogen *Listeria monocytogenes. Photochemistry and photobiology, 88(4),* 985–91.

Losi, A., Polverini, E., Quest, B., & Gärtner, W. (2002). First evidence for phototropin-related blue-light receptors in prokaryotes. *Biophysical Journal, 82*(5), 2627–34.

Losi, A., Quest, B., & Gärtner, W. (2003). Listening to the blue: the time-resolved thermodynamics of the bacterial blue-light receptor YtvA and its isolated LOV domain. *Photochemical & Photobiological Sciences, 2*(7), 759.

Luksiene, Z., Buchovec, I. & Paskeviciute, E. (2010). Inactivation of several strains of *Listeria monocytogenes* attached to the surface of packaging material by Na-Chlorophyllin-based photosensitization. *Journal of Photochemistry and Photobiology. B, Biology, 101(3),* 326–31.

Luksiene, Z. & Paskeviciute, E. (2011). Microbial control of food-related surfaces: Na-Chlorophyllin-based photosensitization. *Journal of photochemistry and photobiology. B, Biology, 105(1),*69–74.

Marc, Y. Le et al., (2009). Modelling the photosensitization-based inactivation of *Bacillus cereus. Journal of Applied Microbiology*, 107(3), 1006–11.

Marles-Wright, J., Grant, T., Delumeau, O., Van Duinen, G., Firbank, S. J., Lewis, P. J., Murray, J. W., et al., (2008). Molecular architecture of the "stressosome," a signal integration and transduction hub. *Science (New York, N.Y.), 322*(5898), 92–6.

Marles-Wright, J., & Lewis, R. J. (2010). The stressosome: molecular architecture of a signalling hub. *Biochemical Society transactions, 38*(4), 928–33.

Martinez, L., Reeves, A., & Haldenwang, W. (2010). Stressosomes formed in *Bacillus subtilis* from the RsbR protein of *Listeria monocytogenes* allow σ^B activation following exposure to either physical or nutritional stress. *Journal of Bacteriology*, *192*(23), 6279–86.

Möglich, A., & Moffat, K. (2007). Structural Basis for Light-dependent Signaling in the Dimeric LOV Domain of the Photosensor YtvA. *Journal of Molecular Biology*, *373*(1), 112–126.

Miller, R. V. et al., (1999). ASM News Vol 65, Num 8, August 1999, Bacterial Responses to Ultraviolet Light. *ASM News*, 65(8).

Murdoch, L. E., Maclean, M., Endarko, E., MacGregor, S. J., & Anderson, J. G. (2012). Bactericidal Effects of 405 nm Light Exposure Demonstrated by Inactivation of *Escherichia, Salmonella, Shigella, Listeria,* and *Mycobacterium* Species in Liquid Suspensions and on Exposed Surfaces. *The Scientific World Journal*, *2012*, 1–8.

Murray, J. W., O. Delumeau, & R. J. Lewis (2005). Structure of a nonheme globin in environmental stress signaling. *Proc. Natl. Acad. Sci. USA,* 102,17320–5.

Nadon, C. A., Bowen, B. M., Wiedmann, M., & Boor, K. J. (2002). Sigma B Contributes to PrfA-Mediated Virulence in *Listeria monocytogenes*. *Infection and Immunity*, *70*(7), 3948-3952.

Nitzan, Y. et al., (2004). ALA induced photodynamic effects on gram positive and negative bacteria. *Photochemical & photobiological sciences : Official Journal of the European Photochemistry Association and the European Society for Photobiology*, 3(5), 430–5.

NUI Galway (2013). Monthly Trends. Available at: http://weather.nuigalway.ie/monthly Trends.php.

O'Byrne C.P. & Karatzas K.A. (2008) The role of sigma B (sigma B) in the stress adaptations of *Listeria monocytogenes*: overlaps between stress adaptation and virulence. *Adv Appl Microbiol*. 65, 115-40

Ondrusch, N., & Kreft, J. (2011). Blue and red light modulates SigB-dependent gene transcription, swimming motility and invasiveness in *Listeria monocytogenes*. *PloS One*, *6*(1), e16151.

Panek, H. & Brian, M.R.O. (2002). A whole genome view of prokaryotic haem biosynthesis. *Microbiology*, 148, 2273–2282.

Price, C. W., Fawcett, P., Cérémonie, H., Su, N., Murphy, C. K., & Youngman, P. (2001). Genome-wide analysis of the general stress response in *Bacillus subtilis*. *Molecular microbiology*, *41*(4), 757–74.

Rea, R., Hill, C. & Gahan, C.G.M. (2005). *Listeria monocytogenes* PerR Mutants Display a Small-Colony Phenotype , Increased Sensitivity to Hydrogen Peroxide , and Significantly Reduced Murine Virulence. *Applied and Environmental Microbiology,* 71(12), 8314-8322.

Romanova, N.A. et al., (2003). Assessment of Photodynamic Destruction of *Escherichia coli* O157 : H7 and *Listeria monocytogenes* by Using ATP Bioluminescence. *Applied and Environmental Microbiology*, 69(11), 6393–6398.

Salomon, M., Christie, J. M., Knieb, E., Lempert, U., & Briggs, W. R. (2000). Photochemical and Mutational Analysis of the FMN-Binding Domains of the Plant Blue Light Receptor, Phototropin. *Biochemistry* (12), 9401–9410.

Smirnova, N., Scott, J., Voelker, U., & Haldenwang, W. G. (1998). Isolation and characterization of *Bacillus subtilis sigB* operon mutations that suppress the loss of the negative regulator RsbX. *Journal of Bacteriology*, *180*(14), 3671–80.

Swartz, T. E., Corchnoy, S. B., Christie, J. M., Lewis, J. W., Szundi, I., Briggs, W. R., & Bogomolni, R. A. (2001). The photocycle of a flavin-binding domain of the blue light photoreceptor phototropin. *The Journal of Biological Chemistry*, *276*(39), 36493–500.

Taylor, B. L., & Zhulin, I. B. (1999). PAS domains: internal sensors of oxygen, redox potential, and light. *Microbiology and molecular biology reviews : MMBR*, *63*(2), 479–506.

Tiensuu, T., Andersson, C., Rydén, P., & Johansson, J. (2013). Cycles of light and dark co-ordinate reversible colony differentiation in *Listeria monocytogenes. Molecular microbiology*, *87*(4), 909–24.

Toledo-Arana, A., Dussurget, O., Nikitas, G., Sesto, N., Guet-Revillet, H., Balestrino, D., Loh, E., et al., (2009). The *Listeria* transcriptional landscape from saprophytism to virulence. *Nature*, *459*(7249), 950–6.

Vijay, K., Brody, M. S., Fredlund, E., & Price, C. W. (2000). A PP2C phosphatase containing a PAS domain is required to convey signals of energy stress to the sigmaB transcription factor of *Bacillus subtilis. Molecular Microbiology*, *35*(1), 180–8.

Yang, X., Kang, C. M., Brody, M. S., & Price, C. W. (1996). Opposing pairs of serine protein kinases and phosphatases transmit signals of environmental stress to activate a bacterial transcription factor. *Genes & Development*, *10*(18), 2265–2275.

In: *Listeria monocytogenes*
Editor: Edmund C. Hambrick
ISBN: 978-1-63117-054-6

Chapter 6

SPATIAL DISTRIBUTION OF *LISTERIA MONOCYTOGENES* AND *PSEUDOMONAS FLUORESCENS* IN MIXED BIOFILMS

C. H. Puga*[*], *C. SanJose and B. Orgaz
Department of Nutrition, Food Science and Technology,
University Complutense of Madrid, Spain

ABSTRACT

The idea of primary colonization of surfaces by *Pseudomonas* as a requirement or a decisive step for *Listeria monocytogenes* (Lm) integration in biofilms (Sasahara & Zottola, 1993) has been of primary influence in food industry sanitation. The assumption has been that *L. monocytogenes* is not capable of sufficient EPS production to create biofilms on its own and that *Pseudomonas* (*fluorescens, putida, fragi* or other species) are clearly superior in this respect and can provide a generous matrix able to host unobtrusive guests such as *Listeria*. For further insight into this classic 20-year-old topic, mixed biofilms with several strains of Lm and *P. fluorescens* (Pf) were obtained at room and refrigeration temperatures to be studied with confocal microscopy. Differential labeling of both species allowed identification of their respective location in biofilm structure. Both types of cells were separately quantified by plating in specific media. Co-culture benefits for Lm were confirmed in biofilms, particularly at 4°C, although not in planktonic cells. Besides, *Listeria* cells appeared preferentially located in the deeper, more anaerobic layers, with *Pseudomonas* cells placed on top of them. It is well known that bacteria in biofilms are better protected against various antimicrobial agents than planktonic cells; the protection is increased for cells in mixed biofilms and even more so when the cells are located in their deeper layers. The observed distribution pattern may guide the development of new and more effective cleaning and disinfection strategies to be used in food processing associated surfaces.

Keywords: *Listeria monocytogenes, Pseudomonas,* mixed biofilms, spatial distribution, layering, CLSM

[*] E-mail: chpuga@pdi.ucm.es.

INTRODUCTION

Biofilms are communities of microorganisms embedded in a matrix of extracellular polymeric substances (EPS) produced by themselves (Kjelleberg & Givskov, 2007; Karunakaran,., 2011). This matrix is an adhesive gel that attaches to a surface and provides protection to the biofilms inhabitants (Flemming & Wingender, 2010). In the food industry context, this protection stands against sheer forces of flowing liquids and chemical action of tensoactives and biocides during cleaning and disinfection operations (Chmielewski & Frank, 2003; Fratamico et al., 2009; Srey et al., 2013; Valderrama & Cutter, 2013). When these operations do not achieve efficient biofilm elimination, the biotransfer of shed cells or biofilm fragments may contaminate raw materials or foods (Verran et al., 2008). This is more dangerous when the embedded cells are pathogens, as it is the case for Lm, when the contaminated food can support their growth during the storage and commercialization lifespan and when the foods do not receive further hygienic treatment, as in Ready To Eat (RTE) foods. When biofilm inhabitants are very active food spoilage agents, as it is the case of Pf, they tend to be widely widespread and abundant in food processing plants, having an impact not only on the product's organoleptic and preserving qualities, but also on different aspects of the plant's technological performance (Fratamico et al., 2009) and on costs of cleaning and disinfection. So, even if biofilms are pathogen-carrying or not, their eradication from food associated surfaces is a challenge for the food hygienist.

Pseudomonas sp. is a versatile and fast psychrotroph known to dominate the spoilage pace and pattern of protein-rich refrigerated foods (Huis in't Veld J. H. J., 1996). Pf is most prominent in milk dairy products, while *P. putida* and *P. fragi* are more commonly associated with meat products (Liao, 2006). *P. aeruginosa* has minor importance as a food pathogen, but it is obviously the best known *Pseudomonas* species (Silby et al., 2011) because of its role in cystic fibrosis; it is thus a reference for the whole gender. *Pseudomonas* biofilms have been extensively studied, not just for their clinical and food related relevance (Mettler & Carpentier, 1998; Mann & Wozniak, 2012), but as a convenient biofilm model (Lopez et al., 2010) due to rapid development, considerable thickness and rich structural complexity.

Pseudomonas interactions with *Listeria* were already studied in planktonic cultures in milk (Farrag & Marth, 1989) and special media (Buchanan & Bagi 1999). *Pseudomonas* protective role on surfaces was described very early (Breen et al., 1982). Observation of mixed biofilms of Lm strain Scott A and *P. fragi* served to draw the inspiring idea of *Listeria*'s utilization of a primary colonizer to enhance its attachment to food related surfaces (Sasahara & Zottola, 1993). *P. putida* and Lm have been found to actually coincide in real biofilms in the food industry (Hassan et al., 2004) and similar partnerships are likely to be common in the refrigerated food processing context.

Human listeriosis owes its "emergence", in the eighties of the twentieth century, to advances in food industry sanitation (Ryser, 2007). Before that, Lm was not able to achieve significant figures due to its poor competitive skills to face the resident microbiota, traditionally kept at much higher levels than now. Interestingly enough, some of the antagonistic mechanisms used by other organisms, such as lactic acid bacteria, are now attempted for Lm control without the use of biocides (Nilsson et al., 1999; Mariani et al., 2011). Additionally, certain food preservation conditions, such as refrigeration or multiple-hurdle strategies, have offered a unique chance for Lm to occupy "new" niches. Thus, from the ecological point of view, the upsurge and persistence of this pathogen in the food chain is a remarkable case of manmade situation (Carpentier & Cerf, 2011; Valderrama & Cutter, 2013).

Many Lm strains have been tested in their ability to form monospecies biofilms in experimental conditions (Moretro 2004, Nilsson et al., 2011; da Silva et al., 2013). Their structure is very different from the more frequently described biofilm models, which are mushroom-like, such as those of *Pseudomonas* (Costerton, 2007). The structures of Lm are thin and sparse, sometimes described as scattered colonies or knitted chains (Rieu et al., 2008). Their matrix composition is also rather different and particularly rich in protein and extracellular DNA (Harmsen et al., 2010; Combrouse et al., 2013). *Listeria*'s interactions with various bacteria in mixed biofilms have been previously investigated (Jeong & Frank, 1994; Carpentier & Chassaing, 2004).

In real environments of food processing Lm is thought to be present in biofilms in commensal association with compatible partners (Elias & Banin, 2012) one of which seems to be *Pseudomonas*. As multispecies biofilms are more resistant to cleaning and disinfection agents than monospecies biofilms (Simoes et al., 2009) better knowledge of their structural features may help to develop more efficient eradication alternatives.

This chapter focuses on three issues: 1) Global outcome of the interactions between Lm and Pf on both biofilm and planktonic cell numbers. 2) Comparison of the results at 20 and 4°C, as low temperatures are very relevant in the food industry context. 3) Spatial distribution of the two species in biofilms, as seen by CLSM. One psychrotrophic Pf collection strain was used for these aims, together with six different Listerias: the reference strain of Lm, Scott A, four persistent strains of the same species, repeatedly sampled in a meat product processing plant (Ortiz et al., 2010) and one strain of *L. innocua*.

MATERIALS AND METHODS

Bacterial Strains

Pf ATCC 948^{TM}, *Listeria innocua* ATCC 33090^{TM} and five strains of *Lm*: 4 persistent strains, S1 (serotype 1/2a; lineage II), S2 (1/2a; lineage II), S4-1 (1/2b; lineage I), and S10-1 (1/2a; lineage II), isolated from an Iberian pig slaughterhouse and a processing plant by Ortiz et al., (2010) and *Lm* Scott A (4b; lineage I), were selected as former biofilm organisms. All the strains were stored at −20°C in Tryptone Soya Broth (TSB, Oxoid) with 15% glycerol. Preinocula were obtained after incubation for 24h under shaking (80 rpm) at 20°C in TSB and reaching up to mid-exponential phase. Cells were harvested by centrifugation at 4000 × g for 10 min, washed twice in sterile TSB and their OD_{600} adjusted in order to reach 10^4 $CFU{\cdot}mL^{-1}$ of each bacteria after inoculation, whether in single or binary cultures.

Experimental System

Biofilms were developed on single-use 22 × 22 mm thin, borosilicate commercial microscope glass coverslips. As described in Orgaz et al., (2011), 16 coverslips were held vertically by marginal insertion into the narrow radial slits of a Teflon carousel platform (6.6 cm diameter). The platform and its lid were assembled by an axial metallic rod for handling and placed into a 600 mL beaker. The whole system, i.e., coverslips, carousel and the covered 600 mL beaker, were heat-sterilized as a unit, before aseptically introducing 60 mL of inoculated TSB. In the case of multispecies biofilms both bacteria were inoculated at the same initial level, and their corresponding monospecies biofilms were used as controls. Incubation was carried out at 20°C/96h or 4°C/20 days, in a rotating shaker at 80 rpm. Under these conditions, the biofilm growth occupied about 70% of the coverslip's surface.

Cell Recovery and Counting

For sampling biofilm cells, glass coverslips were withdrawn with tweezers and were carefully rinsed in sterile 0.9% NaCl to discard weakly attached cells. Then, attached cells were removed by swabbing both coverslip faces and then dispersed into tubes containing 1.5 mL peptone water. These were then vigorously stirred in a vortex to break up cell aggregates. Sampling of the planktonic phase was also carried out in parallel by withdrawing 100 μL of the culture medium at each time. Biofilms and planktonic cells were appropriately diluted in peptone water and the pertinent dilutions were plated according to the drop method described by Hoben & Somasegaran (1982). Briefly, 3 drops of 20μL were deposited into plates of selective media, PALCAM (Oxoid) and *Pseudomonas* Agar Base (Oxoid) for counting *Listeria* spp. *and Pseudomonas* spp., respectively. Colonies were counted after 48 h incubation at 37°C or 30°C, for *Lm* or *Pf*, respectively.

Biomass Determination

To evaluate the attached biomass (cells plus matrix) two coverslips of each carousel were dried and stained for 2 min in a 1‰ Coomassie Blue solution in acetic acid/methanol/water (1:2.5:6.5). Once rinsed and dried again, coverslips were scanned in a Bio-Rad GS-690 densitometer, with Molecular Analyst software (Bio-Rad Inc.) for image analysis. The average optical density of the whole glass coupon surface was obtained, using a green filter corresponding to a 520–570 nm wavelength.

Apparent Attached-cell Generation Times

Apparent generation times of the attached counts over the exponential phase were calculated. *Apparent* means that a combination of the overall proliferation and the balance between attached and detached cells is integrated in each result of viable attached cells. *Apparent* generation times were calculated for monospecies biofilms of *Lm* Scott A and binary biofilms of *Lm* Scott A and *Pf* ATCC 948^{TM}, at 20ºC and 4ºC, using the following equations:

$$\mu = (\text{Log } N_2 - \text{Log } N_1)/(t_2 - t_1) \quad (1)$$

$$g = \log 2/\mu \quad \mathbf{(2)}$$

being N_2 the attached cell density at the end of the exponential growth phase; N_1 the attached cell density at the beginning of the exponential growth phase; t_2 and t_1 were the length of time to attain N_2 and N_1, respectively.

Statistical Analysis

Analysis of the association effect was performed on planktonic and biofilm counts of *Pf* and *Lm* both in mono and binary cultures. Data were analyzed using Statgraphics Plus 5.0 software (Statistical Graphics Corporation, Rockville, Md., USA) and ONE-way ANOVA analysis of variance. Mean comparisons were carried out to determine significant differences at a 95.0% confidence level ($p < 0.05$).

Confocal Laser Scanning Microscopy (CLSM)

For CLSM observation, biofilms developed in glass coverslips were rinsed with sterile 0.9% NaCl and then stained with ViaGram™ Red+ Bacterial Gram Stain and Viability kit (V7023, Invitrogen). This kit includes DAPI, which stains all the cells and Texas Red-X, which selectively binds surfaces of Gram-positive bacteria. Thus, the blue cells here correspond to *Pf*, whereas cells that exhibit red surfaces correspond to *Lm*. CLSM images from 0.12 x 0.12 mm coupon areas were obtained using a TCS SP2 microscope (Leica Lasertechnik, Heidelberg, Germany) with a water immersion objective lens 63X. Three-dimensional projections (Maximum Intensity Projection, MIP) and extended section views of biofilm structures were reconstructed from z-stacks using IMARIS® 7.6 software (Bitplane AG, Zúrich, Switzerland). In order to analyze the spatial distribution of the *Lm* population within the binary species biofilms, a parameter here called biovolume was calculated using the MeasurementPro module of Imaris 7.6. For this, the whole image was first segmented and analyzed to obtain the total volume occupied by cells (that is, blue cells plus red cells); then the image was divided into two sections, the top layer corresponding to the upper 50% of the biovolume and the bottom layers to the inferior 50%. *Listeria* spatial distribution was thus calculated as a percentage of red in each of the above-mentioned sections.

RESULTS

Counts in Planktonic Phase and Biofilms from Monospecies and Dual-species Cultures

In the planktonic phase of the dual cultures with Pf and Lm strain Scott A, the net outcome of species interaction for Lm, in terms of viable cells, was practically neutral at both temperatures (Figure 1A and B). In other words, Lm growth was not affected by the presence of Pf. At some intermediate culture stages at 4°C, however, the interaction outcome was even negative for Lm. Pf cell counts, planktonic or in biofilms, were not affected by the presence of Lm (data not shown).

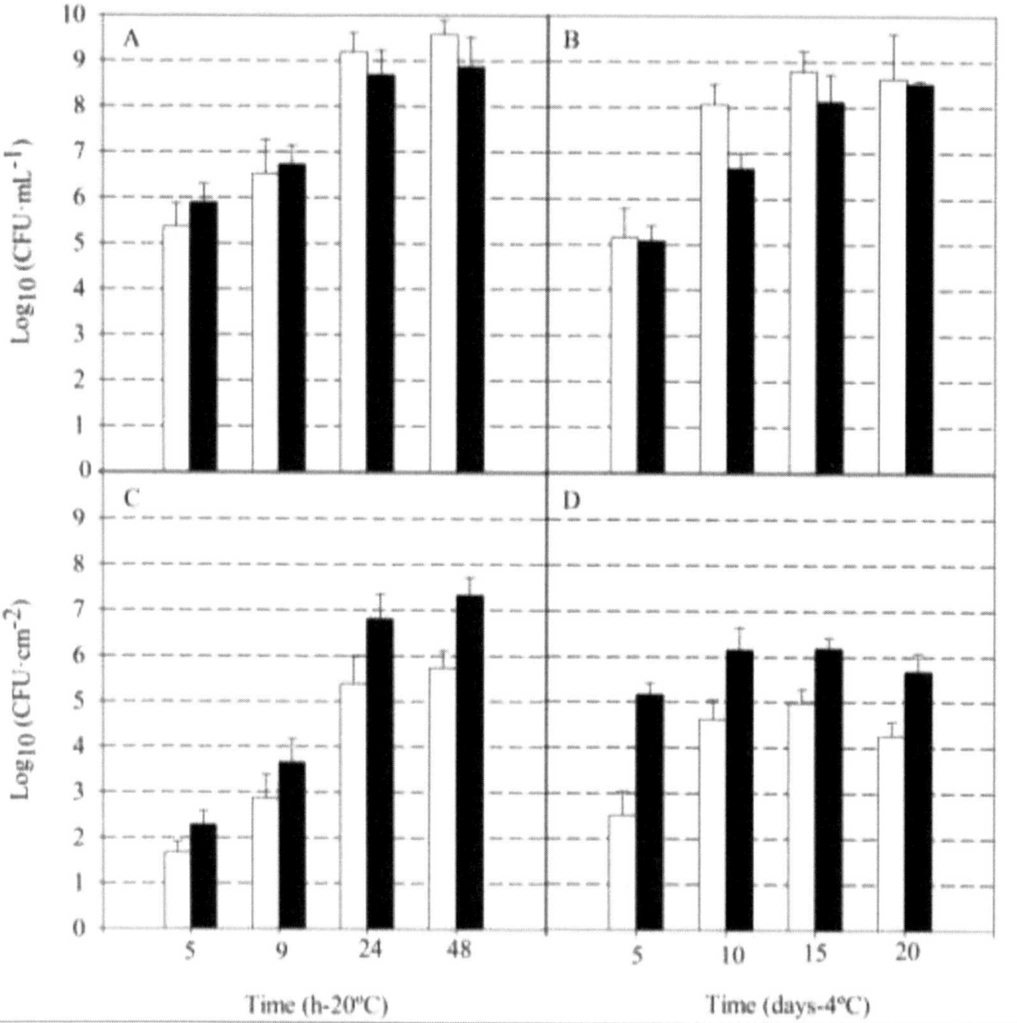

Figure 1. Counts of Lm Scott A in monospecies (white bars) or in dual cultures with Pf (black), incubated at 20ºC (left) or 4ºC (right). Planktonic cells above, biofilm cells below. Error bars represent the confidence interval for the mean (n = 4). Biofilm counts in mono- and multi-species cultures were all significantly different ($P < 0.01$).

At least with the inocula sizes used in these experiments (yielding 10^4 CFU ml^{-1} of each species at the start of the culture) all the strains, tested here, of Lm produce monospecies biofilms. Therefore, no absolute dependence for *Pseudomonas* primary colonization was observed. However, in the dual cultures biofilm phase (Figure 1C and D) the combination of Pf and Lm Scott A was always beneficial for Lm (at least 1 log extra growth) and even more at 4ºC. There were up to 200-fold more Lm cells per cm^2 in 5-day biofilms at 4ºC with *Pseudomonas* than without. The apparent growth rate for Lm Scott A in biofilms, i.e., the figure combining attached cell multiplication rate and new cell attachment, was 0.05 h^{-1} at the monospecies and 0.09 h^{-1} at the dual-species biofilms at 4ºC.

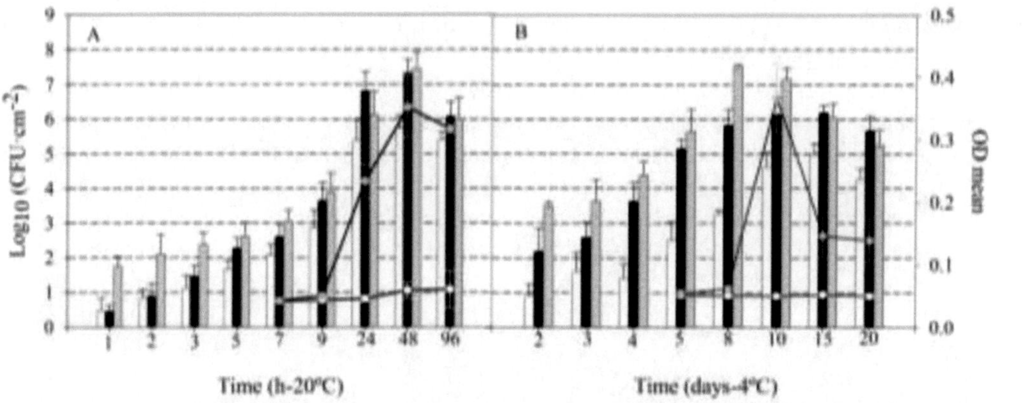

Figure 2. Counts of Lm Scott A in monoespecies (white bars) or dual species (black) biofilms and Pf counts in the dual species biofilms (lined bars), along incubation time at 20ºC (A) or 4ºC (B). Lines represent biomass OD in pure (white circles) or dual-species cultures (gray circles). (n=5). Differences between Lm counts in pure and dual-species biofilms were all significantly different ($P < 0.02$), except for the cases of 1h, 2h and 3h at 20ºC.

Figure 2 presents count levels of the two species in mixed biofilms, from the early stages. The proportion between them was quite large initially; one Lm for every 10 or more Pf, but figures of the two species became closer later on. It could also be noted that during the first 3h at 20ºC, Lm adhered at the same pace with or without Pf. At 4ºC, however, Lm attachment was helped by Pf from the early stages.

Overall biofilm biomass (cells plus matrix) had a significantly different development pattern, depending on temperature and the biofilm being mono- or dual species. Matrix development in pure Lm biofilms was very scarce from the point of view of optical density and was practically unchanged even when the cell density reached 10^7 CFU cm^{-2}. In mixed biofilms the matrix was poorly developed until it rather suddenly started to grow. This sharp start took place at 20ºC when Pf reached 10^4 CFU cm^{-2} in biofilms, but appeared delayed to a density of 10^7 CFU Pf cm^{-2} when at 4ºC. Over time, the biomass quantitative profile also varied with temperature (Figure 2 A and B); at 4ºC, a larger decrease of biomass in old cultures was observed.

CLSM Images of Biofilms with Pf and Listeria

In monocultures of Lm Scott A cells look more aggregated at 20ºC than at 4ºC (Figure 3), but in both cases, biofilms were sparse and thin (3μm thick). In the zenital images of dual species with similar Lm cell density, known from selective media plate counting (Figure 3B and D), red cells (Lm) were less visible and apparently much fewer at 4ºC than at 20ºC. The hypothesis then was that a masking effect of Pf was hiding most of Lm cells underneath.

An effect of blanketing or layering, to confirm that hypothesis, could be observed in the images of Figure 4, showing profiles of dual species biofilms with Pf and Lm Scott A, obtained at different temperatures and times. A layering distribution with Lm underneath Pf was observed in all cases, except after 15 days at 4ºC, where few Pf were visible. It must be remembered that images such as those in Figure 4 do not correspond to the same individual biofilm through time, but to randomly taken zones of different coupons taken from the same culture at different times of incubation.

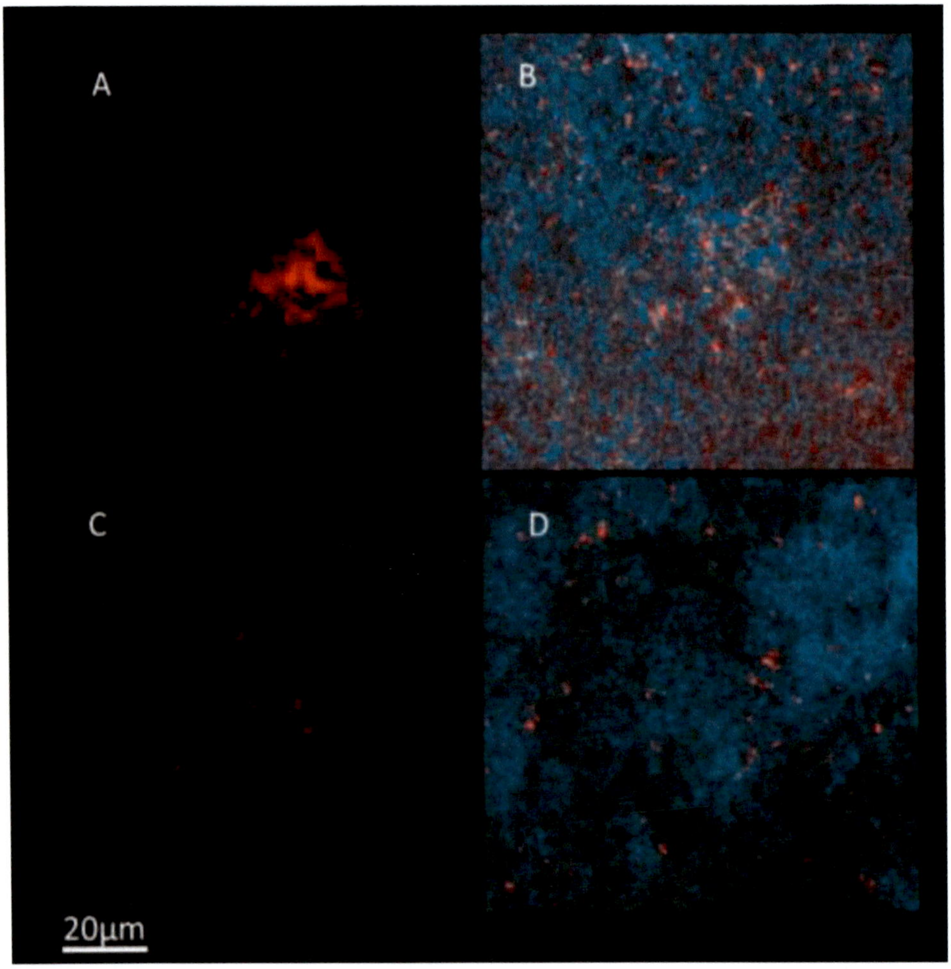

Figure 3. CLSM zenital images of biofilms such as those in Figure 1, with Lm Scott A in red and Pf in blue. A) Monospecies of Lm, after 24h at 20ºC (CFU·cm^{-2}: 2.4 x 10^5; thickness ≤ 3μm). B) Dual species, 24h at 20ºC (CFU·cm$^{-2:}$ 6.4 x 10^6 Lm and 1.3 x 10^6 Pf; thickness 20 μm). C) Monospecies of Lm, after 10 days at 4ºC (CFU·cm^{-2}: 9.7 x 10^4; thickness ≤ 3μm). D) Dual species, 10 days at 4ºC (CFU·cm^{-2}: 1.5 x 10^6 Lm and 1.1 x 10^6 Pf; thickness 14 μm).

In order to confirm this pattern further, four Lm persistent strains obtained from a food industry facility were incubated with the same Pf for 10 days at 4ºC. The resulting biofilm profiles are shown in Figure 5. Two types of structure could be there observed: a thin and compact one, with clear layering of the two species, shown by consortia containing Lm strains S1 and S10, and a spongy, fluffy and more complex profile, with less clear layering, in the case of biofilms with S2 and S4-1. Besides, whereas in most cases biofilm profiles show mixtures of single cells, in those of consortium Pf/Lm S2 the local accumulation of microcolonies of just one of the species was visible.

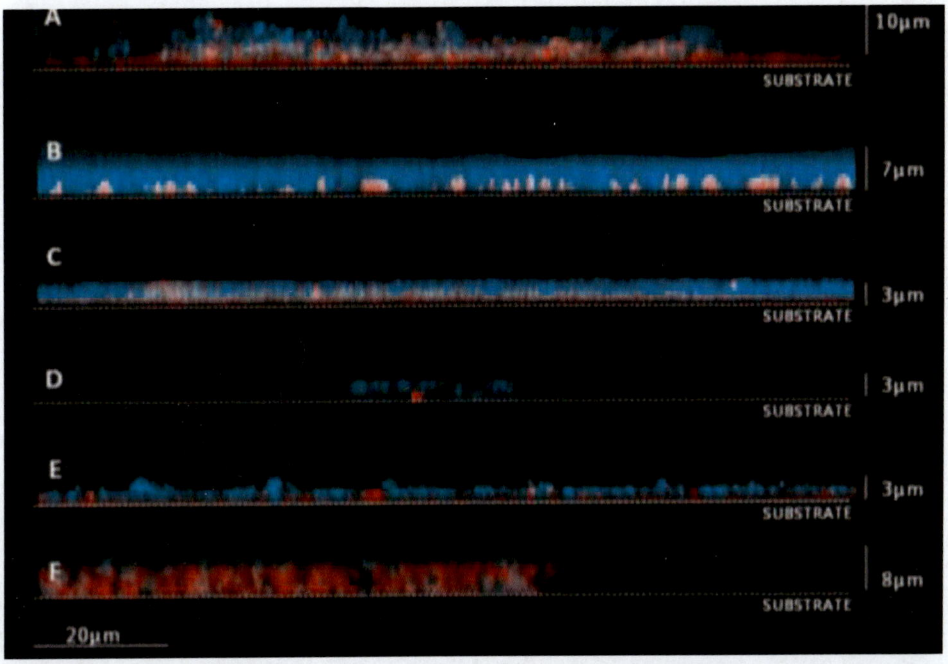

Figure 4. CLSM transversal images of dual species biofilms with Lm Scott A in red and Pf in blue. A) After 48h at 20ºC (CFU·cm^{-2}: 2.14 x 10^7 Lm and 2.93 x 10^7 Pf). B) 72h at 20ºC (CFU·cm^{-2}: 2.87 x 10^7 Lm and 2.93 x 10^8 Pf). C) 96h at 20ºC (CFU·cm^{-2}: 1.17 x 10^6 Lm and 1.01 x 10^6 Pf). D) 4 days at 4ºC (CFU·cm^{-2}: 4.27 x 10^3 Lm and 2.41 x 10^4 Pf). E) 8 days at 4ºC (CFU·cm^{-2}: 6.73 x 10^5 Lm and 3.12 x 10^7 Pf). F) 15 days at 4ºC (CFU·cm^{-2}: 1.53 x 10^6 Lm and 1.10 x 10^6 Pf).

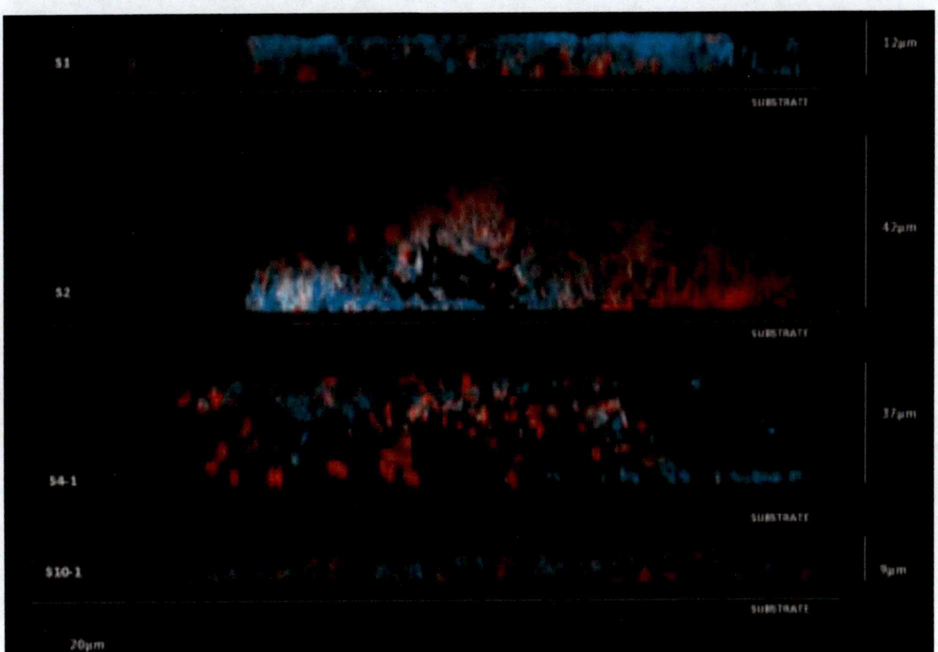

Figure 5. CLSM transversal images of dual species biofilms with four different persistent Lm strains in red, and Pf in blue, after 10 days at 4ºC. S1) CFU·cm^{-2}: 5.48 x 10^5 Lm and 2.82 x 10^7 Pf. S2) CFU·cm^{-2}: 1.74 x 10^5 Lm and 5.01 x 10^7 Pf. S4-1) CFU·cm^{-2}: 2.45 x 10^5 Lm and 3.39 x 10^7 Pf. S10-1) CFU·cm^{-2}: 2.24 x 10^5 Lm and 5.01 x 10^7 Pf.

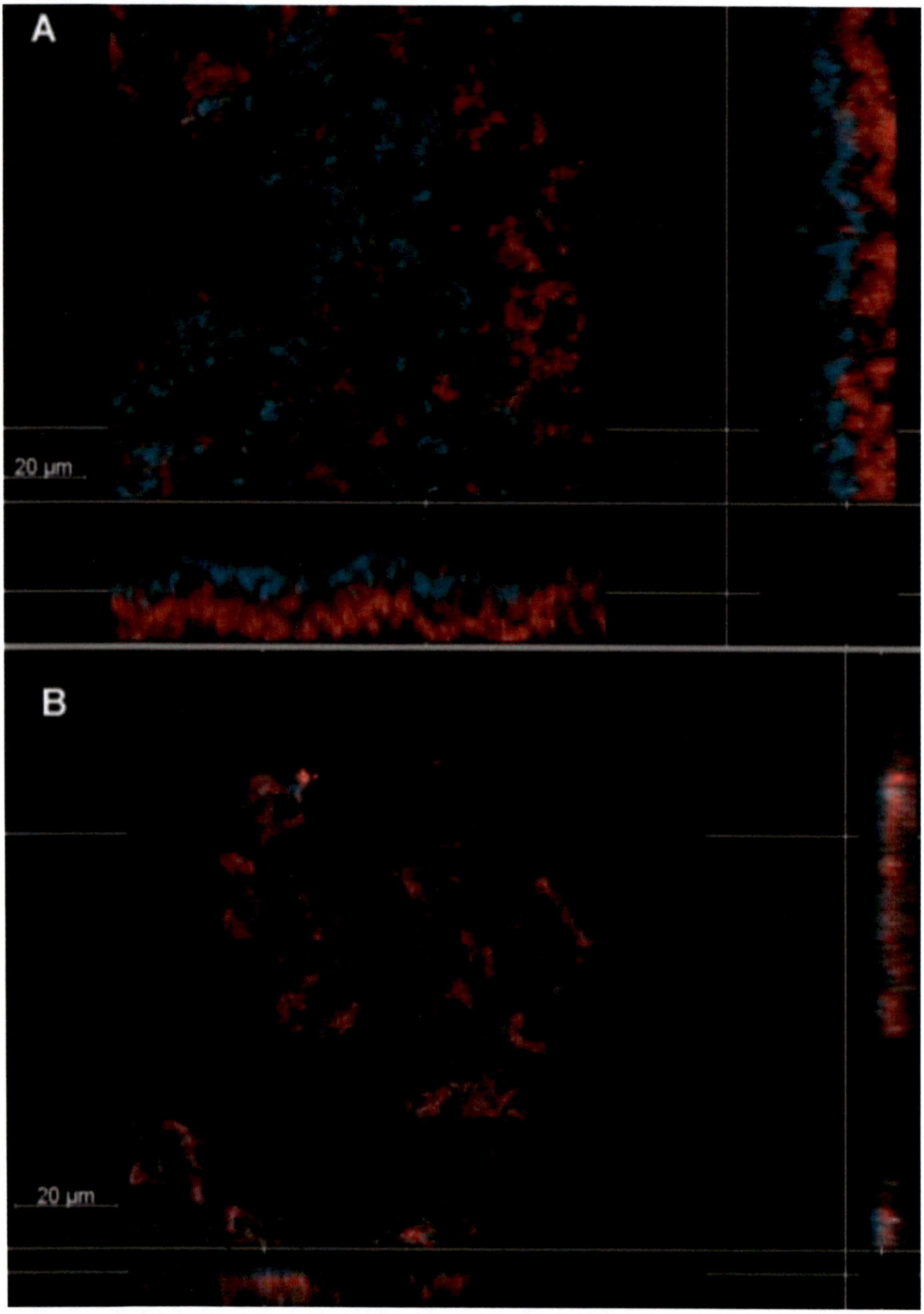

Figure 6. CLSM extended section view (z-y and z-x planes) of dual species biofilms with *Listeria innocua* ATCC® 33090™ in red and Pf in blue. A) after 96h at 20ºC (CFU·cm^{-2}: 1.27 x 10^8 *L. innocua* and 6.16 x 10^8 Pf; Thickness 28 µm). B) 20 days at 4ºC (CFU·cm^{-2}: 2.02 x 10^5 *L. innocua* and 4.86 x 10^7 Pf; Thickness 8 µm).

In consortia of Pf with *L. innocua* (Figure 6), species layering in the biofilms was similar to that observed for Lm, with a spongy but neat stratification after 96h at 20ºC and a denser picture, apparently with less EPS, in the 20-day at 4ºC case, probably corresponding to an old biofilm.

Listeria's percentage distributions between the top half and bottom half of the profile of each dual species biofilms of Pf with *Listeria* studied here are shown in Table 1. The figures represent the % of *Listeria* color label present in the top and bottom halves of the biofilm biovolume, the bottom being the substratum's surface. Of the twelve cases checked for quantifiable layering, nine presented more cells in the "deep compartment" or bottom layer of the biofilms, although in two of those cases only 54% of the total *Listeria* cells were there. The more clear-cut layered distribution could be observed at early and intermediate incubation times. *L. innocua* dual biofilms with Pf also presented layering in the images from cultures at 20ºC. No relationship between serotype or lineage of Lm strains and their spatial distribution in dual biofilms was detected (Valderrama & Cutter, 2013).

Table 1. Layering of *Listeria* cells in the dual-species biofilms shown in figures 4 to 6

					Distribution (%)	
Species	Strain	Serotype (lineage)	T (°C)	Time	Top Layer	Bottom Layer
L m	Scott A	4b (I)	20	48h	19	81
L m	Scott A	4b (I)	20	72h	45	54
L m	Scott A	4b (I)	20	96h	22	78
L m	Scott A	4b (I)	4	4days	19	81
L m	Scott A	4b (I)	4	8days	11	89
L m	Scott A	4b (I)	4	15 days	45	54
L m	S1p	1/2a (II)	4	10 days	53	47
L m	S2p	1/2a (II)	4	10 days	34	66
L m	S4-1p	1/2b (I)	4	10 days	63	37
L m	S10p	1/2a (II)	4	10 days	40	60
L. innocua	ATCC® 33090™	-	20	96h	14	86
L. innocua	ATCC® 33090™	-	4	20 days	72	28

DISCUSSION

Outcome of the Interactions between Pf and Lm Scott A on the Planktonic and Biofilm Phases of Dual Cultures

In the conditions used, there was a striking contrast between the neutral relationship in the planktonic phase of the co-culture and the positive relationship (at least for Lm) in biofilms. This was more intense at low temperatures, where the specific cold adaptation physiology of both species apparently meet or are introduced in a way that gives an advantage to Lm (Chan & Wiedmann, 2009).

Many interactions are likely to take place between these two species, their outcome depending on strains, environment, time and relative populations. In an attempt to find a specific Pf isolate to limit growth of Lm, Buchanan & Bagi (1999) worked with co-cultures of another (non-siderophore producing) Pf strain on Scott A Lm's planktonic growth kinetics. They concluded that all ecologica outcome possibilities, that is, no effect, stimulation or inhibition of Lm, could be found with those co-cultures depending on environmental conditions and initial population levels. Buchanan achieved a remarkable reduction in Maximal Population Density of Lm (over 3-log) in spite of a decrease in generation time, as a result of the coculture at 4°C, pH 7 and 5% NaCl in BHI medium. Inhibition of Lm by *Pseudomonas* spp. had been related on other occasions to the production of siderophores or bacteriocins, factors that were checked out in Buchanan's work.

Other authors, however, have described the stimulation effects on planktonic growth of Lm by various species and strains of *Pseudomonas* (Farrag & Marth, 1989; Gram, 1993). One of the reasons frequently suggested to account for positive effects on Lm was *Pseudomonas* ability to produce extracellular proteinases that could mobilize essential amino acids, particularly in media such as milk. Extracellular enzymes such as proteinases, lipases and other hydrolases are very often produced by *Pseudomonas* (Sorhaug & Stepaniak, 1997) and at least some of them are particularly well expressed at low temperatures (Jaspe et al., 1995) and could contribute to nutrient commensalism.

Published antecedents on biofilm consortia involving Lm have reported the influence of strain precedence in contact with the substratum surface and the partners respective initial population; in general, reports on *Pseudomonas* stimulation tend to dominate those on inhibition of Lm attachment (Breen et al., 1982; Sasahara & Zottola, 1993; Carpentier & Chassaing, 2004; Hassan et al., 2004; Mellefont & Ross, 2007). Ecological interaction mechanisms involving cell-to-cell communication are likely to be more effective inside biofilms than in the planktonic environment (Elias & Banin, 2012) as the smaller distance between cells allows for higher concentrations of autoinducers, metabolic end-products that may act as precursors or inhibitors, enzymes, etc., playing a syntrophic role (Morris et al., 2013).

EPS and Layering in Dual Biofilms

The main and more frequently proposed factor for Pf cooperation with Lm in biofilms has always been EPS, large amounts of *Pseudomonas* EPS that would physically fix, embed and protect Lm. According to our results, however, dual Pf/Lm biofilms are not particularly thick and do not have much EPS. Some mutual benefit between Pf and Lm (nutrient or signal-based) may be holding or reinforcing their interaction at the expense of reduced EPS production. This appeared to be regulated by a QS switch mechanism, having a lower cell density threshold at 20° and at 4°C.

The most characteristic feature of these dual Pf/Lm biofilms was layering; a characteristic type of spatial distribution that may be a key to decipher the specific interactions between these two species when living in biofilms. Lm appeared to be located at the bottom layers of the dual biofilms, but at early stages Lm attached cells were fewer than those of Pf. There were many more Pf than Lm cells to cover the substratum surface for quite a long time, particularly at 4°C, so Lm cells probably have to make their way towards the biofilm bottom across the Pf matrix. It may be that the extracellular chitinase activity, that Tirumalai & Prakash (2011) have described for Lm, helps their cells get through the entangled Pf's EPS network. Amino sugars, acetylated or not, have been described as components of Pf's extracellular polysaccharides (Kives et al., 2006). In our experience we found an exception to layering with *Listeria* (Lm and L. innocua) where at the bottom were old biofilms grown in batch; spontaneous or physiological shedding or detachment of superficial Pf cells located at upper layers could account for it (Kaplan, 2010; McDougald et al., 2012).

A bottom location in the biofilm structure is, so to say, not comfortable but safe. Survival in anaerobiosis (Lungu et al., 2009) and through starvation stress (Lungu et al., 2010) has been described for Lm. Depth and very slow doubling time would contribute to the highly increased resistance to antimicrobials observed for Lm in multispecies biofilms by different authors (Bourion & Cerf, 1996; Fatemi & Frank, 1999; Simoes et al., 2009; Lourenco et al., 2011; Ibusquiza et al., 2012).

Layering is a structural setup fit for cooperation (Elias & Banin, 2012) or syntrophy (Morris et al., 2013). When its outcome is negative for the covered partner, it is called blanketing (Rendueles & Ghigo, 2012). One such case involving Lm in coculture at 25°C has been described (Habimana et al., 2011); also in the bottom layers, Lm was overtaken by *Lactococcus lactis*. Further, Almeida et al. (2011) reported a clear case of layering in biofilms at 21°C involving Lm at the bottom layers together with *Salmonella enterica*, being covered by fast growing *Escherichia coli.*

Is Pseudomonas a Primary Surface Colonizer Needed for Lm Attachment?

Pseudomonas early colonization of surfaces is beneficial for Lm attachment but not indispensable. Diversity in Lm independent biofilm forming ability according to strain and environment has been previously reported (Borucki et al., 2003; Moretro & Langsrud, 2004; Moltz & Martin, 2005). Lm in TSB can form monospecies biofilms reaching rather high cellular densities, around 10^5 CFU cm^{-2} at 20ºC and slightly less at 4ºC. When Pf was present however, Lm cells per cm^2 in biofilms increased about 1 log at 20ºC and 2 log at 4ºC; not needed, right, but very convenient.

The primary colonization concept may be complemented by the subsequent layered structure: the first surface colonizer (Pf) could be later colonized itself (its matrix) by a guest (Lm) that may work out a thorough restructuration of its initial shelter. Pf could be somehow compensated by Lm, possibly providing the host with some useful service, such as acting as a sink for a self-inhibitory end product that would otherwise be dumped outside.

Dual Biofilms at Low Temperatures

Pf/Lm co-culture appeared to be more beneficial for Lm at 4ºC than at 20ºC. Shortening of the lag phase, which is rather long for Lm at low temperatures (Chan & Wiedmann, 2009) seems to be one of the mechanisms; the other presumably being the migration to the bottom layers, later on. This late advantage would disappear when Pf cells get detached, as the biofilm reaches a dispersion or detachment stage (McDougald et al., 2012).

In the food chain context, refrigeration of foods and equipment represent a selective niche for psychrotrophs such as Pf and Lm (Valderrama & Cutter, 2013). There is therefore a need to know the conditions or factors that allow or enhance Lm growth, survival or cell transfer in the cold. Biofilms favor all that, and the presence of *Pseudomonas* favors higher levels of Lm in biofilms. Elimination of *Pseudomonas* and layered Pf and Lm-carrying biofilms are therefore adequate targets to reduce Lm prevalence and improve food safety.

Other saprophytic, psychrotrophic and good biofilm former bacteria different from *Pseudomonas,* which could form layered biofilms the strain of Pf used here. And maybe not all Lm strains are fit to dive into the biofilm's bottom layers; they may stay in more accessible locations, to the advantage of the food plant hygienist. In summary, cleaning procedures aimed at detaching the top biofilm layers occupied by *Pseudomonas* or a similar host, or penetration through them, could be a good strategy for Lm eradication.

ACKNOWLEDGMENTS

The authors thank J.V. Martinez-Suarez for kindly providing persistent Listeria monocytogenes strains, CLSM services of the University Complutense for their skillful assistance, and the Spanish Ministry of Economy and Competition for funding project AGL2010-22212-C02-01.

REFERENCES

Almeida C, Azevedo NF, Santos S, Keevil CW, Vieira MJ. Discriminating Multi-Species Populations in Biofilms with Peptide Nucleic Acid Fluorescence In Situ Hybridization (PNA FISH). *Plos One.* 2011; 6(3).

Borucki MK, Peppin JD, White D, Loge F, Call DR. Variation in biofilm formation among strains of *Listeria monocytogenes. Applied and Environmental Microbiology.* 2003; 69(12):7336-42.

Bourion F, Cerf O. Disinfection efficacy against pure-culture and mixed-population biofilms of *Listeria innocua* and *Pseudomonas aeruginosa* on stainless steel, Teflon(R) and rubber. *Sciences Des Aliments.* 1996; 16(2):151-66.

Breen J, Barnes D, Ganguly R. Protective effect of *Pseudomonas* slime polysaccharide against *Listeria monocytogenes. International Journal of Immunopharmacology.* 1982; 4(4):262-268.

Buchanan RL, Bagi LK. Microbial competition: effect of *Pseudomonas fluorescens* on the growth of *Listeria monocytogenes. Food Microbiology.* 1999; 16(5):523-9.

Carpentier B, Chassaing D. Interactions in biofilms between *Listeria monocytogenes* and resident microorganisms from food industry premises. *International Journal of Food Microbiology.* 2004; 97(2):111-22.

Carpentier B, Cerf O. Review - Persistence of *Listeria monocytogenes* in food industry equipment and premises. *International Journal of Food Microbiology.* 2011; 145(1):1-8.

Chan YC, Wiedmann M. Physiology and Genetics of *Listeria monocytogenes* Survival and Growth at Cold Temperatures. *Critical Reviews in Food Science and Nutrition.* 2009; 49(3):237-53.

Chmielewski RAN, Frank JF. Biofilm Formation and Control in Food Processing Facilities. *Comprehensive reviews in food science and food safety.* 2003; 2:22-32.

Combrouse T, Sadovskaya I, Faille C, Kol O, Guerardel Y, Midelet-Bourdin G. Quantification of the extracellular matrix of the *Listeria monocytogenes* biofilms of different phylogenic lineages with optimization of culture conditions. *Journal of Applied Microbiology.* 2013; 114(4):1120-31.

Costerton, J.W. Toward a unified biofilm theory. In *The Biofilm Primer* (1st ed.); Heidelberg, CE, Ed.; Springer: Berlin, Germany. 2007.

Da Silva EP, Pereira De Martinis EC. Current knowledge and perspectives on biofilm formation: the case of *Listeria monocytogenes. Applied Microbiology and Biotechnology.* 2013; 97(3):957-68.

Elias S, Banin E. Multi-species biofilms: living with friendly neighbors. *Fems Microbiology Reviews.* 2012; 36(5):990-1004.

Farrag SA, Marth EH. Growth of *Listeria monocytogenes* in the presence of *Pseudomonas fluorescens* at 7-degrees-c or 13-degrees-c in skim milk. *Journal of Food Protection.* 1989; 52(12):852-5.

Fatemi P, Frank JF. Inactivation of *Listeria monocytogenes/Pseudomonas* biofilms by peracid sanitizers. *Journal of Food Protection.* 1999; 62(7):761-5.

Flemming H-C, Wingender J. The biofilm matrix. *Nature Reviews Microbiology.* 2010; 8(9):623-33.

Fratamico PM, Annous BA, Gunther IV NW. *Biofilms in the food and beverage industries.* CRC Press, Woodhead Pub: Boca Raton, Oxford. 2009.

Gram L. Inhibitory effect against pathogenic and spoilage bacteria of *Pseudomonas* strains isolated from spoiled and fresh fish. *Applied and Environmental Microbiology.* 1993; 59(7):2197-203.

Habimana O, Guillier L, Kulakauskas S, Briandet R. Spatial competition with *Lactococcus lactis* in mixed-species continuous-flow biofilms inhibits *Listeria monocytogenes* growth. *Biofouling.* 2011; 27(9):1065-72.

Harmsen M, Lappann M, Knochel S, Molin S. Role of Extracellular DNA during Biofilm Formation by *Listeria monocytogenes. Applied and Environmental Microbiology.* 2010; 76(7):2271-9.

Hassan AN, Birt DM, Frank JF. Behavior of *Listeria monocytogenes* in a *Pseudomonas putida* biofilm on a condensate-forming surface. *J Food Prot.* 2004; 67(2):322-7.

Hoben HJ, Somasegaran P. Comparison of the pour, spread, and drop plate methods for enumeration of *Rhizobium* spp in inoculants made from pre-sterilized peat. *Applied and Environmental Microbiology.* 1982; 44(5):1246-7

Huis in 't Veld JH. Microbial and biochemical spoilage of foods: an overview. *International journal of food microbiology.* 1996; 33(1):1-18.

Ibusquiza PS, Herrera JR, Vazquez-Sanchez D, Cabo ML. Adherence kinetics, resistance to benzalkonium chloride and microscopic analysis of mixed biofilms formed by *Listeria monocytogenes* and *Pseudomonas putida. Food Control.* 2012; 25(1):202-10.

Jaspe A, Oviedo P, Fernandez L, Palacios P, Sanjose C. Cooling raw-milk - change in the spoilage potential of contaminating *Pseudomonas*. *Journal of Food Protection.* 1995; 58(8):915-21.

Jeong DK, Frank JF. Growth of *Listeria monocytogenes* at 10-degrees-c in biofilms with microorganisms isolated from meat and dairy processing environments. *Journal of Food Protection.* 1994; 57(7):576-86.

Kaplan JB. Biofilm Dispersal: Mechanisms, Clinical Implications, and Potential Therapeutic Uses. *Journal of Dental Research.* 2010;89 (3):205-18.

Karunakaran E, Mukherjee J, Ramalingam B, Biggs CA. "Biofilmology": a multidisciplinary review of the study of microbial biofilms. *Applied Microbiology and Biotechnology.* 2011; 90(6):1869-81.

Kives J, Orgaz B, SanJose C. Polysaccharide differences between planktonic and biofilm-associated EPS from *Pseudomonas fluorescens* B52. *Colloids and Surfaces B-Biointerfaces.* 2006; 52(2):123-7.

Kjelleberg S, Givskov M. *The Biofilm mode of life. Mechanisms and adaptations.* Horizon Bioscience: Norfolk, UK. 2007

Liao CH, (2006). *Pseudomonas* related genera. In *Food Spoilage Microorganisms.* Blackburn CV, Ed.; CRC Press: Boca Raton, Fl. 2006: 507-540.

Lopez D, Vlamakis H, Kolter R. Biofilms. *Cold Spring Harbor Perspectives in Biology.* 2010; 2(7).

Lourenco A, Machado H, Brito L. Biofilms of *Listeria monocytogenes* produced at 12 degrees C either in Pure Culture or in Co-Culture with *Pseudomonas aeruginosa* showed Reduced Susceptibility to Sanitizers. *Journal of Food Science.* 2011; 76(2):M143-M8.

Lungu B, Ricke SC, Johnson MG. Growth, survival, proliferation and pathogenesis of *Listeria monocytogenes* under low oxygen or anaerobic conditions: A review. *Anaerobe.* 2009; 15(1-2):7-17.

Lungu B, Saldivar JC, Story R, Ricke SC, Johnson MG. The Combination of Energy-Dependent Internal Adaptation Mechanisms and External Factors Enables *Listeria monocytogenes* to Express a Strong Starvation Survival Response During Multiple-Nutrient Starvation. *Foodborne Pathogens and Disease.* 2010; 7(5):499-505.

Mann EE, Wozniak DJ. *Pseudomonas* biofilm matrix composition and niche biology. *Fems Microbiology Reviews.* 2012; 36(4):893-916.

Mariani C, Oulahal N, Chamba JF, Dubois-Brissonnet F, Notz E, Briandet R. Inhibition of *Listeria monocytogenes* by resident biofilms present on wooden shelves used for cheese ripening. *Food Control.* 2011; 22(8):1357-62.

McDougald D, Rice SA, Barraud N, Steinberg PD, Kjelleberg S. Should we stay or should we go: mechanisms and ecological consequences for biofilm dispersal. Nature *Reviews Microbiology.* 2012; 10(1):39-50.

Mellefont LA, Ross T. Effect of potassium lactate and a potassium lactate-sodium diacetate blend on *Listeria monocytogenes* growth in modified atmosphere packaged sliced ham. *Journal of Food Protection.* 2007; 70(10):2297-305.

Mettler E, Carpentier B. Variations over time of microbial load and physicochemical properties of floor materials after cleaning in food industry premises. *Journal of Food Protection.* 1998; 61(1):57-65.

Moltz AG, Martin SE. Formation of biofilms by *Listeria monocytogenes* under various growth conditions. *Journal of Food Protection.* 2005; 68(1):92-7.

Moretro T, Langsrud S. *Listeria monocytogenes*: biofilm formation and persistence in food-processing environments. *Biofilms.* 2004; 1:107-121.

Morris BEL, Henneberger R, Huber H, Moissl-Eichinger C. Microbial syntrophy: interaction for the common good. *Fems Microbiology Reviews. 2013*; 37(3):384-406.

Nilsson L, Gram L, Huss HH. Growth control of Listeria monocytogenes on cold-smoked salmon using a competitive lactic acid bacteria flora. *Journal of Food Protection.* 1999; 62(4):336-42.

Nilsson RE, Ross T, Bowman JP. Variability in biofilm production by *Listeria monocytogenes* correlated to strain origin and growth conditions. *International Journal of Food Microbiology.* 2011; 150(1):14-24.

Orgaz B, Lobete MM, Puga CH, SanJose C. Effectiveness of Chitosan against Mature Biofilms Formed by Food Related Bacteria. *Int J Mol Sci.* 2011; 12(1):817-28.

Ortiz S, Lopez V, Villatoro D, Lopez P, Davila JC, Martinez-Suarez JV. A 3-Year Surveillance of the Genetic Diversity and Persistence of *Listeria monocytogenes* in an Iberian Pig Slaughterhouse and Processing Plant. *Foodborne Pathogens and Disease.* 2010; 7(10):1177-84.

Rendueles O, Ghigo J-M. Multi-species biofilms: how to avoid unfriendly neighbors. *Fems Microbiology Reviews.* 2012; 36(5):972-89.

Rieu A, Briandet R, Habimana O, Garmyn D, Guzzo J, Piveteau P. *Listeria monocytogenes* EGD-e biofilms: no mushrooms but a network of knitted chains. *Appl Environ Microbiol.* 2008; 74(14):4491-7.

Ryser ET, Marth EH. *Listeria, listeriosis and food safety* (3rd ed.). CRC Press: Boca Raton, FL. 2007.

Sasahara KC, Zottola EA. Biofilm formation by *Listeria monocytogenes* utilizes a primary colonizing microorganism in flowing systems. *Journal of Food Protection.* 1993; 56(12):1022-8.

Silby MW, Winstanley C, Godfrey SAC, Levy SB, Jackson RW. *Pseudomonas* genomes: diverse and adaptable. *Fems Microbiology Reviews.* 2011; 35(4):652-80.

Simoes M, Simoes LC, Vieira MJ. Species association increases biofilm resistance to chemical and mechanical treatments. *Water Research.* 2009; 43(1):229-37.

Sorhaug T, Stepaniak L. Psychrotrophs and their enzymes in milk and dairy products: Quality aspects. *Trends in Food Science & Technology.* 1997; 8(2):35-41.

Srey S, Jahid IK, Ha S-D. Biofilm formation in food industries: A food safety concern. *Food Control.* 2013; 31(2):572-85.

Tirumalai PS, Prakash S. Expression of chitinase and chitin binding proteins (CBP's) by *Listeria monocytogenes* J0161 in biofilm and Co-culture broths. *African Journal of Microbiology Research.* 2011; 5(29):5188-93.

Valderrama WB, Cutter CN. An Ecological Perspective of *Listeria monocytogenes* Biofilms in Food Processing Facilities. *Critical Reviews in Food Science and Nutrition.* 2013; 53(8):801-17.

Verran J, Airey P, Packer A, Whitehead KA. Microbial retention on open food contact surfaces and implications for food contamination. In *Advances in Applied Microbiology.* Laskin AI, Sariaslani S, Gadd GM, Eds.; 2008; 64:223-46.

In: *Listeria monocytogenes*
Editor: Edmund C. Hambrick
ISBN: 978-1-63117-054-6

Chapter 7

LISTERIA MONOCYTOGENES AND *SALMOSALAR* (LINNAEUS, 1758): A REVIEW ON THE MAIN HYGIENIC ASPECTS OF PROCESSING AND MARKETING

Domenico Meloni
Department of Veterinary Medicine, University of Sassari,
Via Vienna, Sassari, Italy

ABSTRACT

The Atlantic salmon (*Salmosalar L.*) is the most significant aquaculture species in Europe, both in terms of production biomass and economic value, with Norway, followed by Scotland and Ireland as the three major European producers.

The market for "processed" salmon subjected to one or more treatments sensibly changing the original characteristics of the products, such as salting, marinating and smoking is constantly and quickly growing. Nowadays, thanks to modern production technologies and the large scale retail trade, the smoked salmon market has gained great appreciation across Europe.

This product, normally cold-smoked and vacuum packed, with a refrigerated shelf-life of 3-8 weeks, is considered a risk product for human listeriosis: several studies demonstrated that is an excellent substrate for *Listeria monocytogenes*. The processing plants environment is the major source of finished-product contamination for cold-smoked salmon. It is possible that considerable contamination occurs during processing, in particular during evisceration, fish handling and packaging as a result of inadequate hygiene.

The growth of this microorganism in smoked salmon has been reported most of all during the storage period suggesting that the stress applied during the technological processes may extend the lag phase or reduce the growth rate, but is not able to affect the vitality of the cells. In the following, through an up-to-date review of (*personal and non*) published data, the main hygienic aspects of processing and marketing will be discussed.

INTRODUCTION

The Atlantic salmon (*Salmosalar*) is the most significant aquaculture species in Europe, both in terms of production biomass and economic value. It is an important economic resource for a number of countries worldwide: Norway is the main producer (53%) followed by Chile (23%) and Scotland (10%). In the EU, over 60% of salmon is sold fresh (Anon, 2010): Norway, Scotland and Ireland are the three major producers. In the late years, the market for "processed" salmon subjected to one or more treatments sensibly changing the original characteristics of the products (*e.g.,* salting, marinating and smoking) is constantly and quickly growing. Up to a few years ago, the consumers saw them as very high quality niche products to be bought and eaten only at special occasions and festivities, also considering their high price. Nowadays, thanks to modern production technologies and the many more trading solutions, first of all the large scale retail trade, in Europe the smoked salmon market has gained great appreciation of a new group of consumers from every social class and all year long. At present, smoked salmon (cold or hot) comprises > 30% of the "processed" product market (Anon, 2010). This product, normally cold-smoked and vacuum packed, with a refrigerated shelf-life of 3-8 weeks, is a raw ready to eat food which is considered a risk product for human listeriosis. Two main forms of listeriosis have been described in humans: febrile gastroenteritis in healthy individuals and life-threatening invasive infections with a high mortality rate (20-30% of total cases) in vulnerable groups such as young, old, pregnant and immune-compromised, the so called "YOPI"(De Cesare et al., 2007). Since the beginning of the 1980's *Listeria monocytogenes* has been recognized as an emerging food-borne pathogen after several sporadic and epidemic cases of listeriosis associated with smoked fish, including cold smoked rainbow trout (Ericsson et al., 1997, Miettinen et al., 1999) and cold smoked salmon (Garrido et al., 2008). As a consequence, the smoked salmon supply chain has repeatedly had concerns with *Listeria monocytogenes* contamination of the final products (Rotariu et al., 2013). Several studies (Rorvik et al., 1991; Farber, 1991; Jemmi and Keusch, 1992; Nilsson et al., 1997; Rørvik, 2000; Wagner et al., 2007; Meloni et al., 2009; Miya et al., 2010) indicated that the smoked salmon is an excellent substrate for *Listeria monocytogenes*. Several authors (Ben Embarek, 1994, Medrala et al., 2003, Vitas and Garcia-Jalon, 2004, Garrido et al., 2008) reported that the manufacture of smoked salmon is complex and that the processing plants environment is the major source of finished-product contamination (*e.g.,* evisceration, fish handling and packaging as a result of inadequate hygiene). Moreover, the growth of this microorganism during the storage period of smoked salmon has been repeatedly reported suggesting that the stress applied during the technological process may extend the lag phase or reduce the growth rate, but is not able to affect the vitality of the cells. Therefore, understanding both the process of contamination and identification of potential *Listeria* niches within the plant are a prerequisite for designing successful plans to reduce the pathogen burden (Rotariu et al., 2013).

MAIN CHARACTERISTICS OF THE ATLANTIC SALMON (*SALMOSALAR*)

The distinctive characteristics of the Atlantic salmon are represented by a spindle-shaped body, stretched, with thin tail peduncle and a total length of over than 125 cm. The coloration is blue, green or dorsally brown, silver on the sides, white-silvery on the abdomen in particular in the sea specimen (Manzoni e Tepedino, 2008). On the body above the side line, are present some small black spots usually "x" shaped. The Atlantic salmon is an anadromous species and is not found in Mediterranean waters, despite some naturalization attempts (Fasolato et al., 2008). Salmon farming has developed in north European countries and at present is the most significant aquaculture species: it has been estimated that approximately 60% (2 mil. tonnes/year) of salmon production is farm-cultured (Rotariu et al., 2013). The species was also introduced in Chile, New Zealand and Australia, and also along the Pacific coasts of Canada and the USA. In the North West Atlantic the species is present from north Portugal and Gascoigne Gulf to Iceland, to the Barents Sea, to the Kara Sea, to the White Sea and to the Baltic Sea (Fasolato et al., 2008). The Atlantic salmon is a species of primary importance, appreciated all over the world, daily present on every European market. The growing offer of this product is linked to the very quick development of the salmon farming in north European countries. At present, the Atlantic salmon reaching the markets is almost only from farming plants. There can be attempts to pass off farmed subjects as "wild" that means from fishing activity at sea or rivers as that have a definitely higher commercial quality (Fasolato et al., 2008).

THE PROCESSING AND MARKETING OF SMOKED SALMON

The production of smoked salmon includes several technological steps that require a lot of handling by workers (Rørvik, 2000). The production phases worth considering because they can compromise the quality and safety of the final product are the salting and the smoking (Fasolato et al., 2008).

Raw Material Arriving at a Processing Plant

In general, the raw material used for the production of smoked salmon is farmed or fished in North Europe or South America and arrives at a processing plant in either a chilled or a frozen state. The salmons are kept refrigerated until processing: defrosting is the first technological step and is carried out on salmons of North European origin (mostly chilled) and on the Salmons from South America arriving quick-frozen in Europe (Fasolato et al., 2008). The most used method requires a drinking water shower at + 15/18 °C or the submerging of the salmons in stainless steel running drinking water powered tanks.

Filleting

The following phase is filleting. In the small processing plants(fewer than 5 employees) this operation is carried out manually (Rotariu et al., 2013); in the large processing plants is mostly made by using specific machinery. The salmons go through a tunnel where they are first cut and finally gutted also using a specific aspirator (Fasolato et al., 2008). The proper filleting step is entirely machinery-controlled, while the elimination of the spine is carried out both automatically and by hand.

Salting

After filleting, salmons are subjected to salting: the diffusion of the salt inside the muscular tissues helps drain part of the water content to bind the residual internal one lowering the a_w. The salting step can be dry or wet (by injection of/or soaking in brine).For dry salting, the salmons are put in stainless steel tanks and sprinkled with sea-salt to try to obtain an even absorption from the flesh (Fasolato et al., 2008). The fillets are covered in salt and manually stacked onto wire racks which allowed the free drainage of water drawn from the salmon flesh by the sea-salt. The positive aspects of dry salting are the following: fast penetration in the tissues and consequent salmon dehydration; lower final salmon moisture level; higher stabilization of the salmon towards the alteration of the microbial flora the contrary, dry salting can give a superficial oily look to the salmon caused by a too accelerated dehydration and the proteins coagulation (Fasolato et al., 2008). In this case, the deep diffusion of the sea salt becomes slower. At present, dry salt is seldom used because it gives a poor penetration. One of the two wet salting methods with brine, involved the soaking of salmons in more or less concentrated sodium chloride solutions (Fasolato et al., 2008), usually at 80° salinometers (21.1 g NaCl/100 ml of brine). The soaking time varies according to the salt concentration, the salmon sizes and their centesimal composition (Fasolato et al., 2008). This salt concentration is an indicative compromise to prevent a possible recrystallization of the salt on the surface of the product after cooking and smoking, and reach the desired salinity in an acceptable time (Arcangeli, Baldrati and Pirazzoli, 2003). If all circumstances are equal, the salt diffusion inside the flesh will be as slower as the lipids content of the treated salmons are higher. On the contrary, the final moisture level of the salmon is usually higher than the dry treated one (Fasolato et al., 2008). In salting by brine injection, a salt level enough to guarantee the desired "salt point" is reached by injecting the brine in the salmon flesh, in pressure and with the aid of needles (Fasolato et al., 2008). This method allows a quick diffusion of the salt inside the tissues and most of all increases the technical performance (up to 9% in weight in the case of smoked salmons compared to a salting with dry salt). On the contrary, a higher a_w level implies a higher dehumidification difficulty during the desiccation phase and a lower shelf-life of the salmons (Fasolato et al., 2008).

Moreover, the risk of a chain cross-contamination of the fillet can't be avoided, in particular when the process is continuous and with recycling of brine (Fasolato et al., 2008). Various factors can affect the good outcome of the salting: in particular, the brine concentration with a salt level of over 60 tend to drain water from the flesh and to give a not always even diffusion (Fasolato et al., 2008). The salting duration is also important: the so-called "fully cured salmon" is when the salmon reaches a balance with a saturated solution (17%) and is firm to the touch, but moist on the surface (Fasolato et al., 2008).Very often, salmons treated in this way are washed for many hours before being smoked. The brine/flesh ratio should be always 1:1. Regarding the temperature, is not necessary to standardize a system that requires the use of low temperatures (<+4°C) to control the development of microorganisms (Fasolato et al., 2008).

Smoking

Before the smoking process, the fillets are allowed to dry in special ventilated dryers at a temperature over 30 °C for up to 24 hours before smoking (Rotariu et al., 2013). The smoke is obtained from combustion, without flame, or friction of wood (including sawdust) or all wooden plants, except the drenched, coloured, glued, painted or similarly treated ones (Fasolato et al., 2008). The term "smoke" also includes all the by-products obtained by condensation or absorption of smoke in a liquid medium suitable for food use. Liquid smoke when properly diluted can be used to give a smoke flavour to the product or to give natural flavours of hickory smoke, vinegar, molasses, caramel and salt (Fasolato et al., 2008).The fillets can be placed horizontally on grills or hanged vertically on appropriate trolleys. The smoke is brought in contact with the salmons at room temperature and left for weeks (low density smoke, long lasting smoking method) or for a few days or hours (short lasting smoking). The product does not show any proteins heat coagulation signs (Fasolato et al., 2008).The antiseptic action performed by the smoke is very light and limited to the surface of the product that undergoing this treatment has most of all its colour, flavour and taste changed (Fasolato et al., 2008). There are two general methods of smoking salmons: cold-smoking and hot-smoking. The smoking process can be "cold" when the smoke temperature is between + 17 and + 30°C (Rotariu et al., 2013). Cold smoking is the most frequently used method in fishery products preservation. The consumption of these products, classified as lightly preserved, varies considerably according to geographic areas: they are widely consumed in the United States and Europe and thus are of considerable economic importance for the seafood market (Hoffmann et al., 2003; Cardinal et al., 2004). In "hot" smoking, in order to obtain a heat coagulation of proteins in all the product, the temperature is between + 60 and + 100°C with a duration from 20 minutes to 3 hours. The most common combinations of time and temperature are + 65 °C for 20 minutes and +60 °C for 45 minutes (Heintz and Johnson, 1998). These products are classified as mildly heat-preserved seafood products (Arcangeli, Baldrati and Pirazzoli, 2003).

Cutting and Reconstruction of the Smoked Fillet

The smoked fillet, if is not marketed as it is ("side") but has to be sliced, is cut with high speed operated blades. Afterwards is possible to proceed to its reconstruction retrieving the skin as well or to package the slices in appropriate trays or bags of different sizes (Fasolato et al., 2008). At the end of the process, before being packed, the salmons are cooled to about 0°C. However, if the salmon should be vacuum packed and stored under freezing, it is advisable to bagging at room temperature. This is to avoid condensation inside the packages and the inevitable "snow effect" (Arcangeli, Baldrati and Pirazzoli, 2003).

Bagging and Packaging

Both pre-sliced salmon slices and whole fillets are put in appropriate trays with the introduction of particular gas mixtures (modified atmosphere packaging), or in plastic film bags, passing through an air sucking machine in order to vacuum-package the salmon slices (Fasolato et al., 2008).

Marketing

The smoked salmon is an unstable product. It is necessary to constantly keep it refrigerated at temperatures between 0 and + 2 °C (Fasolato et al., 2008). The storage period can vary from 4-5 weeks to 3 months according to the health-hygienic measures taken during the processing phases (Fasolato et al., 2008).

Hygienic Aspects and Critical Points Related to the Presence of *Listeria monocytogenes* in Smoked Salmon

Listeria monocytogenes can contaminate smoked salmons at many levels of processing chain (Vogel et al., 2001b). Contamination of smoked salmon occurs mainly during processing rather than originating from raw materials, even if raw salmon can be a contamination source of the processing environment (Hoffman et al., 2003; Gudmundsdottir et al., 2005; Katzav et al., 2006).

Raw Material Arriving at a Processing Plant

Regarding the raw material quality, the best pre-slaughtering fast time should be two weeks exactly because a shorter time would cause the presence of a bigger quantity of water in the flesh and this would affect the whole production process (Fasolato et al., 2008). Salmons should arrive at a processing plant stored under appropriate temperatures. Usually they are covered with ice flakes (2.5-3.0 Kg of ice for 10 kg of salmon) in perforated containers to allow the drainage of the water formed by the melted covering ice. Before processing, salmons have to be stored in a cell at a temperature of ≤ 3 °C (Fasolato et al., 2008). Defrosting should be carried out at refrigerating temperatures to avoid inappropriate multiplications of *Listeria monocytogenes*. As a matter of fact, the raw materials are often contaminated at negligible levels, from zero (Ben Embarek et al., 1997) to about 30% (Hoffman et al., 2003; Miettinen and Wirtanen, 2005; Di Ciccio et al., 2012) while a massive contamination during processing due to contact with contaminated surfaces or during the storage of the finished products often occurs (Eklund et al., 1995).

Filleting

The filleting phase needs appropriate working conditions: all the cleaning and processing of the salmons should be done in places of work divided in departments and air-conditioned at 12 °C (Fasolato et al., 2008). The hygiene regime and the cleaning and sanitation practices applied during smoked salmon production represent key factors in reducing the *Listeria* burden in smoked products: it will be imperative to set up an accurate washing ad disinfection plan for these areas (Rotariu et al., 2013). Several authors found evidence that the environment is the major source of smoked salmons contamination (Jorgensen and Huss, 1998; Autio et al., 1999; Rørvik, 2000) and that considerable contamination occurs during evisceration, handling and filleting of salmons as a result of inadequate hygiene (Katzav et al., 2006).It was shown (Johansson et al., 1999) that the same *Listeria monocytogenes* strains found in the smoked salmons were also isolated from the plant drains, the skinning and filleting equipment and conveyor belts. Another study (Autio et al., 1999) concluded that plant-resident strains of *Listeria monocytogenes* represented important sources of contamination for the smoked salmons and demonstrated that effective decontamination of the plant environment could be achieved using hot (80 °C) water, steam and hot (80 °C) air ovens. In a recent study, Di Ciccio et al., (2012) found an overall prevalence rate of 16% among environmental samples from a salmon processing plant. The most contaminated niches of the plant were the working tables (43%) and the slicing machine (37%). Previous studies have shown that the main risk factor is represented by the rotation job that can be associated with the movement of smoking plant employees from one department to another of the processing plant without the adoption of appropriate precautionary measures (Rørvik et al., 1995). Smoking plant employees can act as fomites for *Listeria monocytogenes* (Tompkin et al., 1999).

Salting

It is important to use the right amount of salt or brine to avoid, in case of excessive quantities, compromising the final taste of the salmon or, in case of insufficient dosage, compromising the perfect preservation of the product (Fasolato et al., 2008). A light salting treatment can explicate most of all a firming action on the flesh, prevents their discoloration, gives the right flavour, but does not explicate an efficient and lasting *Listeria monocytogenes* inhibition. The risk of a contamination of the fillet can't be avoided when the process is continuous and with recycling of brine (Fasolato et al., 2008). The contamination level of brine containers is usually 17%, while brine is generally contaminated around at 36% (Rørvik et al., 1995).

Smoking

Both hot and cold smoking are usually not capable of preventing the growth of *Listeria monocytogenes* to high numbers, posing an increased risk of causing listeriosis has a high lipid content which may protect microorganisms from a possible thermal treatment or from freezing and cold-smoking below 20°C, cannot be considered a listericidal treatment (Serio et al., 2011). Cold smoking processing does not eliminate the possibility of contamination of raw materials and refrigeration or vacuum storage are not able to prevent the growth of *Listeria monocytogenes* (Duffes, 1999). In a recent study, *Listeria monocytogenes* has been isolated from raw salmon samples and final products highlighting that the pathogen can be reduced during cold smoking, although this process does not include a listericidal heat treatment (Di Ciccio et al., 2012). The temperatures and conditions required to denature salmon protein during hot smoking are generally sufficient to cause the complete kill of *Listeria monocytogenes* present on the surface of the salmon (Poysky et al., 1997). In a 1992 survey, Jemmi and Keush have shown that after experimental inoculation of *Listeria monocytogenes* in fresh salmon that had to be hot-smoked, the microorganism would not survive the process with applied temperatures of >70°C. This led to thinking that hot smoked salmon contamination might be mainly post-process. This treatment alone cannot guarantee the total safety of the product during marketing or a satisfactory shelf life. Therefore, a preliminary salting treatment and a strictly controlled refrigeration when the product is marketed are both necessary. In hot-smoked salmons, the prevalence of *Listeria monocytogenes* never exceed 9.12% (Jemmi and Keush, 1992; Ben Embarek, 1994, Jorgensen et al. 1998; Rørvik, 2000).

Cutting and Reconstruction of the Smoked Fillet

The cutting and reconstruction of the fillet requires a very high attention towards the compliance to the hygienic regulations by the smoking plant employers. The hygiene regime and the cleaning and sanitation practices applied to surfaces and work equipment during smoked salmon production represent key factors in reducing the *Listeria* monocytogenes contamination (Fasolato et al., 2008; Rotariu et al., 2013).

Bagging and Packaging

Vacuum packed smoked salmons are ready to eat products mostly eaten without any further cooking (Eklund et al., 1995; Gudmundsdóttir et al., 2005). Their recontamination with *Listeria monocytogenes* at packaging stage represents a health hazard, although the presence of salt and the refrigeration can limit the risk of development of the microorganism.

Marketing

For smoked salmons, the need to meet the consumers taste forces the producers to limit salt use (Fasolato et al., 2008). Moreover, physical and chemical parameters of smoked salmon are usually within values permitting the growth of *Listeria monocytogenes*, obtaining in this way, preparations that have to be kept compulsorily in strictly controlled refrigerating conditions during their life on the market (T < 3.0 °C). The growth of *Listeria monocytogenes* in vacuum packaged smoked salmon stored at T <5 °C can be inhibited by the use of a competitive lactic acid bacteria flora (Nilsson et al., 1999). Again, the development of *Listeria monocytogenes* during storage is definitely inhibited by the use of bacteriocins: nisin in combination with modified atmosphere packaging seems to have a decisive role on the control of *Listeria monocytogenes* in smoked salmon (Szabo and Cahill, 1999). Other studies have shown that the presence of bacteriocins producing strains can inhibit the growth of *Listeria monocytogenes* (Duffes et al., 1999). Nevertheless, *Listeria monocytogenes* is often isolated from smoked salmon, albeit in low numbers (Katzav et al., 2006) and the prevalence has been documented repeatedly varying between 10% and 40% (Jemmi and Keusch, 1992;

Hudson et al., 1992; Rørvik et al., 2000; Vogel et al., 2001a; Dauphin et al., 2001; Aguado et al., 2001; Autio et al., 2003; Nakamura et al., 2004; Azevedo et al., 2005; Miettinen and Wirtanen, 2005). In recent studies performed in Italy, Di Pinto et al. (2010) reported a prevalence of 34%, Mureddu et al., (2011) and Di Ciccio et al., (2012) 12% while Gambarin et al., (2011) of 16%. The high prevalence of *Listeria monocytogenes* in these products could be due to the low smoking temperature (20 °C) applied during the cold-salmon processing, as this condition would be ideal for the proliferation of *Listeria monocytogenes*. As reported above, the characteristics of the product should allow the growth of the pathogen at levels dangerous to humans (Gombas et al., 2003, Cornu et al., 2006). When the refrigeration temperatures are observed throughout the storage period, the development of *Listeria monocytogenes* is generally moderate (<100 cfu / g), however the long period of shelf life may allow a significant increase in the level of contamination, because of the absence of competitive microflora (Johansson et al., 1999; Little et al., 2007). Temperature abuse may be a factor in growth to high numbers unless prolonged storage has occurred (ILSI, 2005). Although it is rather uncommon, contamination levels up to 10^4cfu/g have been reported in samples stored at temperatures of thermal abuse between 8 and 10°C (Loncarevic et al., 1996, Cortesi et al., 1997, Jorgensen et al. 1998). For these reasons, consumer education to keep cold chain management during the shelf life becomes very important to minimize the growth of the pathogen. Shortening of the best-before date is sometimes useful to reduce the risk of causing food-borne listeriosis (Marklinder et al. 2004; ILSI, 2005; Garrido, 2008) and to respect the food safety criteria provided by the Commission Regulation (EC) N° 2073/2005 on microbiological criteria for foodstuffs.

REFERENCES

Aguado V., Vitas A.I., García-Jalón I. (2001). Random amplified polymorphic DNA typing applied to the study of cross-contamination by *Listeria monocytogenes* in processed food products. *Journal of Food Protection*, 64, 716–720.

Anon V. (2010). *The Marine Harvest Salmon Industry Handbook*, Available athttp://marineharvest.com/PageFiles/1296/2013%20Salmon%20Handbook%2027-04-13.pdf1-72.

Arcangeli G., Baldrati G., Pirazzoli P. (2003). La trasformazione dei prodotti della pesca: tecnologia, controllo e igiene di lavorazione. Ed. Stazione Sperimentale per l'Industria delle Conserve Alimentari, *Litografica Faenza*, 2003.

Autio T. (1999). Sources of *Listeria monoctytogenes* contamination in a cold-smoked rainbow trout processing plant detected by pulsed-field gel electrophoresis typing. *Applied Environmental Microbiology* 65:150-155.

Autio T., Keto-Timonen R., Biorkroth J. and Korkeala H. (2003). Characterization of persistent and sporadic *Listeria monocytogenes* strains by pulsed-field gel electrophoresis (PFGE) and amplified fragment length polymorphism (AFLP). *Systematic and Applied Microbiology* 26, (4), 539.

Azevedo I., Regalo M., Mena C., Almeida G., Carneiro L., Teixeira P.,Hogg T.,Gibbs P. (2005*). Incidence of Listeria spp. in domestic refrigerators in Portugal Food Control,* 16:121-4.

Ben Embarek P.K. (1994). Presence, detection and growth of *Listeria monocytogenes* in seafoods: a review. *International Journal of Food Microbiology* 23: 17.

Ben Embarek P.K., Hansen L.T., Enger O. and Huss H.H. (1997). Occurrence of *Listeria spp.* in farmed salmon and during subsequent slaughter: comparison of Listertestk Lift and the USDA method. *Food Microbiology* 14: 39.

Cardinal M., Gunnlaugsdottir H., Bjoernevik M., Ouisse A., Vallet J.L. (2004). Sensory characteristics of cold-smoked Atlantic salmon (Salmosalar) from European market relationships with chemical, physical and microbiological measurements. *Food Research International* 37, 181-93.

Cornu M., Beaufort A., Rudelle S., Laloux L., Bergis H., Miconnet N., Serot T., Ignette-Muller M.L. (2006). Effect of temperature, water-phase salt and phenolic contents on *Listeria monocytogenes* growth rates on cold-smoked salmon and evaluation of secondary models. *International Journal of Food Microbiology*, 106,159–168.

Cortesi M.L., Sarli T., Santoro A., Murru N., Pepe T. (1997). Distribution and behavior of *Listeria monocytogenes* in three lots of naturally-contaminated vacuum-packed smoked salmon stored at 2 and 10°C. *International Journal of Food Microbiology*, 37, 209–214.

Dauphin, G., Ragimbeau, C., Malle, P. (2001). Use of PFGE typing for tracing contamination with *Listeria monocytogenes* in three cold-smoked salmon processing plants. *International Journal of Food Microbiology* 64, 51-61.

De Cesare A., Mioni R., Manfreda G. (2007). Prevalence of Listeria monocytogenes in fresh and fermented sausages and ribotyping of contaminating strains. *International Journal of Food Microbiology* 120, (1-2), 124.

Di Ciccio, P.L., Meloni D., Festino A.R., Conter M., Zanardi E., Ghidini S., Vergara A., Mazzette R., Ianieri A. (2012). Longitudinal study on the sources of Listeria monocytogenes contamination in cold-smoked salmon and its processing environment in Italy. *International Journal of Food Microbiology*, 158, 79-84.

Di Pinto A., Novello L., Montemurro F., Bonerba E., Tantillo G. (2010). Occurrence of *Listeria monocytogenes* in ready-to-eat foods from supermarketsin Southern Italy. *New Microbiologica*, 33, 249-252.

Duffes F. (1999). Improving the control of *Listeria monocytogenes* in cold smoked salmon. *Trends in Food Science and Technology* 10: 211.

Ericsson I.L., Eklow A., Danielsson-Tham M.L., Loncarevic S., Mentzing O., Persson I., Unnerstad H., Tham W. (1997). An outbreak of listeriosis suspected to have been caused by rainbow trout. *Journal of Clinical Microbiology*, Nov (1997), 2904–290.

European Commission (EC) 2005. Regulation (EC) No 2073/2005 of 15 November 2005 on microbiological criteria for foodstuffs. *Off. J. Eur. Union L* 338:1.

Eklund, M. W., Poysky, F. T., Paranjpye, R. N., Lashbrook, L. C., Peterson, M. E., & Pelroy, G. A. (1995). Incidence and sources of *Listeria monocytogenes* in coldsmoked fishery products and processing plants. *Journal of Food Protection*, 58,502-508.

Farber J.M., Peterkin P. (1991). *Listeria monocytogenes*, a food-borne pathogen. *Microbiology and Molecular Biology Reviews* 55, (3), 476.

Farber J. M. (1991). *Listeria monocytogenes* in fish products. *Journal of Food Protection,* 54, 922-924.

Fasolato L., Ferito J.C., Mirisola M., Tepedino G., Tanzi E., Barbuti S. (2008). Pro e contro del salmone affumicato. *Eurofishmarket*, 2, 2008, 18-53.

Gambarin P., Magnabosco C., Arcangeli G., Losio M., Pavoni E. (2011). *Listeria monocytogenes* in semiconserve ittiche e potenziali rischi per il consumatore. *Industrie Alimentari*, 19-27.

Garrido V., Torroba L., García-Jalón I., Vitas A.I. (2008). Surveillance of listeriosis in Navarre, Spain, 1995-2005--epidemiological patterns and characterisation of clinical and food isolates. *Euro Surveillance Dec* 4;13(49). pii: 19058.

Gombas D.E., Chen Y., Clavero R.S., Scott V.N. (2003). Survey of *Listeria monocytogenes* in ready-to-eat foods. *Journal of Food Protection* 66:559.

Gudmundsdóttir, S., Gudbjörnsdóttir, B., Lauzon, H.L., Einarsson, H., Kristinsson, K.G., Kristjánsson, M. (2005). Tracing *Listeria monocytogenes* isolates from cold-smoked salmon and its processing environment in Iceland using pulsed-field gel electrophoresis. *International Journal of Food Microbiology* 101, 41-51.

Heintz M.L., Johnson J.M. (1998). The incidence of *Listeria spp.*, *Salmonella spp* and *Clostridium botulinum*in smoked fish and shellfish. *Journal of Food Protection*, 61 (3), 318-323.

Hoffman A., Gall K.L., Norton D.M. and Wiedmann M. (2003). *Listeria monocytogenes* contamination patterns for the smoked fish processing environment and for raw fish. *Journal of Food Protection* 66:52.

Hudson, J. A., Mott, S. J., Delacy, K. M. & Edridge, A. L. (1992). Incidence and coincidence of Listeria spp., motile aeromonads and Yersinia enterocolitica on ready-to-eat fleshfoods. *International Journal of food microbiology*, 16(2), 99-108.

Ilsi Research Foundation-Risk Science Institute - Expert Panel on *Listeria monocytogenes* in foods, Achieving continuous improvement in reductions in foodborne listeriosis - a risk-based approach. (2005). *Journal of Food Protection* 68, (9), 1932.

Jemmi T., Keusch A., (1992). Behaviour of *Listeria monocytogenes* during processing and storage of experimentally contaminated hot-smoked trout. *International Journal of Food Microbiology,* 15, 339–346.

Johansson T., Rantala L., Palmu L., Honkanen-Buzalski T. (1999). Occurrence and typing *of Listeria monocytogenes* strains in retail vacuum packed fish products and in a production plant. *International Journal of Food Microbiology* 47:111.

Jørgensen L.V., Huss H.H. (1998). Prevalence and growth of *Listeria monocytogenes* in naturally contaminated seafood. *International Journal of Food Microbiology*, 42, 127–131.

Katzav, M., Hyvönen, P., Muje, P., Rantala, L., von Wright A. (2006). Pulsed-Field Gel Electrophoresis typing of *Listeria monocytogenes* isolated in two Finnish fish farms. *Journal of Food Protection* 69, 1443-1447.

Little C.L., Taylor F.C., Sagoo S.K., Gillespie I.A., Grant K. and McLauchlin J. (2007). Prevalence and level of *Listeria monocytogenes* and other *Listeria* species in retail pre-packaged mixed vegetable salads in the UK. *Food Microbiology* 24:711.

Loncarevic S., Tham W., Danielsson-Tham M.L. (1996). Prevalence of *Listeria monocytogenes* and *Listeria spp.* in smoked and «gravad» fish. *Acta Veterinaria Scandinava,* 37, 13–18.

Manzoni P., Tepedino V. (2008). Grande Enciclopedia illustrate dei pesci.Ed. *Euro fish market*, Trebbo di Reno (BO).

Marklinder I.M., Lindblad M., Eriksson L.M., Finnson A.M., Lindqvist R. (2004). Home storage temperatures and consumer handling of refrigerated foods in Sweden. *Journal of Food Protection*, 67, 2570–2577.

Medrala, D., Dabrowski, W., Czekajlo-Kolodziej, U., Daczkowska-Kozon, E., Koronkiewcz, A., Augustynowicz, E., Manzano, M. (2003). Persistence of Listeria monocytogenes strains isolated from products in a Polish fish-processing plant over a 1-year period. *Food microbiology*, 20, 715-724.

Meloni D., Galluzzo P., Mureddu A., Piras F., Griffiths M., Mazzette R. (2009). *Listeria monocytogenes* in RTE foods marketed in Italy: prevalence and automated EcoRIribo typing of the isolates. *International Journal of Food Microbiology* 129, (2), 166.

Miettinen M.K., Björkroth K., Korkeala H.J. (1999). Characterization of Listeria monocytogenes from an ice cream plant by serotyping and pulsed-field gel electrophoresis. *International Journal of Food Microbiology*, 46, 187–192.

Miettinen H., Wirtanen G. (2005). Prevalence and location of *Listeria monocytogenes* in farmed rainbow trout. *International Journal of Food Microbiology* 104:135.

Miya, S., Takahashi, H., Ishikawa, T., Fujii, T., Kimura, B. (2010). Risk of Listeria monocytogenes contamination of raw ready-to-eat seafood products available at retail outlets in Japan. *Applied and Environmental Microbiology*, 76(10), 3383-3386.

Nakamura, H., Hatanaka, M., Ochi, K., Nagao, M., Ogasawar, J., Hase, A., Kitase, T., Haruki, K., Nishikawa, Y. (2004). *Listeria monocytogenes* isolated from cold-smoked fish products in Osaka City, Japan. *International Journal of Food Microbiology* 94, 323-8.

Mureddu A., Conter M., Meloni D., Piras F., Ianieri A., Mazzette R. (2011). Molecular characterization of *L. monocytogenes* isolated from ready to eat seafood in Italy. *Italian Journal of Food Science,* 23, 106-117.

Nilsson, L., Huss, H. H., & Gram, L. (1997). Inhibition of *Listeria monocytogenes* on cold-smoked salmon by nisin and carbon dioxide atmosphere. *International Journal of Food Microbiology,* 38, 217-227.

Nilsson, L., Huss, H. H., & Gram, L. (1999). Growth control of Listeria monocytogenes on cold smoked salmon using a competitive lactic acid bacteria flora. *Journal of Food Protection* 62, (4), 336-342.

Poysky, F. T., Paranjpye, R. N., Peterson, M. E., Pelroy, G. A., Guttman, A. E., Eklund, W.M.(1997).Inactivation of Listeria monocytogenes on hot-smoked salmon by the interaction of heat and smoke or liquid smoke. *Journal of food protection*, 60, 649-654.

Rotariu, O., Thomas, D.J.I., Goodburn, K.E., Hutchison, M.L., Strachan, N.J.C. (2013). Smoked Salmon Industry Practices and their association with Listeria monocytogenes, *Food Control*, doi: 10.1016/j.foodcont.2013.07.015.

Rørvik, L. M., Yndestad, M., Skjerve, E., (1991). Growth of Listeria monocytogenes in vacuum packed, smoked salmon, during storage at 4 degrees C. *International Journal of food microbiology,*534 14(2), 111-117.

Rørvik, L.M., Caugant, D., Yndestad, M. (1995). Contamination pattern of Listeria *monocytogenes* and other *Listeria spp.* in a salmon slaughterhouse and smoked salmon processing plant. *International Journal of Food Microbiology* 25, 19-27.

Rørvik, L. M. (2000). *Listeria monocytogenes* in the smoked salmon industry. *International Journal of Food Microbiology*, 62, 183-190.

Serio A., Chaves-López C., Paparella A. (2011).*Listeria monocytogenes* isolated from the smoked salmon industry: Growth potential under different environmental conditions. *Food Control* 22 (2011) 2071-2075.

Szabo E.A., Cahill M.E. (1999). Nisin and ALTA 2341 inhibit the growth of Listeria monocytogenes on smoked salmon packaged under vacuum or 100% CO_2. Letters in *Applied Microbiology*, 28:373-377.

Tompkin, R. B., Scott, V.N., Bernard, D.T., Sveum, W.H., &Gombas, K.S. (1999). Guidance toprevent post processing contamination from Listeria monocytogenes. *Dairy, food and environmental sanitation*, 19, 551-561.

Vitas, A. I., Garcia-Jalon, V.A. (2004). Occurrence of Listeria monocytogenes in fresh andprocessed foods in Navarra (Spain). *International Journal of food microbiology,* 90(3), 349-356.

Vogel, B. F., Huss, H. H., Ojneiyi, B., Ahrens, P., & Gram, L. (2001[a]). Elucidation of *Listeria monocytogenes* contamination routes in cold-smoked salmon processing plants detected by DNA-based typing methods. *Applied and Environmental Microbiology*, 67, 2586-2595.

Vogel, B.F., Jorgensen, L.V., Ojeniyi, B., Huss. H.H., Gram, L. (2001[b]). Diversity of *Listeria monocytogenes* isolates from cold-smoked salmon produced in different smokehouses as assessed by Random Amplified Polymorphic DNA analyses. *International Journal of Food Microbiology* 65, 83-92.

Wagner, M., B. Auer, C. Trittremmel, I. Hein, & Schoder, D. (2007). Survey on the Listeria contamination of ready-to-eat food products and household environments in Vienna, Austria. *Zoonoses and public health*, 54(1), 16-22.

In: *Listeria monocytogenes*
Editor: Edmund C. Hambrick

ISBN: 978-1-63117-054-6

Chapter 8

DETECTION OF *LISTERIA SPP.* AND *LISTERIA MONOCYTOGENES* IN FOOD AND FEED PRODUCTS

Elodie Barbau-Piednoir[1], Jacques Mahillon[2], Nancy H. Roosens[1] and Nadine Botteldoorn[1]
[1]Scientific Institute of Public Health, Brussels, Belgium
[2]Université Catholique de Louvain,
Faculty of Bioscience Engineering, Earth and Life Institute,
Louvain-la-Neuve, Belgium

ABSTRACT

Listeria monocytogenes is one of the most important foodborne pathogens because of the high mortality rate of this pathogen infection. Thus, its presence in food along the food chain is intensively controlled. The detection of Listeria in food samples is traditionally performed by culture-based methods using selective enrichment and plating media. The isolated strains should then be confirmed by biochemical characterization. This method, described in the ISO 11290-1, is the gold standard but is time consuming and may not be suitable to test food with short shelf life or in the case of outbreaks. Faster alternative methods based on different technologies have been developed and validated, they are antibodies-based (ELISA, ELFA) or acid nucleic based (RNA or DNA hybridization, PCR, RT-PCR, qPCR) technologies. In this chapter, the standard method for detecting *L. monocytogenes* in food and feed products as well as the different alternative technologies currently available are described in detail. The advantages and drawbacks of each technology are presented as well as a time scale comparison between the different technologies.

1. INTRODUCTION

The official controls of food and feed in Europe is performed according to the regulation n°882/2004 [1], which is carried out to comply with feed and food laws. It consists of controls at all production stages and in all sectors. In the European regulation 2073/2005 [2], the limit of acceptance for each foodborne pathogen is defined for each type of food category. In the case of *Listeria* spp., it is recommended to keep the concentration of *Listeria monocytogenes* in food below 100 CFU/g during the entire shelf life. More stringent rules are applied for ready-to-eat foods intended for infants and for special medical purposes since they should not contain any *Listeria* in 25 g during their entire shelf life. Consequently, the food business operators, responsible for the manufacture of the product, must conduct studies in order to investigate compliance with these criteria. In particular, this involves ready-to-eat food products that are able to support the growth of *L. monocytogenes* and may therefore be a risk for public health.

ISO11290-1 is the culture-based International Standard for the detection of *L. monocytogenes* in food and feed [2; 3]. Like other culture-based methods, this technique is relatively easy to implement and cheap in consumables but is time consuming and labor intensive, i.e., at least three days are necessary to obtain a presumptive (positive or negative) result, followed by two additional days for confirmation.

As an alternative, several approaches have been developed to accelerate the detection of *Listeria* spp. and *L. monocytogenes* in food. In this chapter, besides the standard method for *L. monocytogenes* detection in food and feed products [2; 3], an overview of alternative technologies currently available is provided.

2. Reference *Listeria* Detection Method

The standard culture method ISO 11290-1:1996 [Microbiology of food and animal feeding stuffs - Horizontal method for the detection and enumeration of L. monocytogenes - Part 1: Detection method] [3] is a horizontal method, meaning that it can, theoretically, be applied to all food categories and feed matrices. It has been modified in 2005 by the amendment ISO 11290-1/A1:2005 [4].

As shown in Figure 1, the method includes four successive steps:

1. Primary enrichment in moderately selective liquid medium
2. Secondary enrichment in fully selective liquid medium
3. Isolation and identification
4. Confirmation

A. Enrichments

In brief, the primary and secondary enrichments are performed in Half-Fraser and Fraser media, respectively. These selective media for *Listeria* spp [5] contains, in addition to growth components, aesculin, ferric ions, lithium chloride, nalidixic and acriflavine. All *Listeria* spp. hydrolyse aesculin to aesculetin, which reacts with ferric ions and results in blackening of the medium. Blackening of the medium is therefore a presumptive evidence for the presence of *Listeria*. However, non-blackening cultures can contain *Listeria*. Ferric ions enhance the growth of *L. monocytogenes* [6] and lithium chloride is included in the medium to inhibit the growth of enterococci that can also hydrolyze aesculin.

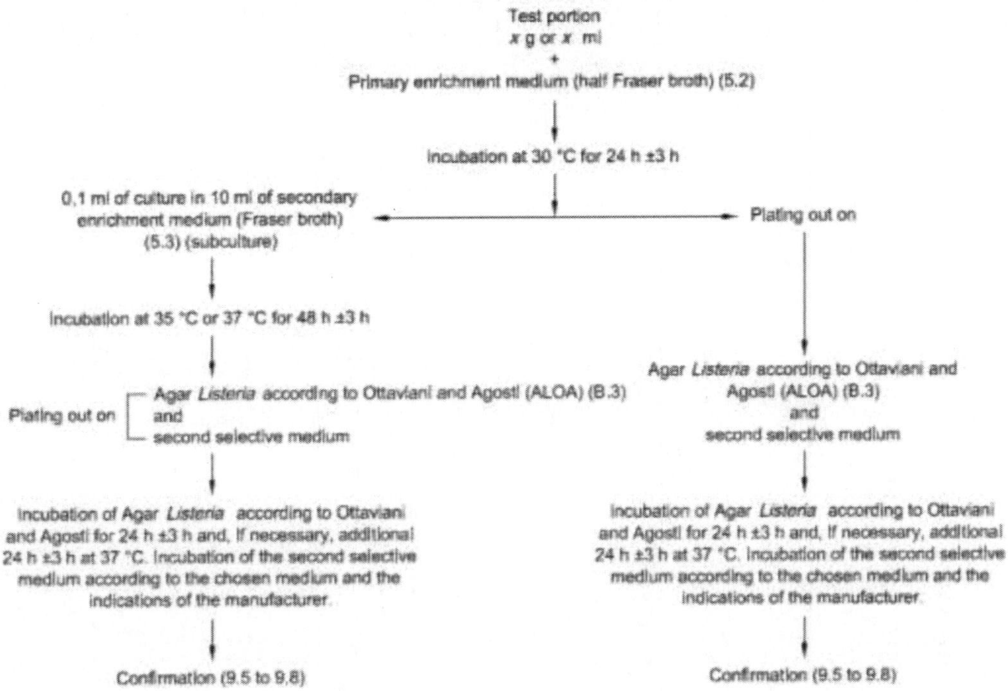

Figure 1. Flow diagram for the detection of *L. monocytogenes* using the ISO 11290-1 (4).

Nalidixic acid inhibits growth of Gram-negative organisms while acriflavine inhibits many Gram-positive bacteria. The last two compounds are present at half concentration in the Half-Fraser medium.

B. Isolation

From the primary and the secondary enrichments, dilution and plating out are made on two different selective agar media, such as to get well-separated colonies. According to the Amendment 1 of ISO 11290-1 [7], the selective media to use are Agar *Listeria* according to Ottaviani and Agosti (ALOA), combined with another selective medium at the choice of the laboratory (e.g., Oxford, PALCAM or MOX). A list of all commercially available AFNOR-validated media and their exclusivity and inclusivity results are shown in Table 1. It is, for instance, difficult to have a *Listeria* spp. specific medium able to discriminate *L. monocytogenes* from *Listeria ivanovii*. Only the Rapid *L'mono* from Bio-Rad allows this discrimination, but with this selective medium, some false positives have been recorded with other bacterial genera (Table 1).

Antibiotics against Gram-negative bacteria provide the selectivity of all of these media. The chromogenic differentiation of *Listeria* spp. from the other bacteria is based on the β-D-glucosidase activity present in all *Listeria* species. The second discrimination is done via the phosphatidylinositol phospholipase C (PIPL-C) activity, which is present in *L. monocytogenes* and *L. ivanovii.*

Table 1. List of available selective media for isolation of *Listeria spp.* and *L. monocytogenes* validated by AFNOR

			From AFNOR validation dossier	
Name	Company	Specificity	Inclusivity	Exclusivity
ALOA one day	AES Chemunex	*Listeria monocytogenes*	Ok	Ok
CHROMagar *Listeria*	CHROMagar		Ok	Ok
CHROMID *Lmono*	Biomérieux		Ok	False positive *L. ivanovii* (= *L. monocytogenes*)
Listeria Precis	OXOID		*L. mono* 3a et 4e: false negative	Ok
Rapid *Listeria* spp	Bio-Rad	*Listeria spp.*	one *L. ivanovii* false negative	False positive *Enterococcus faecalis, E. faecium, E. durans, Jonesia denitificans, Bacillus*
AL Detection	Bio-Rad	*Listeria* spp. and discrimination *L. monocytogenes*	Ok	False positive *L. ivanovii* (= *L. monocytogenes*)
ALOA	Biomérieux		Ok	False positive *L. ivanovii* (= *L. mono*) and among *Bacillus* & *Enterococcus* spp.
Compass *Listeria* agar	Solabia SAS		Ok	False positive *L. ivanovii* (= *L. monocytogenes*)
Rapid *L'mono*	Bio-Rad		Ok	False positive among *Bacillus, Enterococcus faecium, Oekskovia xanthineolytica, Gardnerella vaginalis, Lactobacillus*(*)

*This genera contains phosphatidylinositol phospholipase C activity; Inclusivity: test of more than 50 targets strains, Ok: all give typical colonies; exclusivity: test of more than 30 non-target strains, Ok: all give no typical colonies.

Finally in most selective media (such as Rapid *L'mono*), *L. monocytogenes* and *L. ivanovii* are discriminated from each other by the xylose fermentation activity, which is present in *L ivanovii*, but not in *L. monocytogenes*. The advantages of these chromogenic media are that they are simple to use, cost effective, simple to interpret and allow the isolation of the bacterial strain. However, although these selective/chromogenic plates are robust for common strains, atypical strains lacking one of the enzymatic activities used for the chromogenic discrimination, will give rise to false negative results.

C. Confirmation

After the isolation on selective plates, typical colonies should be confirmed to be *L. monocytogenes*, or not, via morphological, physiological and biochemical tests [3;7]: catalase reaction, Gram staining, motility test, Christie, Atkins, Munch-Petersen (CAMP) test, hemolytic activity and fermentation of xylose/rhamnose. Up to now, all these tests have to be performed and have to give the features of *L. monocytogenes* (Table 2). In the future, the confirmation step will be simplified, through a current ISO 11290-1 revision.

3. Immunological *Listeria* Detection Methods

Immunological methods, based on the antibody-antigen affinity, are popular because they are simple, sensitive, and accurate and can be carried out directly on enrichments from food or feed samples without cumbersome sample preparation. However, due to the antigenic heterogeneity of microorganisms, the specificity of these immunological assays is not always acceptable. To solve this drawback, alternative assays were recently developed using the bacteriophage binding protein domain instead of antibodies [8;9]. Many immunological assays are commercially available and are AOAC- or AFNOR-validated. In Table 3, a comparison of most of the available kits is reported. Positive results obtained with immunoassays should always be confirmed by isolation from the enrichment broth on *Listeria* selective plating media since both dead and alive bacteria can be detected by immunological tests. Once isolated, confirmation by morphological, physiological and biochemical tests is also requested.

A. Enzyme-Linked Immunosorbent Assay (ELISA)

ELISA methods used immobilized antibodies in a microtitre well (plate) or on a nitrocellulose membrane as a single line (strip). If *Listeria* is present in the enriched sample, they will bind to the anti-*Listeria* antibodies. For the plate tests, the sample is added into the well. Then unbound sample is removed by washing. A second antibody coupled with an enzyme (or other label) is added, followed by another washing step to remove unbound antibodies. The staining is done by the addition of a substrate, which displays color after an enzymatic reaction. After a final washing step, this color signal is measured for each sample.

Table 2. Biochemical characteristics of *Listeria* species

		L. monocytogenes	*L. innocua*	*L. ivanovii*	*L. seeligeri*	*L. welshimeri*	*L. grayi*
	Motile	+	+	+	+	+	+
	Hemolysin	+	-	+	(+)	-	-
	Catalase	+	+	+	+	+	+
	Oxidase	-	-	-	-	-	-
Fermentation of:	L-Rhamnose	+	+/-	-	-	+/-	+/-
	D-Xylose	-	-	+	+	+	-
CAMP test	*S. aureus*	+	-	-	(+)	-	-
	R. equi	-	-	+	-	-	-

Table 3. Commercially available Immunological tests

Application	Concentration/Separation of target	Detection of target																
Test	Pathatrix® Pooling *Listeria* species Test system	VIDAS *Listeria*	VIDAS *Listeria* Species Xpress	VIDAS UP *Listeria*	RAYAL *Listeria*	Tecra Ultima *Listeria*	Transia Plate *Listeria*	*Listeria* Rapid Test	Solus *Listeria* spp. ELISA	Reveal® for *Listeria* *one-Step* Test system	VIDAS *Listeria* DUO	VIP® (Visual Immunoprecipitate Assay) gold *Listeria*	TRANSIA strip *Listeria*	*Listeria*- TEK Elisa	Transia Plate *Listeria monocytogenes*	VIDAS *Listeria monocytogenes* 2 (enrichissement à 30°C)	VIDAS *Listeria monocytogenes* 2 (enrichissement à 37°C)	VIDAS *Listeria monocytogenes* Xpress
Manufacturer	Matrix microSciences Limited	Biomérieux			Rayal	Tecra International	BioControl Systems	OXOID	Solus Scientific	Neogen corp.	Biomérieux	BioControl Systems		Organon Teknica corp.	BioControl Systems	Biomérieux		
Specificity	***Listeria spp.***	***Listeria* spp.**						*L. spp excluding L. grayi*			***Listeria* spp. & *L. monocytogenes***			***Listeria monocytogenes***				
Matrices	Pasteurized milk, raw ground beef, Coocked sliced ham, milk powder, orange juice, ground black pepper, chocolate, soft cheese, lettuce, raw fish, and lasagna read to eat meal	all food and feed matrices	dairy, meat and vegetal products	all food matrices	all food matrices	all food and feed matrices	all food and feed matrices	all food and feed matrices	all food matrices	Pasteurized liquid eggs, ice cream, parmesan cheese, 2% milk, deli turkey, hot dogs, frozen coocked hamburgers, pepperoni, pasteurized crab meat, smoked salmon	all food and feed matrices	Dairy foods, red meats, pork, poultry and poultry products, seafood, fruits, vegetables, nutmeats, pasta, chocolate, eggs, bone meal	all food matrices	meat, meat product, fish, dairy product	all food matrices	all food matrices (excpt raw products)	all food matrices	all food matrices
Type of test	Immunocapture	ELFA			ELISA			ELISA		Chromatographique ELISA (strip)	ELFA	Chromatographique ELISA (strip)		ELISA		ELFA		
Validation dossier number	AOAC-RI#090201B	AFNOR BIO-12/2-06/94	AFNOR BIO-12/12-07/04	AFNOR BIO-12/33-05/12	AFNOR RAY-32/03-07/10	AFNOR TEC-24/4-09/04	NordVal 002	AFNOR UNI-03/02-04/95	AFNOR SOL-37/02-06/13	AOAC-RI#960701. and AOAC-RI#070601	AFNOR BIO-12/18-03/06	AOAC-OMA#997.03	AFNOR TRA-02/10-12/07 (2)	AOAC-OMA#994.03	AFNOR TRA-02/11-03/08	AFNOR BIO-12/9-07/02	AFNOR BIO-12/11-03/04	AFNOR BIO-12/27-02/10
Limit of detection (cfu/25g (ml)	Report not found	0.4-1.9	0.2-0.9	0.2-1.1	0.2-1	0.1-4	1-10	0.2-1.3	0.2-1.2	1-10	0.3-2.5	Report not found	0.2-1.3	Report not found	0.2-2.8	0.3-1.3	0.3-1.8	0.2-1.8
Inclusivity		Ok	Ok	Ok	Ok	Ok	Ok	Ok (except *L. grayi*)	Ok (except *L. grayi*)	Ok (except *L. grayi*)	Ok		Ok		Ok	Ok	Ok	Ok
Exclusivity		Ok	Ok	Ok	Ok	false positive with *Lactobacillus casei* & *Lactobacillus plantarum*	Ok	Ok	Ok	Ok	Ok		Ok		Ok	Ok	Ok	false positive with *Bacillus sphaericus*, *Staphylococcus intermedius* & *Staphylococcus aureus*
Practicability positif results (days)		5	3-5	2-6	5-7	5-6	nf	3-4	3-4	nf	3-5		2-3		2-3	3-10	3-10	2-8
Practicability negatif results (days)		2	2	1	2	2-5	nf	2	2	nf	2		2-3		2-3	2	2	1

nf: data not found.

Table 4. Commercially available nucleic acid-based Listeria detections kits and comparison of performances

Type of test	Target Capture*, Transcription Mediated Amplification* (TMA), and Hybridization Protection Assay* (HPA)	RNA Hybridization		DNA hybridation		PCR				
							Double-Stranded probes	SYBR®Green	TaqMan®	TaqMan®
Test	Atlas® *Listeria* detection assay	Accuprobe *Listeria monocytogenes*	Lumiprobe 24 Listeria monocytogenes	GeneQuence® *Listeria* spp.	GeneQuence® *Listeria monocytogenes*	Veriflow™ *Listeria monocytogenes*	IQ-Check *Listeria* spp	BAX *Listeria* 24E	GeneDisc *Listeria* spp.	MicroSEQ *Listeria* spp.
Manufacturer	Roka biosciences	Biomérieux	Lumiprobe SA	Neogen corp.	Neogen corp.	Invisible Sentinel	Bio-Rad	DuPont	Pall	Life Sciences®
Matrices	Pasteurized whole milk, ice cream, brie cheese, hot dogs, cured ham, deli chicken, chicken salad, cold-smoked salmon, romaine lettuce	all food matrices	all food matrices excepte "cantal" and "salers" cheeses	Parmesan cheese, brie cheese, pasteurized milk, ice cream, cottage cheese, mayonnaise, ground beef, deli turkey, hot dogs, ground pork, deli ham, lettuce, mixed vegetables, alfalfa sprouts, raw shrimp, pasteurized crab meat, smoked salmon.	nf	Beef hot dogs, deli turkey	all food and feed matrices	all food and feed matrices	all food matrices	all food matrices
Validation dossier number	AOAC-RI#011201	AFNOR BIO-12/4-02/95	AFNOR EUR-15/3-12/05	AOAC-RI#010403	AOAC-RI#120501	AOAC-RI#051303	AFNOR BRD-07/13-05/07	AFNOR QUA-18/06-07/08	AFNOR GEN-25/07-07/10	AFNOR ABI 29/04 - 12/11
Specificity	*Listeria* spp.	*Listeria monocytogenes*		*Listeria* spp.	*Listeria monocytogenes*	*Listeria monocytogenes*	*Listeria* spp			
Limit of detection (cfu/25g (ml)	nf	0.1-5.2	0.6-2.2	1-5	1-5	nf	0.2-1.5	0.3-1.4	0.1-2	0.3-1.6
Inclusivity	Ok	Ok	Ok	Ok	nf	Ok	Ok	Ok	Ok for *Listeria monocytogenes* but false negative among *L.* spp.	Ok except *L. grayi*
Exclusivity	Ok	Ok	Ok	Ok	nf	Ok	Ok	Ok	Ok	Ok
Practicability positif results (days)	nf	3-6	2	2-3	2-3	nf	2	2-3	2	3-4
Practicability negatif results (days)	nf	2-4	1	1-2	1-2	nf	1	1	1	2

qPCR										
qPCR	IMS + qPCR	Double-Stranded probes	SYBR®Green	Molecular beacon	TaqMan®	TaqMan®	qPCR	TaqMan®	fast qPCR (PikoReal 24)	IMS + qPCR
InstantLabs® *Listeria* species food safety kit	Assurance GDS *Listeria* spp Test Kit	IQ-check Listeria monocytogenes II	BAX *Listeria monocytogenes* 24E	Adiafood *Listeria monocytogenes*	GeneDisc *Listeria monocytogenes*	MicroSEQ *Listeria monocytogenes*	InstantLabs® *Listeria monocytogenes* food safety kit	foodproof® *Listeria monocytogenes* Detection Kit	SureTect™ *Listeria monocytogenes* PCR assay	Assurance GDS *L. monocytogenes* Test Kit
InstantLabs Medical Diagnostics Corp.	BioControl Systems	Bio-Rad	DuPont	AES Chemunex	Pall®	Life Sciences®	InstantLabs Medical Diagnostics Corp.	BIOTECON®	Thermo Fisher Scientific	BioControl Systems
Hot dogs, raw shrimp, cheddar cheese	Liquid pasteurized milk, mexican soft cheese, frankfurter, deli turkey, raw fish, raw green beans	all food matrices	all food matrices (excepted smoked fish and raw charcuteries)	all food matrices	all food matrices	all food matrices	Hot dogs, deli turkey, romaine lettuce, raw shrimp, cheddar cheese, vanilla ice cream, pasteurised whole milk	all food matrices	Raw ground beef, pork frankfurters, salami, coocked sliced turkey, fresh bagged spinach, cantaloupe, processesd cheese, ice cream, smoked salmon, coocked prawns	Liquid pasteurized milk, mexican soft cheese, frankfurter, deli turkey, raw fish, raw green beans
AOAC-RI#041304	AOAC-RI 070701	AFNOR BRD-07/10-04/05	AFNOR QUA-18/05-07/08	AFNOR AES-10/08-12/09	AFNOR GEN-25/08-07/10	AFNOR ABI 29/05-12/11	AOAC-RI#051302	NordVal 025	AOAC-RI#061302 (AFNOR validation on going)	AOAC-RI 070702
		Listeria monocytogenes								
nf	nf	0.2-1.6	0.2-1.3	0.3-2.4	0.1-2	0.2-1.6	nf	1-10	nf	nf
nf	nf	Ok	Ok	Ok	Ok	Ok	nf	Ok	Ok	nf
nf	nf	One false positive: *Enterococcus faecium*	Ok	Ok	Ok	Ok	nf	Ok	Ok	nf
nf	nf	2	2-3	3	2	3-4	nf	nf	nf	nf
nf	nf	1	1	1	1	1-2	nf	nf	nf	nf

nf: data not found.

For the strip tests, the ELISA is performed on a nitrocellulose membrane (the stationary phase). The loading extremity of the strip is immersed into an aliquot of the enriched medium (the mobile phase). Gold colloid labeled anti-*Listeria* antibodies, free on the membrane, are dissolved into the mobile phase and can form complexes with the *Listeria,* if present. These complexes migrate across the membrane by capillarity and are trapped by anti-*Listeria* antibodies fixed at the test line position to develop a reddish purple line by the gold colloid. Thus, whether the sample solution contains *Listeria* or not can be determined by visually inspecting the test line. Irrespective of the presence/absence of *Listeria* in the sample, the excess gold colloid labeled antibodies further migrates across the membrane and are then trapped by specific antibodies of the free antibodies, fixed at the control line position to develop a reddish purple line. It can be confirmed that the sample solution has migrated across the membrane by the development of the control line.

The Enzyme Linked Fluorescent Assay (ELFA), which combines the ELISA method with a final fluorescence reading, is mainly used in Vitek Immuno Diagnostic Assay System assays (VIDAS®, Biomérieux). It is the same principle as the ELISA, described above, except that the antibodies are coated on a Solid Phase Receptacle® (SPR®) that serves as stationary phase and pipetting device. The second antibody is conjugated with the alkaline phosphatase and binds to the *Listeria* captured on antibody immobilized on the wall of the SPR®. Finally, the enzyme catalyses the conversion of 4-methyl-umbelliferyl phosphate into the fluorescent 4-methly-umbelliferone phosphate [10]. The optical scanner of the instrument measures the intensity of fluorescence in each sample. VIDAS® assays are commercially available and are AFNOR/ISO16140-validated [7]. VIDAS® have several advantages: it is fast, automated and uses a general protocol for all food products. However, the VIDAS® system also has a drawback: the apparatus is not versatile and can only be used with the VIDAS® detection kits. Several additional kits exist for the specific detection of *L. monocytogenes*, the discrimination between *L. monocytogenes* and *Listeria* spp. or for a general detection of *Listeria* spp. (Table 3).

B. Immuno-Magnetic Separation (IMS)

Immuno-magnetic separation is based on antibodies that are immobilized on magnetic beads that are mixed with the enriched sample. If *Listeria* is present, it will bind to the anti-*Listeria* antibodies. A magnetic separation will allow isolation of the specific bacteria from the background flora and PCR inhibitors. Consequently, downstream steps (e.g., isolation on selective plates, immunoassays or nucleic acid based assays) will be simpler, faster and will give less false negative results. This technology is used for example into the Dynabeads® Anti-*Listeria* and Pathatrix® *Listeria* spp. kits both from Applied Biosystems® (Table 4).

4. Nucleic Acid-Based Detection Systems

The detection of *L. monocytogenes* and/or *Listeria* spp. from food samples via molecular methods is gaining recognition since it is also fast, accurate, specific and sensitive. Moreover, unlike biochemical and RNA-based detection methods, the DNA-based detection methods are not influenced by the expression of specific biochemical features. A list of commercially available molecular kits for the detection of *L. monocytogenes* and/or *Listeria* spp. from food samples is given in Table 4. The type of technology used and technical parameters obtained during AFNOR/AOAC validations are reported for each kit, for comparison purposes. As for the immunological assays, positive results obtained with molecular assays should always be confirmed by isolation from the enrichment broth on *Listeria* selective plating media. Once isolated, confirmation by morphological, physiological and biochemical tests is also required.

A. DNA/RNA Extraction

A critical pre-analytical step for nucleic acid-based assays is nucleic acid extraction. It is important to use an extraction protocol that is suitable with the tested matrix to get the best yield and less PCR inhibitors [11-13]. Indeed, some matrices with high background flora could inhibit the growth of the target bacteria. In this case, extraction methods giving high yield will be needed. For matrices containing a high level of PCR inhibitors, a highly selective extraction method will be needed. However, more selective and more efficient extraction methods are also more expensive and time-consuming.

This nucleic acid extraction consists in lysis of the cell to free RNA and DNA, which are then separated from the background sample by affinity membranes, organic extraction, filter membranes or magnetic particles. After this separation, RNA and DNA are eluted into an appropriated buffer. If only RNA or DNA have to be isolated, the DNase or RNase step will be included, respectively.

B. DNA or RNA Hybridization

DNA or RNA hybridization is the simplest molecular method for detection of *Listeria* spp. in food samples. The presence of the target is detected using a single stranded oligonucleotide probe complementary to the RNA or DNA target sequence. Probes are immobilized in a well and are labeled on one or both ends by radioactive isotopes, biotin, digoxygenin or fluorescent dyes. DNA or RNA hybridization could be used on enriched samples, on isolated colonies or after PCR amplification (14). The reading is done according to the dye selected. Four commercial kits (Table 4) are using this technology: Accuprobe (Biomérieux), Luminprobe (Europrobe), GeneQuence® *Listeria* spp. and GeneQuence® *L. monocytogenes* (Neogen corp.).

C. Reverse Transcriptase

The reverse transcriptase is a step performed before the Polymerase Chain Reaction (PCR) when the target of the detection is RNA. This step consists of reverse transcription of RNA to a complementary DNA (cDNA). This is done with the use of the reverse transcriptase (RT) enzyme. This step is necessary since RNA cannot be used as a template for a PCR reaction and is highly sensitive to degradation by RNases. When reverse transcription is performed in the same tube as the PCR amplification, this is called an RT-PCR.

D. Polymerase Chain Reaction

PCRs were developed in the 1980's by Mullis et al., [15]. Prior to a PCR analysis, the DNA should be extracted from the enriched sample. PCR is an enzymatic reaction amplifying *in vitro* a specific DNA region. Amplification is performed by a thermostable DNA polymerase using specific oligonucleotides, complementary to the target sequence, as primers for polymerization. PCR is a cycled thermal process: DNA denaturation (90-95°C), hybridization of oligonucleotides (annealing: 50-65°C) and DNA polymerase elongation (synthesis of DNA). This reaction is performed in a thermocycler. Usually PCR reaction is made of ca. 35 cycles. Each cycle leads to the synthesis of one new copy of the targeted DNA region from one target DNA strand, and each created DNA strand serves as a template in the following step. Thus, it is an exponential amplification that leads to billions of copies of the DNA target at the end of the reaction. This DNA amplification is visualized on agarose gel with ethidium bromide staining.

E. Real-Time PCR

The real-time or quantitative PCR (qPCR), developed by Heid et al., [16] in the 1990's, is an evolution of the classical PCR technology. In qPCR assays, the DNA polymerization is not measured post-amplification, as in classical PCR, but all along the process (hence "real-time"). At each cycle of amplification, the amount of polymerized DNA is measured via fluorescent dyes. The result of a qPCR assay is expressed as a threshold cycle (Ct), crossing point (Cp), take-off point (TOF) or quantification cycle (Cq) value, all different names of the same measure. Note that the Cq abbreviation has now been recommended [17]. The Cq-value represents the time-point at which the PCR amplification reaches the prior-set threshold level for the reaction. The threshold level takes into consideration the background level during the early cycle and the start of measurable amplification [18;19]. As such, the lack of measurable Cq-value in a qPCR assay is to be interpreted as the absence of amplification above the background level.

qPCR is more sensitive than PCR and has been widely accepted as a faster alternative to culture-based methods for the detection of pathogenic bacteria in food [20]. The different qPCR technologies are explained in detail below and schematized in Figure 2. qPCR could be performed as simplex (one target per well) or multiplex assays (several targets per well). Most commercially available qPCR detection assays are single target, using only one target to detect *Listeria* spp. or *L. monocytogenes*.

It is important to note that the use of only one target could lead to false negative results if the strain present in the sample is mutated or deleted in the targeted sequence, as illustrated in Barbau-Piednoir et al., 2013 [21]. A multi-level detection system could also be applied, allowing discrimination of *L. monocytogenes* from the other *Listeria* species [22]. Since to perform a qPCR detection, DNA from the enriched sample should be extracted, most commercially available *Listeria* qPCR detection systems provide/advise a specific DNA extraction kit.

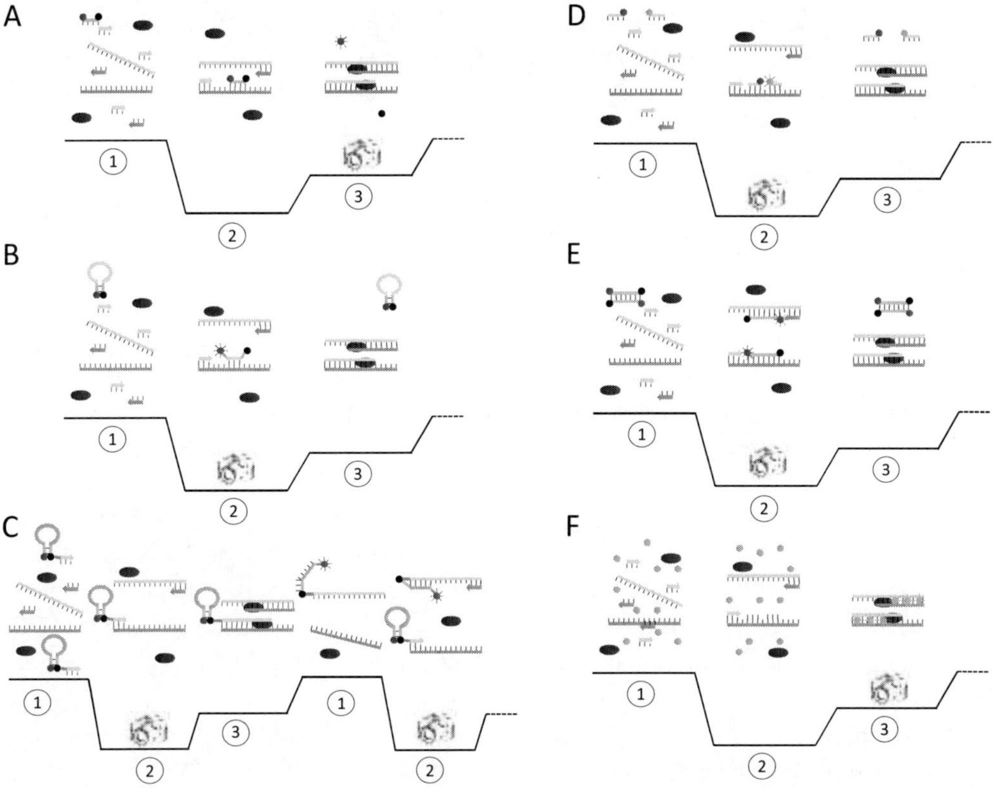

A.5' nuclease probe (TaqMan®), B. Molecular beacons, C. Scorpion, D. Fluorescent resonance energy transfer, E. Double-stranded probes. F. Intercaling agent (SYBR®Green). 1. Denaturation, 2. Annealing and 3. Elongation. : Taq-polymerase, : Forward and reverse primers, : fluoresecnt dyes, : probe

Figure 2. Schemes of the different qPCR technology mechanisms.

1. Probe-Based qPCR Methods

In the probe-based qPCR methods, in addition to the primers, one or two labeled probe(s) is/are used. The advantages of the probe-based assay are the extra specificity provided by the probe, and the possibility of multiplex assays. However, the drawbacks reside in the risk of false negatives when the target sequence is polymorph, and the price of the labeled probes.

a) 5'-Nuclease Probe (TaqMan®)

The 5'-nuclease probe technology, also named TaqMan® technology, was the first qPCR technology to be developed [16] and is the most used technology for qPCR analysis. This technology is used in Life Sciences, Biotecon and Pall kits (Table 4). A TaqMan® probe is a short oligonucleotide containing a 5'-fluorescent dye and 3'-quenching dye. The quenching dye inhibits the fluorescence of the fluorescent dye when they are close to each other. To generate a signal, the fluorescent dye should be separated from the quenching dye by hydrolysis of the probe by the DNA polymerase (5'-3' exonuclease activity) during the elongation step. The fluorescence signal is then measured at the elongation step of the qPCR reaction (Figure 2A). However, since the probe is hydrolyzed during the elongation step no melting curve analysis is possible.

Most of the kits available with the TaqMan® technology are performed in regular qPCR thermocyclers, except for the GeneDisc kits (Pall), delivered as ready-to-use discs usable into the GeneDisc® Cycler devise only. This system is not modular but is very user-friendly.

b) Molecular Beacons

Molecular beacons were described for the first time in 1996 [23]. This technique is based on one oligonucleotide hybridization probe, called molecular beacon, that is hairpin-shaped thanks to complementary regions at both extremities. The loop is an 18-30 base region complementary to the target sequence, while the dsDNA stem corresponds to the two hybridized ends. This probe is 5'-end labeled with a reporter fluorescent dye and 3'-end labeled by a suppressor fluorescent dye (quencher). No fluorescence is emitted from the probe at the hairpin shape stage. During denaturation, the probe is linearized. During the following annealing step, it either returns to its original hairpin shape if no complementary sequence is present in the sample (no fluorescence), or it hybridizes to its complementary sequence of the target DNA (emission of fluorescence as quencher and reporter are separated) (Figure 2B). Contrary to TaqMan® probes, this probe will not be hydrolyzed by the DNA polymerase, and the fluorescence will be measured at the annealing step. As the probe is not hydrolyzed during the elongation step, a melting curve analysis is possible. This technology is used in Adiafood *L. monocytogenes* qPCR detection kit and AES chemunex (Table 4).

c) Scorpion

The scorpion technology is a variant of the molecular beacons . The hairpin structure is dyed at both ends as a molecular beacon. The main difference is that the hairpin probe is coupled with one of the primers. The primer is linked to the hairpin structure via the PCR blocker hexaethylene glycol phosphoramidite (HEG). The polymerization of the target DNA occurs, but the PCR blocker stops the polymerization to avoid the probe being extended. At the next denaturation step, the probe is linearized (giving a scorpion tail shape) then during the annealing step the probe is able to hybridize to the polymerized DNA from the last amplification by a turnaround, which allows the reporter dye to emit fluorescent by distancing from the quencher dye. The fluorescence at each cycle is measured during the annealing step (Figure 2C).

d) Fluorescent Resonance Energy Transfer: FRET

The FRET technology uses two probes, named FRET hybridization probes or LightCycler® probes. One probe is a 3'-end labeled with a fluorescent dye; the other is a 5'-end labeled by an acceptor dye. Probes are designed to hybridize on the target sequence, separated by 1 to 5 bases. When alone, the fluorescent dye emits only a background signal. When both dyes are very close, the proximity allows the fluorescent resonance energy transfer from the donor fluorescent dye to the acceptor dye that emits a strong fluorescent signal. The fluorescence is measured at every annealing step; during the elongation step the probes are not hydrolyzed by the polymerase but only detached (Figure 2D). The two probes encompass a region of 40 to 50 bases, providing a larger specificity than the other probe-based qPCR technology. As the probes are not hydrolyzed during the elongation step, a melting curve analysis is possible.

e) Double-Stranded Probes

The double-stranded (ds) probe technology is specific to the iQ-check test kits of Bio-Rad® [24]. Ds-probes are labeled on both strands in the 5'-end with a fluorescent dye and in the 3'-ends with a quencher. When the probes are double-stranded, no fluorescence is emitted since the quencher inhibits the fluorescence of the dye. When the probes are hybridized to their specific targets, the fluorescent dye can be emitted and measured (Figure 2E). This technology has the advantage over other probe-based technologies to double the fluorescent signal per target as the probe is labeled on both strands.

2. Intercaling Agent-Based qPCR Assays

Intercaling agent-based qPCR uses intercaling dyes to monitor the polymerization of the target sequence in real-time. This technology was first described in the early 1990's by Higuchi et al, using ethidium bromide [25]. However, ethidium bromide is known to be a carcinogen and was replaced by SYBR®Green in the late 1990's [26]. Indeed, SYBR®Green is both safer to work with (less carcinogen) and free from the complex waste disposal issues. SYBR®Green is an asymmetric cyanine dye [27] intercaling dsDNA [28]. In the SYBR®Green qPCR assay, the reaction is performed as a classical PCR reaction, except that SYBR®Green is added into the mix. SYBR®Green is very lightly fluorescent when it is free and highly fluorescent when bound into dsDNA. During the annealing and the elongation steps, SYBR®Green is intercaled into the dsDNA and gives a fluorescent signal. At each cycle the amount of polymerized DNA increases and thus the SYBR®Green fluorescent intensity increases as well (Figure 2F). This fluorescent signal is reversible as SYBR®Green is freed at each denaturation step. This property allows a post amplification check of the end product (amplicon) specificity. So the SYBR®Green qPCR assay, in addition to the Cq-value, gives a second value: the melting temperature of the amplicon (Tm). This Tm-value represents the temperature at which 50% of the SYBR®Green fluorescence is lost due to denaturation (separation of the strands) of the amplicon. The Tm-value is a physical parameter specific and inherent to the sequence of the amplified product (especially GC content) and is influenced by chemical factors affecting the dsDNA stability (e.g., salt concentration or DMSO) (29). The other advantages of SYBR®Green technology is the easiness of its implementation and its low cost, since no probes are needed. The drawback of this technique is that multiplexing is less easy and effective than with probe-based qPCR assays. Indeed, several amplicons obtained from a SYBR®Green qPCR assay would give a common Cq-value since SYBR®Green is not sequence specific and the discrimination of each amplicon would be done based on the Tm-values recorded. Thus, a multiplex with more than two targets seems difficult to perform. This technology is the one used in BAX® kits, DuPont Qualicon (Table 4), and in the CoSYPS *Listeria* recently developed [22].

3. Important Controls to Perform with qPCR Detection Systems

As with every detection method, controls are needed to validate the results of qPCR analysis. With qPCR detection assays, in addition to the classical positive and negative controls, an inhibition control is advised [30]. These controls avoid occurrence of false negative or false positive results due to technical problems. They verify the efficiency of the PCR reaction (positive control), the non-contamination of the PCR mix (negative control) and the quality of the extract DNA (inhibition control). Common inhibitors are numerous and include various constituents of bacterial cells, non-target DNA, food constituents (e.g., organic and phenolic compounds, glycogen, cellulose, fats, and calcium), components of enrichment media and DNA extraction kits (e.g., detergents, lysozyme, NaOH, alcohols), components of body fluids (e.g., hemoglobin, urea, and heparin), environmental compounds (e.g., pollen, phenolic compounds, humic acids, and heavy metals), and laboratory items (e.g., glove powder, and laboratory plastic ware) [11;31;32]. The choice of a DNA extraction kit appropriate to the tested food matrices/enrichment media is important to reduce the presence of inhibitors into the extracted DNA sample. The presence of inhibitors into a sample could be tested with an external (not performed in the same well as the detection reaction) or an internal (performed in the same well as the detection reaction) inhibition control. Most commercial kits include an internal inhibition control. This inhibition control could be a housekeeping gene (gene present in the sample) or an heterologous gene (addition of a target such as plasmid containing a target gene different from the detection gene) or a homologous gene (addition of a target such as plasmid containing a target gene amplified by the same primers as the detection gene but with a different internal sequence (annealing of the probe)).

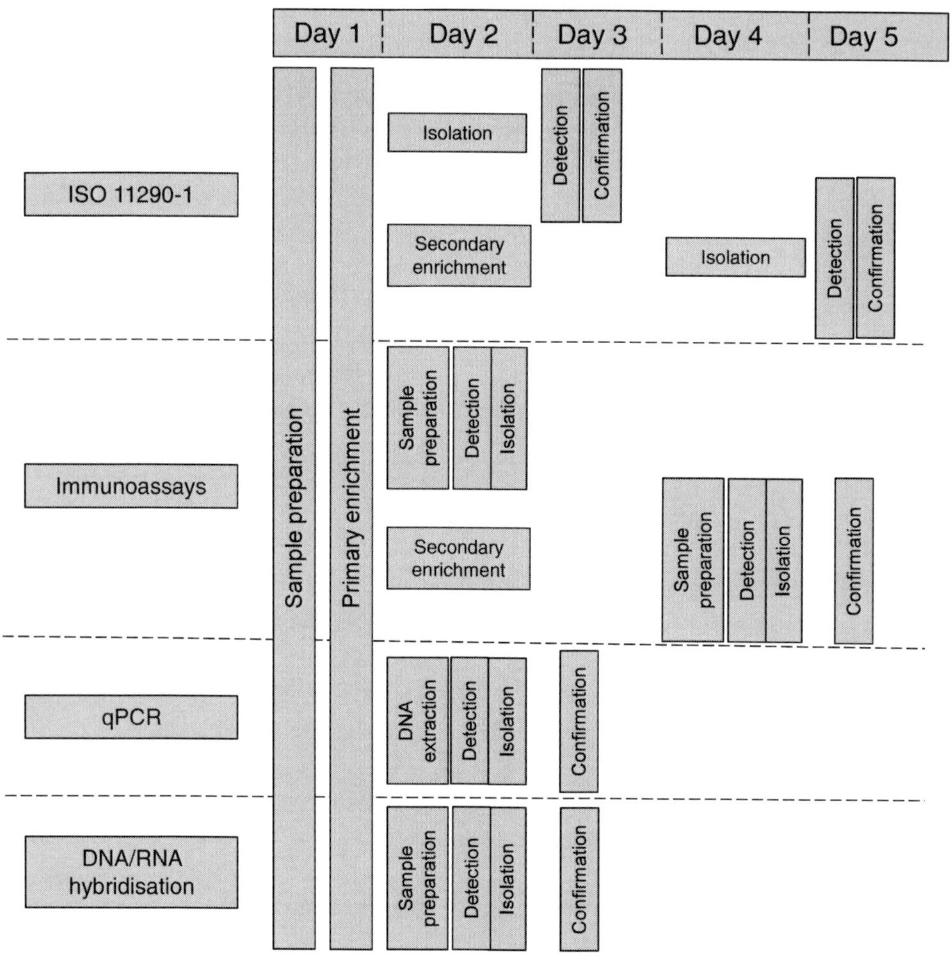

Figure 3. Diagram of time needed for the detection of *Listeria spp.* from feed and food samples via different technologies.

CONCLUSION AND PERSPECTIVES

Since listeriosis is the most severe food-borne human disease in term of hospitalization and fatal cases [33], its rapid and sensitive detection in food products is an important challenge to protect the consumer. For this purpose, several detection methods have been developed and are in constant improvement. The choice of a detection system is based on different criteria such as specificity, cost and time of analysis.

In Table 5, advantages and drawbacks of the different methodologies for the detection of *Listeria* spp. and *L. monocytogenes* are presented. Since there is a 'zero tolerance' for the presence of *L. monocytogenes* in certain foodstuffs in Europe [2], it is important to swiftly obtain the information on the potential presence of a suspected pathogen. This rapidity is of particular importance in the case of food poisoning investigations and for food business operators that need to release their products as fast as possible, especially for those with a short shelf life.

Figure 3 schematically presents the time needed for each methodology to confirm the presence of *Listeria* into a food sample. As all available detection methodologies are performed after an enrichment step, a possible way to reduce the time of analysis is the replacement of this enrichment step by a shorter step. Substitution technologies have been recently developed: filtration [34-36], buoyant density gradient centrifugation [36;37] and propidium monoazide (PMA) [38].

Table 5. Advantages and drawbacks of different technologies for detection of *Listeria* spp. and *Listeria monocytogenes* in food and feed products

	Bacteriological method	Immunological tests	Nucleic acid based tests
Advantages	Specific Easy to perform Sensitive (depends on the matrix) Cheap an isolate is available for further characterizations	Fast Specific (often false + results) Sensitive Not dependent of expression of biochemical characteristics Low influence of annexe flora	Fast Specific Sensitive Not dependent of expression of biochemical characteristics Low influence of annexe flora and analyzed matrix
Drawbacks	Time-consuming Dependent of expression of biochemical characteristics Large influence of analyzed matrix Large influence of annexe flora	Need of specific apparatus Cost Influence of the matrix Isolation has to be performed after positive result	Need of specific apparatus Cost Risk of contamination (false + results) Isolation has to be performed after positive result Influence of inhibitors (false - results)

Filtration and buoyant density gradient intend to separate and concentrate the targeted bacteria from the food product to allow its detection. Buoyant is even able to separate dead from living cells. PMA is a molecule theoretically able to discriminate living cells from dead cells without the enrichment step by binding DNA from dead cells, removing its "amplificability" by PCR or qPCR assays [39]. Buoyant density gradient could be used to replace enrichment since it is faster in separating dead from living cells and concentrate them, however it is a highly complex system to perform.

Filtration is able to concentrate bacteria but does not discriminate living cells from dead cells. PMA discriminates living cells from dead cells but does not concentrate them, which is crucial in food pathogen detection as the contamination is usually very low.

References

[1] Official journal of the European Union (2004). Regulation (EC) No 882/2004 of the European parliament and of council of 29 April 2004 on official controls performed to ensure the verification of compliance with feed and food law, animal health and animal welfare rules.

[2] European Commission (2005). Commission Regulation (EC) No 2073/2005 of 15 November 2005 on microbiological criteria for foodstuffs.

[3] International Organization for Standardization (ISO) (1996). ISO 11290-1:1996 Microbiologie des aliments - Méthode horizontale pour la recherche et le dénombrement de *Listeria monocytogenes* - Partie 1: Méthode de recherche.

[4] International Organization for Standardization (ISO) (2005). ISO 11290-1/A1:2005 Microbiology of food and animal feeding stuffs-Horizontal method for the detection and enumeration of *Listeria monocytogenes-Part 1: Detection method - Amendment 1: Modification of the isolation media, of the haemolysis test and inclusion of precision data.*

[5] Fraser JA, Sperber WH (1988). Rapid detection of *Listeria* spp. in food and environmental samples by esculin hydrolysis., *J. Food Prot* 51,10:762-765.

[6] Cowart RE, Foster BG (1985). Differential effects of iron on the growth of *Listeria monocytogenes*: minimum requirements and mechanism of acquisition, *J. Infect Dis.* 151,4:721-730.

[7] International Organization for Standardization (ISO) (2005). ISO 11290-1/A1:2005 Microbiology of food and animal feeding stuffs-Horizontal method for the detection and enumeration of *Listeria monocytogenes* - Part 1: Detection method - Amendment 1: Modification of the isolation media, of the haemolysis test and inclusion of precision data.

[8] Schmelcher M, Shabarova T, Eugster MR, Eichenseher F, Tchang VS, Banz M, et al., (2010). Rapid Multiplex Detection and Differentiation of *Listeria* Cells by Use of Fluorescent Phage Endolysin Cell Wall Binding Domains, *Appl Environ Microbiol* 76,17:5745-5756. doi:10.1128/AEM.00801-10.

[9] Kretzer JW, Lehmann R, Schmelcher M, Banz M, Kim K-P, Kom C, et al., (2007). Use of High-Affinity Cell Wall-Binding Domains of Bacteriophage Endolysins for

Immobilization and Separation of Bacterial Cells., *Appl Environ Microbiol* 73,6:1992-2000.

[10] Sewell AM, Warburton DW, Boville A, Daley EF, Mullen K (2003). The development of an efficient and rapid enzyme linked fluorescent assay method for the detection of *Listeria* spp. from foods, *International Journal of Food Microbiology* 81,2:123-129.

[11] Hyeon JY, Hwang IG, Kwak HS, Park C, Choi IS, Seo KH (2010). Evaluation of PCR inhibitory effect of enrichment broths and comparison of DNA extraction methods for detection of *Salmonella* Enteritidis using real-time PCR assay, *J. Vet. Sci.* 11,2:143-149. doi:10.4142/jvs.2010.11.2.143.

[12] Heller LC, Davis CR, Peak KK, Wingfield D, Cannons AC, Amuso PT, et al., (2003). Comparison of methods for DNA isolation from food samples for detection of Shiga toxin-producing *Escherichia coli* by real-time PCR, *Appl. Environ Microbiol.* 69,3:1844-1846.

[13] Elizaquivel P, Aznar R (2008). Comparison of four commercial DNA extraction kits for PCR detection of *Listeria monocytogenes*, *Salmonella, Escherichia coli* O157:H7, and *Staphylococcus aureus* in fresh, minimally processed vegetables, *J. Food Prot* 71,10:2110-2114.

[14] Ninet B, Bannerman E, Bille J (1992). Assessment of the Accuprobe *Listeria monocytogenes* culture identification reagent kit for rapid colony confirmation and its application in various enrichment broths, *Appl. Environ. Microbiol* 58,12:4055-4059.

[15] Mullis K, Faloona F, Scharf S, Saiki R, Horn G, Erlich H (1986). Specific enzymatic amplification of DNA in vitro: the polymerase chain reaction, Cold Spring Harb Symp *Quant. Biol.* 51 Pt 1:263-273.

[16] Holland PM, Abramson RD, Watson R, Gelfand DH (1991). Detection of specific polymerase chain reaction product by utilizing the 5'-3' exonuclease activity of *Thermus aquaticus* DNA polymerase, *Proc. Natl. Acad. Sci. USA* 88,16:7276-7280.

[17] Bustin SA, Benes V, Garson JA, Hellemans J, Huggett J, Kubista M, et al., (2009). The MIQE guidelines: minimum information for publication of quantitative real-time PCR experiments, *Clin. Chem* 55,4:611-622. doi:10.1373/clinchem.2008.112797.

[18] Bustin SA (2000). Absolute quantification of mRNA using real-time reverse transcription polymerase chain reaction assays, *J. Mol. Endocrinol* 25,2:169-193.

[19] Tse C, Capeau J (2003). [Real time PCR methodology for quantification of nucleic acids], *Ann. Biol. Clin.* 61,3:279-293.

[20] Postollec F, Falentin H, Pavan S, Combrisson J, Sohier D (2011). Recent advances in quantitative PCR (qPCR) applications in food microbiology, *Food Microbiol* 28,5:848-861. doi:10.1016/j.fm.2011.02.008.

[21] Barbau-Piednoir E, Roosens NH, Bertrand S, Mahillon J, Botteldoorn N (2013). SYBR®Green qPCR *Salmonella* detection system allowing discrimination at the genus, species and subspecies levels, *Applied Microbiology and Biotechnology*. doi:10.1007/s00253-013-5234-x

[22] Barbau-Piednoir E, Botteldoorn N, Yde M, Mahillon J, Roosens NH (2013). Development and validation of qualitative SYBR®Green Real-Time PCR for detection and discrimination of *Listeria* spp. and *Listeria monocytogenes*, *Applied Microbiology and Biotechnology* 97,9:4021-4037. doi:10.1007/s00253-012-4477-2.

[23] Tyagi S, Kramer FR (1996). Molecular beacons: probes that fluoresce upon hybridization, *Nat. Biotechnol* 14,3:303-308. doi:10.1038/nbt0396-303 [doi]

[24] Dagland T, Rieunier F, inventors (2008). Double-Stranded probes for the fluorescent detection of nucleic acids, WO2008044129A2.

[25] Higuchi R, Dollinger G, Walsh PS, Griffith R (1992). Simultaneous amplification and detection of specific DNA sequences, *Biotechnology* (N Y) 10,4:413-417.

[26] Wittwer CT, Herrmann MG, Moss AA, Rasmussen RP (1997). Continuous fluorescence monitoring of rapid cycle DNA amplification, *Biotechniques* 22,1:130-138.

[27] Zipper H, Brunner H, Bernhagen J, Vitzthum F (2004). Investigations on DNA intercalation and surface binding by SYBR Green I, its structure determination and methodological implications, *Nucleic Acids Res* 32,12:e103.

[28] Haugland RP (2002). *Handbook of fluorescence probes and research products*, ninth edition, Molecular Probes, Inc.

[29] Ye J, Parra EJ, Sosnoski DM, Hiester K, Underhill PA, Shriver MD (2002). Melting curve SNP (McSNP) genotyping: a useful approach for diallelic genotyping in forensic science, *J. Forensic Sci* 47,3:593-600.

[30] Hoorfar J, Malorny B, Abdulmawjood A, Cook N, Wagner M, Fach P (2004 May). Practical considerations in design of internal amplification controls for diagnostic PCR assays, *J. Clin. Microbiol.* 42,5:1863-1868.

[31] Wilson IG (1997). Inhibition and facilitation of nucleic acid amplification, *Appl. Environ. Microbiol.* 63,10:3741-3751.

[32] Rossen L, Norskov P, Holmstrom K, Rasmussen OF (1992). Inhibition of PCR by components of food samples, microbial diagnostic assays and DNA-extraction solutions, *International Journal of Food Microbiology* 17,1:37-45.

[33] European Food Safety Authority (EFSA), European Centre for Disease Prevention and Control (ECDC) (2013). The European Union Summary Report on Trends and Sources of Zoonoses, Zoonotic Agents and Food-borne Outbreaks in 2011, *EFSA Journal* 11,4:3129. doi:10.2903/j.efsa.2013.3129.

[34] Wolffs PF, Glencross K, Thibaudeau R, Griffiths MW (2006). Direct quantitation and detection of salmonellae in biological samples without enrichment, using two-step filtration and real-time PCR, *Appl. Environ Microbiol* 72,6:3896-3900.

[35] Murakimi T (2012). Filter-Based Pathogen Enrichment Technology for Detection of Multiple Viable Foodborne Pathogens in 1 Day, *J. Food Protection* 75,9:1603-1610. doi:10.4315/0362-028X.JFP-12-039.

[36] Kim HJ, Cho JC (2010). Simple and rapid detection of *Listeria monocytogenes* in fruit juice by real-time PCR without enrichment culture, *Food Control* 21,10:1419-1423. doi: 10.1016/j.foodcont.2010.04.006

[37] Fukushima H, Katsube K, Hata Y, Kishi R, Fujiwara S (2007). Rapid separation and concentration of food-borne pathogens in food samples prior to quantification by viable-cell counting and real-time PCR, *Appl. Environ. Microbiol.* 73,1:92-100. doi:10.1128/AEM.01772-06.

[38] Nocker A, Sossa-Fernadez P, Burr MD, Camper, A. K. (2007) Use of Propidium Monoazide for Live/Dead Distinction in Microbial Ecology, *Appl. Environ. Microbiol* 73(16): 5111–5117.doi:10.1128/AEM.02987-06.

[39] Fittipaldi M, Nocker A, Codony F (2012). Progress in understanding preferential detection of live cells using viability dyes in combination with DNA amplification, *Journal of Microbiological Methods* 91,2:276-289. doi:10.1016/j.mimet.2012.08.007.

In: *Listeria monocytogenes*
Editor: Edmund C. Hambrick

ISBN: 978-1-63117-054-6

Chapter 9

PREVALENCE AND CONTROL OF *LISTERIA MONOCYTOGENES* IN FOOD PROCESSING ENVIRONMENTS

Snehal Jadhav, Vandana Gulati, Mrinal Bhave and Enzo A. Palombo
Faculty of Life and Social Sciences, Swinburne University of Technology, Hawthorn, Victoria , Australia

ABSTRACT

Listeria monocytogenes is an ubiquitous foodborne pathogen responsible for 'listeriosis' which is one of the most severe foodborne diseases in humans. Although the disease has an overall low incidence, it has a very high mortality rate of 20-30%. A diverse range of food products have been implicated with sporadic and epidemic incidents of listeriosis linked to the consumption of ready-to-eat foods, dairy products, seafood, cooked or frozen meat, pork and fresh produce. The pathogen is a major threat to the food industry since it can survive the harsh conditions encountered in food processing environments including high salinity, acidity and refrigeration temperatures. More importantly, it has the ability to attach to abiotic surfaces and form biofilms that can frequently contaminate food products and the production environment. Such biofilms tend to be more resistant to antimicrobial treatments. This has generated the need to develop reliable control strategies for *L. monocytogenes* (existing either as planktonic or sessile communities) in processing environments. With the recent emergence of bacteria resistant to antimicrobials and the inclination of consumers towards products derived from natural sources, it is important to explore alternatives for disinfection. This review will focus on the prevalence and control of *L. monocytogenes* in food processing environments. In particular, the role biofilms play in persistence of the pathogen in production environments and the use of essential oils as suitable disinfectant agents will be highlighted.

INTRODUCTION

Listeria monocytogenes is a gram positive, intracellular human pathogen responsible for the foodborne disease 'listeriosis'. The genus *Listeria* is composed of two pathogenic species, *L. monocytogenes* (main cause of human infection), *L. ivanovii* (predominantly a ruminant pathogen, rarely responsible for human infection), and eight non-pathogenic species, *L. seeligeri, L. innocua, L. welshimeri, L. grayi, L. marthii, L. rocourtiae, L. weihenstephanensis* and *L. fleischmannii*. Listeriosis was first described in 1926 as the cause of an epizootic outbreak in guinea pigs and rabbits. It was subsequently identified as a human infection in 1929, following which it was recognized as a serious public health problem (Swaminathan et al., 2007). The disease has a lower incidence compared to other foodborne illnesses such as campylobacteriosis and salmonellosis but is considered a major threat due to its high case fatality rate (20-30%) (Allerberger and Wagner 2010). *L. monocytogenes* is widely distributed in the environment including soil, water, sewage, vegetation (e.g., grass, meadows, forests and silage), wild animal faeces, as well as on-farm and in food processing facilities (Kerouanton et al., 2010; Todd et al., 2011). Some common foods associated with the presence of *L. monocytogenes* include uncooked meat products, vegetables, seafood, pork products, salads and soft cheeses (Swaminathan et al., 2007). Contaminated refrigerated ready-to-eat (RTE) foods which can support growth of *L. monocytogenes* are of particular concern since such products typically are not heated prior to consumption. There is a 'zero tolerance' policy for RTE foods in the United States; under this policy any food contaminated with detectable levels of *L. monocytogenes* is deemed adulterated (U.S. FDA, 2013). Serological techniques that detect interactions between somatic (O) and flagellar (H) antigens with the corresponding antisera are used to classify *L. monocytogenes* into 13 serovars. Based on its virulence potential in mammalian hosts, it has been noted that *L. monocytogenes* serovars 4b, 1/2a, 1/2b, and 1/2c are responsible for over 98% of the isolations from clinical cases of human listeriosis (Viswanath et al., 2013).

Listeriosis mostly affects the elderly, neonates, pregnant women and immune-compromised populations. It can result in meningitis, septicaemia, encephalitis and can also cause still births and abortions in pregnant women (Farber and Peterkin 1991; reviewed in Jadhav et al., 2012). In humans, the infection occurs mainly due to consumption of contaminated foods such as dairy, vegetables, meats, fish, chicken and hot dogs. The Centre for Disease Control and Prevention (CDC) estimates that approximately 1600 illnesses and 260 deaths due to listeriosis occur annually in the United States (Anonymous 2013).

Linking foods with outbreaks is particularly challenging, as the incubation period for listeriosis can often vary between 3-70 days (Dawson et al., 2006). Currently, ampicillin plus gentamicin is standard therapy for listeriosis, while trimethoprim-sulfamethoxazole, vancomycin and erythromycin are also used to treat bacteremia and listeriosis in pregnant women (Yan et al., 2010). Listeriosis can exist in invasive or non-invasive forms. The invasive form is recognized as a more serious manifestation of the disease due to a longer incubation period, severity of symptoms and the associated high case-fatality rate. The predominant forms of presentation correspond to disseminated infection (bacteraemia or septicaemia) or to local infection in the central nervous system (meningitis or meningoencephalitis) (Garrido et al., 2008). In contrast, non-invasive listeriosis results in a milder gastroenteritis-type illness with symptoms including fever, diarrhoea and vomiting.

These clinical conditions have non-specific symptoms and therefore remain vastly underdiagnosed. In particular, febrile gastroenteritis may affect perfectly healthy individuals following ingestion of heavily contaminated food. Cutaneous listeriosis is another manifestation that can result in an eczematous skin infection caused by direct exposure of intact skin to *L. monocytogenes*. This form is typically observed in veterinarians exposed to a diseased animal often after a listerial abortion in animals (McLauchlin et al., 1994; Godshall et al., 2013).

FOODBORNE OUTBREAKS OF LISTERIOSIS

Epidemiological surveillance studies conducted in the past have clearly indicated that listeriosis is a global issue (Table 1). Over the years, several large epidemics of human listeriosis have been reported, but its attribution to a food source was first reported in 1981, linked to the consumption of contaminated coleslaw (Linnan et al., 1988). Subsequent to this, a common source of pasteurized milk was found to be associated with an outbreak in Massachusetts in 1983. In 1985, the outbreak in California resulting from consumption of inadequately pasteurized cheese made listeriosis a concern for the food industry (Linnan et al., 1988). Since then, *L. monocytogenes* has been involved in many outbreaks linked to food items such as dairy products, deli meats, and seafood (Gaulin et al., 2012). Over the past two decades, numerous foodborne outbreaks have been reported all over the world including the USA, Canada, Japan, New Zealand, Germany, England, France and other European countries (Warriner et al., 2009). Table 2 gives an overview of the outbreaks that have occurred during the period 2005-2013.

Table 1. Incidences of listeriosis in different countries

Country	Period	Incidence of listeriosis per 100,000	Reference
Australia	2005-2009	0.3	OzFoodNet (2012)
USA	2004-2009	0.25-0.32	Silk et al., (2012)
New Zealand	2009-2010	0.5-0.6	Lim et al., (2012)
Canada	1990-1998	0.18-0.34	Todd and Notermans (2011)
European Union	2000-2005	0.1-0.3	Denny and McLauchlin (2008)
Belgium	2000-2005	0.43-0.86	Goulet et al., (2008)
Denmark	2005-2008	0.52-0.85	Goulet et al., (2008)
England-Wales	1990-2000; 2001-2009	0.21; 0.36	Mook et al., (2011)
Finland	2000-2005	0.35-0.79	Goulet et al., (2008)
Germany	2001; 2005	0.26; 0.62	Koch and Stark (2006)
Netherlands	2002-2005	0.2-0.56	Doorduyn et al. (2006); Goulet et al., (2008)
Sweden	2000-2005	0.44-0.75	Goulet et al., (2008)
Switzerland	2001-2005	0.38-0.98	Goulet et al., (2008)

PREVALENCE OF *LISTERIA* IN FOOD PROCESSING ENVIRONMENTS

L. monocytogenes has been isolated previously from poultry, meat, trout and shrimp processing plants, and ice cream plants (Fonnesbech Vogel et al., 2001). It can be a serious threat to the food industry, as it has the ability to adhere to various abiotic surfaces in the processing environments, forming biofilms that can continuously contaminate the food products. Incoming raw materials are considered an important source of listerial contamination (Viswanath et al., 2013). Some of the other contributors to contamination in food processing environments include continuous presence of water on floors, failure to properly clean and sanitize drains, food contact surfaces and unregulated human traffic patterns (Viswanath et al,. 2013). Common sampling points considered important for monitoring the presence of *L. monocytogenes* are cutting boards, utensils, raw material, drains, floors, conveyor belts, slicers, skinners, work tables, tanks, vats, centrifuges, strainers, sieves, knives, walls, personnel hands, gloves, aprons and boots (Dass et al., 2011; Fallah et al., 2013).

Contamination can either be due to sporadic strains or persistent strains in the plant environment. The sporadic strains may frequently be detected but are not able to establish their niche, while the persistent strains can exist in a particular processing stage for a long period of time, leading to recurrent contamination in that stage of food production. Indeed, recurrent contamination spanning over a decade in the same production environment has been reported (Swaminathan et al., 2007; Takahashi et al., 2010). This persistent type of contamination by *L. monocytogenes* can be attributed to the microorganism's ability to tolerate extreme physiological conditions such as vacuums, freezing temperatures, high salinity, desiccation and extreme acidic and alkaline environments (Vogel et al., 2010). Studies have also suggested that *L. monocytogenes* can exhibit resistance to one or more antimicrobial agents and disinfectants, allowing it to survive in food processing environments (Carpentier et al., 2011; Luz et al., 2012). Hence, most jurisdictions mandate that high risk food processing environments have their own hazard analysis and critical control point (HACCP) program for controlling *Listeria* contamination in the food product as well as the entire facility. These programs require the industries to have systematic and periodic sampling regimes in their production facilities. Subsequent *L. monocytogenes* detections in the product are notified to the relevant regulatory authority and may eventually lead to product recalls. This may not only have serious economic impacts, but can also destroy the confidence of customers in a particular brand.

AREAS OF CONCERN IN THE FOOD PRODUCTION ENVIRONMENT

Product testing alone cannot solve contamination issues in a food processing environment. It is important to perform environmental sampling in order to trace the source of the contamination (Fox et al., 2011). Over the years, researchers have highlighted many problematic areas in the production or processing environment that have been associated with detection of *Listeria* spp. The prevalence of *Listeria* in food processing environments is generally investigated using different molecular subtyping techniques. These techniques are used for 'source tracking' or 'source attribution' wherein the probable origins of contamination in the food processing environments are estimated. By comparing the subtyping profiles of all the isolates collected post-sampling, correlations with their origins can be made and used to associate individual strains with particular areas of the processing environment. A number of molecular subtyping methods have been developed and are commonly used for *L. monocytogenes*, including automated ribotyping, pulsed-field gel electrophoresis (PFGE), multiple-locus sequence typing (MLST) and multiple-locus VNTR analysis (MLVA). A detailed description of the techniques, the advantages and disadvantages associated with each technique has been provided previously by our group (Jadhav et al., 2012) and will not be discussed in the current review. The prevalence of *L. monocytogenes* in some of the major food industries is discussed below.

Table 2. Foodborne outbreaks of listeriosis (2005-2013)

Year	Location	Infected Cases	Food vehicle	Serotype	References
2005	Switzerland	10	Tomme cheese	1/2a	Garrido et al., (2010)
2006-2007	Germany	189	Cheese made from pasteurised milk	4b	Koch et al., (2010)
2006-2007	Belgium	11	NA	4b	Valderrama and Cutter (2013)
2006	Czech Republic	75	Cheese and Salad	1/2b	Garrido et al., (2010)
2006-2007	Germany	16	Scalded sausages	4b	Garrido et al., (2010)
2007	Norway	21	Cheese	NA	Todd and Notermans (2011)
2007	USA	5	Pasteurised Milk	NA	Garrido et al., (2010)
2008	Canada	57	Ready-to-eat meats	1/2a	Garrido et al., (2010)
2008	Chile	119	Brie and Camembert cheese	NA	Garrido et al., (2010)
2008	Austria	19	Jelly Pork	4b	Garrido et al., (2010)
2008	USA	20	Sprouts	1/2a	Cartwright et al., (2013)
2009	Austria and Germany	14	Curd cheese	1/2a	Fretz et al., (2010)
2009	Denmark	8	Beef Meat	NA	Garrido et al., (2010)
2009-2010	Germany, Austria and Czech republic	34	Curd Cheese	1/2a	Garrido et al., (2010)
2010	USA	15	Celery	NA	Valderrama and Cutter (2013)
2011	USA	147	Cantaloupe	1/2a, 1/2b	Lomonaco et al., (2013)
2012	USA	22	Soft Ricotta cheese	NA	CDC (2012)
2012	Spain	10	Latin style fresh cheese	1/2a	de Castro et al., (2012)
2013	USA	6	Cheese	NA	CDC (2013)

A) Smoked fish industry

Several listeriosis outbreaks have been associated with smoked fish. Fish processing facilities are often observed to harbour some at-risk sites for listerial contamination (Rotariu et al., 2014). Other than contaminated raw material, the multiple processing stages including trimming, brining, drying, vacuum storage and refrigeration are major contributors to contamination. In particular, the brining, slicing and skin removing equipment have complicated assemblies making certain areas inaccessible for cleaning (Rørvik et al., 2000). These areas can then become sources of bacterial contamination as they are perfect niches for biofilms. Dripping condensation in chillers and reuse of brine for multiple batches may also be probable reasons for listerial contamination in this processing environment (Rotariu et al., 2014).

In a study by Autio et al., (1999), the source of *L. monocytogenes* contamination was tracked in a cold smoked rainbow trout production plant at various stages of production using PFGE. Sampling was done from the production line, environment and fish at different stages of production. Since brining and post-brining areas were found to be the most contaminated sites, the skinning, brining and slicing machines were disassembled, cleaned and disinfected by heating directly in an oven or by immersing their parts in hot water. Hence, a suitable eradication program was devised for the reduction in *L. monocytogenes* contamination based on this study. In a more recent study conducted by Fallah et al., (2013), sampling of various sites in seafood processing facilities revealed the presence of *L. monocytogenes* on cutting instruments, cutting boards, trimmers, fish fillet holding trays, and on factory personnel. Such investigations help in developing better control strategies for *Listeria* in the processing environment.

B) Dairy industry

Many dairy products such as cheese (soft cheese in particular), pasteurised milk (contaminated post-pasteurisation) and creams have been associated with some of the major listeriosis outbreaks (Kabuki et al., 2004; Borucki et al., 2005). In the case of dairy products, farm environments (contamination of raw milk in bulk tanks or in-line milk filters, silage contamination, drinking water contamination, or mammary contamination) can be major sources of *Listeria* leading to contamination of the processing environment (Borucki et al., 2005; Lattore et al., 2008; Fox et al., 2011). Healthy and infected ruminants shed the organism on a routine basis in their faeces, which can often result in contamination of the raw milk (Borucki et al., 2005). Hence, products made from unpasteurised milk are often at high risk. In contrast, products made from pasteurised milk are mainly contaminated during the post-pasteurisation processing (Cagri-Mehmetoglu et al., 2011). The most likely environmental sources are surfaces that come in contact with the product, such as plastic tube connectors used to connect holding tanks to pasteurisers, ripening room racks in the case of cheese factories, sinks, crates used for holding the food, tables, plastic milk packages, floor drains, and floors. Hence, to minimise listerial contamination in dairy processing environments, it is essential to use clean raw materials, appropriate materials (which can be easily sanitised) for surfaces that are in direct contact with the food, have suitable workflows and maintain hygienic environments (Kabuki et al., 2004; Fox et al., 2011).

In a study conducted by Alessandria et al., (2010), a cheese processing plant was sampled for two consecutive years (2007-2008). Semi-finished and finished products, equipment and the other raw materials were sampled and the genetic relatedness of the isolates obtained was deciphered by using molecular typing methods (RAPD, Rep-PCR, and Sau-PCR). *L. monocytogenes* was detected from the packaging area (which included conveyor belts, portioning knives and packaging machines), from the semi-finished products and final products. Since the pathogen was isolated from different areas, it clearly suggested that it was possible that the same contaminating isolates were carried from one stage of processing to the next, leading to a persistent contamination. In the first year of sampling, significant contamination was observed in the brines and in the equipment. Following implementation of stricter cleaning regimes and replacement of a brine pasteuriser, the level of contamination was lower in the second year of sampling. This clearly shows that source tracking of the contamination and subsequent reduction of the problem in the dairy plant can be achieved by using a correct sampling scheme, adequate cleaning and disinfectant techniques and by application of appropriate molecular and analytical methods.

C) Ready-to-eat meat and poultry processing industries

RTE meat products are considered to fall under the high risk category foods responsible for listeriosis, as mentioned earlier. The US Food and Drug Administration (FDA) allows minimal detectable levels of the pathogen in some foods that do not support its growth. However, a zero tolerance policy is applied to RTE foods, as *L. monocytogenes* can grow to high cell numbers and they are consumed without heating (Kushwaha et al., 2009). Nearly 90% of listeriosis cases in USA are caused by RTE deli meats. An RTE product can be considered to be contaminated with *L. monocytogenes* if it contains the pathogen or if it comes in direct contact with a food contact surface (such as slicers, dicers or packagaing equipment) that has tested positive for *L. monocytogenes* (Kushwaha et al., 2009; Williams et al., 2011). In spite of such stringent regulations, contamination still occurs in RTE meat processing plants. A large outbreak in Canada in 2008 involving deli meat products from the Maple Leaf Company was traced to the slicing equipment in a Toronto-area processing plant. This resulted in 22 deaths out of 57 reported cases of listeriosis and resulted in huge economic loss to the company (Howell et al., 2010; Todd and Notermans 2010). In RTE product processing plants, any resident *L. monocytogenes* biofilms in areas close to food contact surfaces that are a rich source of water and food ingredients and environments that undergo multiple steps of food processing are at potential risk for product contamination (Kovačević et al., 2009; Kovačević et al., 2012).

In the study conducted by William et al., (2011), six meat processing plants were sampled for environmental listerial contamination of non-food contact surfaces for a period of eleven months and *L. monocytogenes* was isolated mainly from raw/in-process areas. Post speciation, the isolates obtained were subtyped using automated *Eco*RI ribotyping and PFGE. Based on these results, strategies focussed on correct personnel flow and rigorous sanitisation of problematic areas were developed to control the detections. Moreover, similar molecular subtypes were detected from the same sites of detection in most of these plants, thus highlighting the importance of sanitation in such non-food contact surfaces to avoid product recalls. Kushwaha et al., (2009) have suggested several ways to minimise detections in the

RTE meat processing environments; regulation on the incoming raw meat to avoid it entering the facility, having sanitizing foot baths, sanitation of drains and air ducts, reduced use of high pressure water hoses during shifts (these can spread the contamination to different areas), designing of instruments with minimal crevices for bacterial protection, sloped surfaces to avoid stagnant water collection, prevention of personnel flow from raw to post processing areas without appropriate sanitation and segregating drainage systems for raw and post processing areas. Thus, with appropriate designs, sanitation programs and trained personnel, *Listeria* contamination can be minimised in RTE meat processing plants.

D) Fresh produce processing industry

Most studies of listerial contamination have carried out an in-depth analysis in seafood, dairy and meat processing facilities. However, very few studies have focussed on fresh produce processing facilities, the rationale being that very few listeriosis outbreaks have been linked to the consumption of contaminated fruits or vegetables. However, fresh produce such as cabbage, lettuce, onions, fennel and mushrooms can support the growth of *Listeria* (Bari et al., 2005). A recent listeriosis outbreak linked to whole cantaloupes led to 33 fatalities (Lomonaco et al., 2013). This indicates the need for more investigations in fresh produce processing facilities to study the prevalence of *Listeria* in the plant environment. Viswanath et al., (2013) carried out an investigation to trace the presence of *Listeria* in a small mushroom production facility. Most of the positive *L. monocytogenes* detections were observed in the raw material composting area. Although no *L. monocytogenes* serotype linked to an outbreak was isolated from that particular facility, the presence of the pathogen suggested that similar investigations of larger mushroom processing facilities are warranted. An increase in such investigations will help decipher the possible sources of contamination in fresh produce processing plants and potentially prevent episodes of listeriosis.

Overall, *L. monocytogenes* does prevail in a range of food processing facilities and molecular subtyping seems to be the key in tracing the sources of contamination in the plant environment. Good manufacturing practises, HACCP plans, suitable work flows and appropriately trained personnel can help in reducing the incidence of *Listeria* in the processing facility. However, the foremost priority of manufacturers should be to prevent *L. monocytogenes* from surviving and residing in their facilities. This requires efficient disinfection and sanitation approaches that can eradicate both planktonic and sessile communities of *Listeria* from the processing environment.

CONTROL OF *L. MONOCYTOGENES* IN FOOD PROCESSING ENVIRONMENTS

In the food industry, different foods require different control strategies. For example, addition of antimicrobials such as potassium lactate or sodium di-acetate during product manufacturing, thermal treatment of the final product with steam, application of high hydrostatic pressure or irradiation or reduction in the product shelf life would be appropriate control measures for RTE meat products (Mataragas et al., 2010). However, alternative strategies would be required for some other foods depending on their physical and chemical properties.

Dealing with a routinely encountered pathogen such as *Listeria* requires a practical approach to achieve the food safety objectives (FSO) in any production environment. It is important to devise and implement various risk management strategies throughout the food chain to reduce the incidence of listeriosis (Walls et al., 2005; Pérez-Rodríguez et al., 2006). In general, minimizing the risk of *Listeria* contamination in large-scale processing environments would require numerous control strategies including good manufacturing practices, sanitation, standard operating procedures, HACCP programs, intensive environmental sampling programs, time and temperature controls throughout the entire processing and storage period and the use of post-packaging treatments to destroy *L. monocytogenes* on products.

BIOFILMS

While developing control strategies for *Listeria* it is crucial to consider the ability of biofilm formation by the pathogen in food manufacturing and processing facilities. In general, biofilms are highly organised cellular arrangements formed through irreversible attachment of cells to various surfaces such as water system pipes, industrial equipment, processing facilities, food handling surfaces, conveyer belts and stainless steel equipment (Gandhi et al. 2007; Takahashi et al., 2010; Xu et al., 2011).

In particular, cells in these sessile communities are difficult to manage as they are known to develop an increased resistance to disinfectants (Chorionopoulas et al., 2008; Sandasi et al. 2008). Thus, the presence of resident biofilms involving a mixture of pathogenic microbes, such as *Listeria*, can sometimes lead to persistent contamination in the facility. In a study conducted by Pan et al., (2006), a five strain mixture of *L. monocytogenes* was exposed to a sanitation regime which would be typically experienced by the bacteria in a food processing environment to evaluate the resistance of biofilms formed on stainless steel and teflon coupons towards various sanitizers such as a hydrogen peroxide-based reagent, chlorine and quaternary ammonium compounds. The biofilms were exposed to conditions of dehydration, starvation and disinfection treatments, conditions they would commonly experience in food processing environments. It was observed that in the first week, the biofilm cell number was reduced under all treatment conditions; however the biofilms subsequently developed resistance towards all the sanitizers (Pan et al., 2006). Thus, there should be an emphasis on using cleaning regimes that reduce the probability of biofilm formation in the processing environments. The resistance of biofilms to disinfectants has been attributed to many factors; a) the inability of disinfectants to penetrate the extracellular polysaccharide layer surrounding the biofilm matrix, b) unlike planktonic cells, cells in a sessile community have a slow growth rate (due to lack of nutrients) and thus are less susceptible to antimicrobials c) differential protein expression under stressed conditions, compared to that in their planktonic counter-parts. The three dimensional arrangements of biofilms under static and flow conditions also play important roles in the resistance of cells to antimicrobials. Generally, under static conditions, a monolayer of cells is observed while in case of biofilms formed under flow conditions, a more complex dense network of cells is observed which hinders the penetration of antimicrobials to the innermost layer of cells (de Oliviera et al., 2012b).

A number of studies aimed at finding effective strategies to eliminate biofilms from food processing environments have been published. Disinfection with commonly used chemical disinfectants such as chlorine, quaternary ammonium compounds, hydrogen peroxide and sodium hypochlorite has been the most popular strategy. In a study conducted by Norwood and Gilmour (2000), exposure to 1000 ppm free chlorine for 20 min was suggested to reduce *L. monocytogenes* biofilms whereas a 10 ppm dose was found to be sufficient for planktonic cells. In another investigation, the effects of aqueous sodium hypochlorite, aqueous chlorine dioxide and gaseous chlorine dioxide on a five strain biofilm formed (containing mixture of *L. monocytogenes* strains) on stainless steel surface were reported. Overall, it was observed that, in comparison to aqueous sodium hypochlorite, low levels of chlorine dioxide gas and aqueous chlorine dioxide were required for the inactivation of listerial biofilms. The authors suggested that, unlike chlorine, chlorine dioxide gas produces less toxic disinfection by-products (the disinfection by-products of chlorine can sometimes be carcinogenic) and is effective over a wide pH range (Vaid et al., 2010). Romanova et al., (2007) showed that killing of live cells of *Listeria* in biofilms required concentrations of benzalkonium chloride (a major component of commonly used industrial sanitizers) higher than 10 mg/ml applied for at least 30 minutes. However, the same concentration was found to be insufficient for removal of biofilms formed on plastic surfaces. One of the problems associated with the use of commercial disinfectants containing hypochlorites, quaternary ammonium compounds and iodophors is their reduced efficacy at low temperatures. Food processing environments generally operate at refrigeration temperatures. As a solution to this problem, Amalaradjou et al., (2009) reported the effect of octenidine hydrochloride (a positively charged bispyridinamine) on free floating and preformed biofilms cells of *L. monocytogenes* at a temperature range of 37°C to 4°C. Overall the non-toxic, non-mutagenic and non-carcinogenic octenidine hydrochloride exhibited significant anti-listerial activity on planktonic and biofilms cells regardless of the temperature or the presence of organic matter or the surface on which, biofilms were formed. Thus, it could be used as a potent disinfectant to avoid the biofilms formation. However, acquired resistance to chemical disinfectants and sanitizers can often create an added delay in the elimination of mature biofilms. In case of such resident biofilms, peracetic acid has been suggested as an efficient disinfectant. Its high oxidizing capacity and low molecular weight facilitates penetration into the interior pockets of mature biofilms that are otherwise inaccessible to many other molecules (Ibusquiza et al., 2011). Other than chemical disinfectants, an alternative approach that has been the used to control *Listeria* and listerial biofilms is the application of lactic acid bacteria (LAB) or bacteriocins as biocontrol agents (Guerrieri et al., 2009). The use of polymers containing natural antimicrobial agents which block or interfere with the initial adhesion and eventual biofilm formation of bacteria is a novel strategy for controlling the problem of biofilms in the food industry, especially for food packaging systems (Romanova et al., 2007; Issepi et al. 2008; Nguyen et al., 2008).

In the food industry, specific 'in-house' microbiota are often used to give the food individual sensory characteristics such as taste, odour or other organoleptic properties. Thus, it is important that the cleaning and disinfection routines remove only the spoilage or pathogenic bacteria without affecting the beneficial microbes (Lebert et al., 2007). Consumers today are more inclined towards the use of products derived from natural sources instead of synthetic chemicals; this tendency has been termed 'green consumerism' (Sandasi et al., 2008; Al-Reza et al. 2009; de Oliveira et al., 2012b). In general, the emergence of highly resistant strains and the reluctance of consumers towards consumption of chemically treated goods have created the need for the use of natural agents for controlling *Listeria*.

ESSENTIAL OILS AS ANTI-LISTERIAL AGENTS

Essential oils (EOs) and their constituents have been explored extensively for their antioxidant, antidiabetic, anticancerous, antiviral, antithrombosis and antimicrobial effects (Djabou et al., 2013). Also called ethereal oils, EOs are volatile secondary plant metabolites with characteristic odours that can be obtained frm flowers, seeds, leaves, fruits, bark and roots (de Oliveira et al., 2012). Although many techniques are available for obtaining EOs, steam distillation is the most widely accepted method (Burt et al., 2004). EOs obtained from plants belonging to similar species often possess varying chemical compositions. The factors altering their chemical composition include the geographical location, physiological conditions of cultivation/growth, time of harvest, the plant organ used for extraction and its developmental stage and the technique used for obtaining the oil (Jordan et al., 2013).

The antimicrobial action of EOs is determined by their composition, which typically includes terpenes, terpenoids, sesquiterpenes and phenylpropanoids (Figure 1) (Carson and Hammer 2011; de Oliveira et al., 2012). EOs, being lipophilic in nature, disrupt the bacterial cell wall and arrangement of the polysaccharide, fatty acid and phospholipid layers in the cytoplasmic membrane, thus altering the membrane permeability. Following exposure to EOs, certain essential enzymes in the cell become inactive and coagulation of the cytoplasm is observed along with a significant reduction in the proton motive force (de Oliveira et al., 2012b). Some EOs have also been associated with inhibition of virulence factors, quorums sensing, toxin synthesis and flagellin production in certain bacteria (Carson and Hammer 2011). The composition of EOs is highly complex. Thus, resistance of bacteria to antimicrobial EOs is less likely as they generally have multiple bioactive components (Delaquis and Stanich 2011). An added advantage is that many of the EO constituents can show synergistic antimicrobial effects when used as mixtures (Dimitrijevic et al., 2007).

Several EOs have been reported to exhibit anti-listerial effects. EOs of clove (*Syzygium aromaticum*), bay (*Laurus nobilis*), cinnamon (*Cinnamomum zeylanicum*), thyme (*Thymus vulgaris*) and pimento (*Pimenta Dioica*) (Owen and Palombo 2007); rosemary *(Rosmarinus officinalis)* and oregano (*Origanum vulgare*) (Dimitrijevic et al., 2007; de Azerado et al. 2012); lemongrass (*Cymbopogan citratus*) (de Oliviera et al., 2010) and cilantro (*Coriandrum sativum*) (Delaquis and Stanich 2011) have exhibited significant anti-listerial properties at very low concentrations. Gutierrez et al., (2008) observed a synergistic anti-listerial effect when a mixture of marjoram (*Majorana hortensis*) and thyme EOs were used in combination with other EOs. Interestingly, the synergy was not observed when a mixture of two strong antimicrobial oils, oregano and thyme, were used in combination with other EOs. Since EOs have similar chemical composition, it is possible that their mixtures exhibit additive rather than synergistic effects. Hence, using EOs in combination with compounds which have a different chemical structure may produce desirable synergistic behaviour.

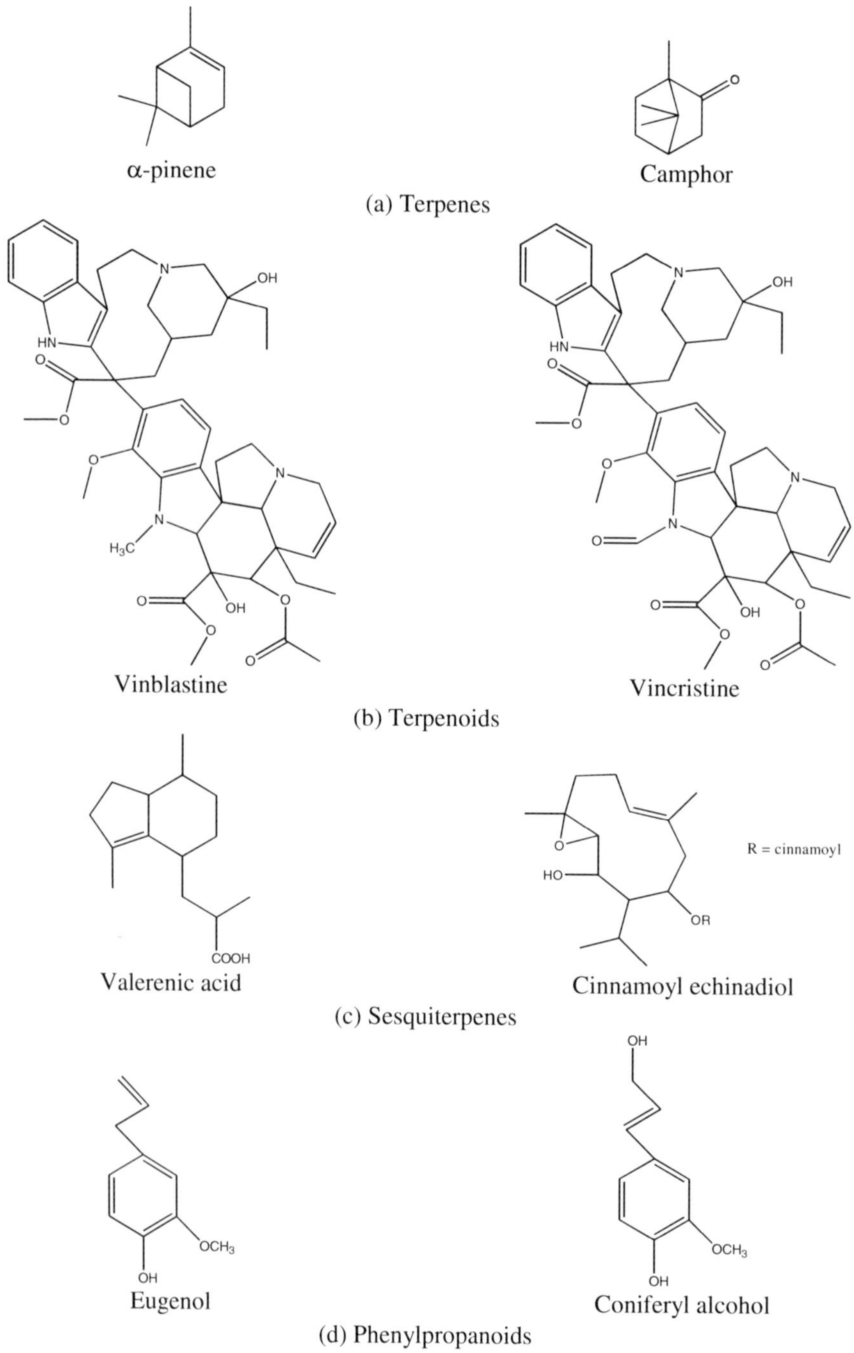

Figure 1. Examples of chemical constituents contributing to the antimicrobial mechanism of essential oils.

The bioactive components in some of these EOs have been elucidated. For instance, the antimicrobial effects of thyme and oregano have been attributed to the presence of phenolic constituents such as carvacrol and thymol that target the bacterial cell membrane altering its structure and function (Dimitrijevic et al., 2007). Certain EO constituents such as carvacrol, thymol, p-cymene and γ-terpinene have been issued GRAS (generally recognised as safe) status by the USFDA (USFDA 2013b) and have also been registered as safe flavouring agents by the European authorities (EFSA 2013).

BIOFILM INHIBITORY EFFECTS OF ESSENTIAL OILS

Although many studies have focussed on the anti-listerial activity of EOs, very few have focused on their biofilm inhibitory effects. Chorianopoulos et al., (2008) provided one of the early reports highlighting the importance of EOs as inhibitors of listerial biofilms. In this study, the disinfecting ability of *Satureja thymbra* EO and its hydrosol fraction against monoculture and mixed culture biofilms was investigated and the effect was also compared to three chemical sanitizers (hydrochloric acid, lactic acid and sodium hydroxide). The authors observed that chemical sanitizers were not as efficient as the EO and its hydrosol fractions in disinfecting listerial biofilms. An acquired adaptation of the cells to maintain the pH fluctuations and the inability of the molecules to diffuse in adequate amounts in the biofilms were suggested as possible explanations for reduced efficacy of the chemical disinfectants. de Oliviera et al., (2010) found disinfectant solutions of *Cymbopogan citratus* (D.C.) Stapf. and *Cymbopogan nardus* (L.) Rendle EOs (used alone or in combination) to be effective against cells of *L. monocytogenes* adhered to stainless steel after a treatment time of 60 min. In an independent study, biofilm inhibitory effects of multiple EOs were investigated by Leonard et al., (2010). EOs of *Syzygium aromaticum, Mentha spicata, Lippia rehmannii* and *Cymbopogon citratus* and their major constituents were studied against listerial biofilms. EOs of *Syzygium aromaticum* and *Mentha spicata* and the pure compounds nerol and citral produced strong biofilm inhibitory effects. It was expected that the major chemical constituents of *S. armaticum* and *M. spicata*, i.e., eugenol and carvone, respectively, would also show the same effect. However, when tested individually, these compounds caused biofilm enhancement rather than inhibition. Thus, it is important to study the actual mechanism of the bioactive components in EOs and also their synergistic or antagonistic behaviours with other EOs or EO components. On the contrary, *L. rehmanni* and *C. citratus* EOs failed to inhibit biofilm formation but their chemical constituent citral showed the same anti-listerial acitivity as the positive control (ciprofloxacin). Although these may seem as conflicting results to the study by de Oliviera et al., (2010), the differences in the concentration of EOs of *C. citratus* used in both studies could account for the discrepancy. *L.* monocytogenes intra-strain diversity in biofilm formation has been the focus of many studies (Soni et al., 2010; Nilson et al., 2011; Kostaki et al., 2012; Kadam et al., 2013). Desai et al., (2012) tested nine different EOs against 21 different strains of *L. monocytogenes* representing all 13 serotypes of the species. Although minor differences in the sensitivities of different strains were observed, all strains were found to be sensitive towards thyme, oregano and carvacrol EOs. The EOs successfully inactivated early and mature phase biofilms formed on stainless and polystyrene surfaces. The strong biofilm inhibitory effect of carvacrol has been previously reported by Pérez-Conesa et al., (2011). The effects of the EOs on biofilm cells is generally evaluated in terms of inhibition of biomass formation (using crystal violet assay/microscopic techniques) or reduction in number of viable cells (plate count assays). In a study conducted by our group, the biofilm inhibitory effects of the EOs obtained from yarrow

(*Achillea millefolium*) were evaluated against biofilms formed by *L. monocytogenes* and *L. innocua* using two assays – crystal violet staining to study inhibition of biofilm formation and 3-[4, 5-dimethylthiazol-2-yl]-2, 5- diphenyltetrazolium bromide (MTT) reduction to study the inhibition of metabolic activity of biofilm cells (Jadhav et al., 2013). Such assays are important as it is possible that, apart from killing cells, EOs may play a role in lowering the metabolic activity of viable cells in a biofilm, which cannot be evaluated using the popular crystal violet assay as it does not differentiate between live and dead cells. Yarrow EO was found to be effective in inhibiting the metabolic activity and initial cell attachment and of listerial cells to polystyrene, stainless steel and high density polyethylene surfaces, even at sub-MIC levels, but was less inhibitory towards preformed (6 hour) biofilms formed on polystyrene surface.

Biofilms generally exist as multispecies communities in food processing environments (Chorianopoulos et al., 2008). de Oliveira et al., (2012) evaluated the effects of EOs of *Cinnamonum cassia* against single and multispecies sessile communities of enteropathogenic *Escherichia coli* (EPEC) and *L. monocytogenes* formed on stainless steel coupons, and compared this effect to that of commercially available chemical sanitizers. In comparison to EPEC, *L. monocytogenes* cells were more resistant to the EO treatment when present as single-species biofilms. However, in mixed-species biofilms, both bacteria showed increased sensitivity to the EO treatment. This emphasizes the importance of studying the antibiofilm effect of EOs against mixed-species biofilms (which may function in synergy or in an antagonistic manner) as they represent the true natural environment. Interestingly, the authors observed *C. cassia* EOs performed as efficiently as the chemical sanitizers (hydrogen peroxide and Diprol containing a quaternary ammonium compound) in killing all the viable bacterial cells from the stainless steel surfaces thus highlighting the potency of EOs as sanitizers. The same group investigated the anti-listerial and biofilm inhibitory activity of EOs obtained from the bark of *Cinnamonum cassia*, and leaves of *Melaleuca alternifolia* and *Cymbopogan flexuosus* and found *C. Cassia* EO to be the most effective anti-listerial agent. More importantly, no synergistic effects were observed when combinations of the three EOs were used (Oliveira et al., 2012b).

Limitations of Essential Oils

Numerous studies have been performed to assess the effects of disinfectants used in industrial environments, including quaternary ammonium compounds, peracetic acids and peroxides. Industrially-used antimicrobials are often hazardous to health and some can even be carcinogenic (Rahnama et al., 2012). In contrast, few studies have explored the use of EOs as surface disinfectants (Chorionopoulas et al., 2008; Ibusquiza et al., 2011). Most studies have focussed on the ability of EOs to be used as food preservatives since some have been given GRAS status (Dimitrijevic et al., 2007; Al-Reza et al., 2009). However, their use as preservatives has also been limited since EOs in foods can possibly alter the organoleptic, textural, odour and other sensory food characteristics (Gutierrez et al., 2008; Ehsani and Mahmoudi 2013). It is also possible that some EOs may be better suited as surface disinfectants rather than food additives in spite of being safe for consumption. For instance, cilantro EO was found to be very strongly anti-listerial (MIC $\leq$ 0.01% v/v) in a study conducted by Delaquis and Stanich (2011). The authors suggested its probable use as an effective sanitizer or disinfectant as its volatile and oxidative properties may alter the sensory characteristics of the foods if used as a food preservative.

Another problem associated with EOs is the variation in chemical composition of the source plant with changes in climate, geographical conditions and/or technique of isolation. Since geographical variation can have a profound impact on the chemical composition of EOs, using the bioactive constituents in formulating sanitizers and disinfectants can provide a more standardised approach (de Oliviera et al., 2012). Jordan et al., (2013) performed an interesting study in which the antimicrobial activity of different chemotypes (abundance of the major chemical constituents defining the oil) of rosemary EOs obtained from multiple climatic locations were compared against different foodborne pathogens including *L. monocytogenes*. The authors concluded that the activity was found to similar for all chemotypes in the case of *L. monocytogenes* but this was not observed for all pathogens.

Using EOs in sanitizing or disinfecting formulations would require considerable standardisation in regards to the optimum concentration, contact time and temperature needed to maximise the effect of the oil and to strategically address the removal of any residual odour. Along with this, toxicity of the oils should also be assessed for possible associated allergen or irritant properties if they are to be applied in industrial environments. Another important factor to be addressed would be the cost involved in manufacturing EOs. This is particularly noteworthy in case of EOs that need to be imported. Using EOs only occasionally or in combination with conventional chemical sanitizers can also prove to be an alternative solution to make their use cost-effective (de Oliviera et al., 2012). However, more investigation is required to understand the efficacy of EOs as antibacterial agents in the presence of other synthetic disinfectants or sanitisers.

Assessing the efficacy of any disinfectant, including EOs, should be performed under conditions which attempt to replicate those found in real-world industrial scenarios. It is also important not to limit antimicrobial studies to planktonic cells as sessile cells may exhibit different behaviours (Leonard et al., 2010). In addition, the efficacy of disinfectants should be tested against mature biofilms as these tend to be more resistant to antimicrobial agents. Structure and arrangement of biofilms have been found to have a major impact on the acquired resistance to disinfectants and sanitizers (Ibusquiza et al., 2011). Most of the studies highlighting the antibacterial properties of EOs are carried out at temperatures suitable for microbial growth. In food processing plants, the operating temperatures are often around refrigeration conditions; hence it is important to test the efficacy of the oils under conditions mimicking specific food processing environments.

Another approach in applying the biofilm inhibitory effects of EOs is their incorporation in the production of materials used in food industries. Nostro et al., (2012) investigated the antibacterial and biofilm inhibitory effects of two EO components with GRAS status, carvacrol and cinnamaldehyde, incorporated in polyethylene-co-vinylacetate (EVA) films. The EVA films with the two constituents showed bacteriostatic effects against *L. monocytogenes* and also inhibited the bacteria's ability to form biofilms, probably by reducing surface hydrophobicity and disturbing its initial attachment process. Although the authors have advocated the use of such films in food packaging, the inclusion of EOs and their constituents into materials used in food processing equipment seems an attractive approach to help control foodborne pathogens and biofilm formation in these environments.

CONCLUSION

New strategies for controlling foodborne pathogens may include the use of plant secondary metabolites or their constituents as disinfectants in the food industry. Recent research in this area has provided many promising results. However, the use of EOs as disinfectants is still limited. The need to decipher the actual antimicrobial mechanism, reduced solubility in the aqueous phase, loss in activity due to reaction with food components and alteration of sensory qualities of the food may be some of the factors contributing to this limited use (Delaquis and Stanich 2011). Further research into understanding the mechanisms and behaviour of EOs in real-world scenarios will provide the knowledge and evidence needed to assist in the development of disinfectant/sanitiser formulations based on these complex products.

ACKNOWLEDGMENT

We are grateful to Rohan Shah for providing us assistance with the figures in the manuscript.

REFERENCES

Allerberger, F. & Wagner, M. (2010). Listeriosis, a resurgent foodborne infection. *Clinical Microbiology and Infection*, *16*, 16-23.

Alessandria, V., Rantsiou, K., Dolci, P. & Cocolin, L. (2010). Molecular methods to assess *Listeria monocytogenes* route of contamination in a dairy processing plant. *International Journal of Food Microbiology*, *141*, S156-S162.

Al-Reza, S. M., Bajpai, V. K. & Kang, S. C. (2009). Antioxidant and antilisterial effect of seed essential oil and organic extracts from *Zizyphus jujuba*. *Food and Chemical Toxicology*, *47*, 2374-2380.

Amalaradjou, M. A. R., Norris, C. E. & Venkitanarayanan, K. (2009). Effect of octenidine hydrochloride on planktonic cells and biofilms of *Listeria monocytogenes*. *Applied and Environmental Microbiology*, *75*, 4089-4092.

Anonymous, (2013). "Statistics". Retrieved 25th September, 2013, from http://www.cdc.gov/listeria/statistics.html.

Autio, T., Hielm, S., Miettinen, M., Sjöberg, A.-M., Aarnisalo, K., Björkroth, J., Mattila-Sandholm, T. & Korkeala, H. (1999). Sources of *Listeria monocytogenes* contamination in a cold-smoked rainbow trout processing plant detected by pulsed-field gel electrophoresis typing. *Applied and Environmental Microbiology*, *65*, 150-155.

Bari, M. L., Nakauma, M., Todoriki, S., Juneja, V. K., Isshiki, K. & Kawamoto, S. (2005). Effectiveness of irradiation treatments in inactivating *Listeria monocytogenes* on fresh vegetables at refrigeration temperature. *Journal of food protection*, *68*, 318-323.

Silk, J. B., Date, K. A., Jackson, K. A., Pouillot, R., Holt, G. K., Graves, L. M., Ong, K. L., Hurd, S., Meyer, R., Marcus, R., Shiferaw, B., Norton D. M., Medus, C., Zansky, S. M., Cronquist A. B., Henao, O. L., Jones, T. F., Vujia, D. J., Farley, M. M. & Mahon, B. E.

(2012). Invasive listeriosis in the foodborne diseases active surveillance network (FoodNet), 2004-2009: Further targeted prevention needed for higher-risk groups. *Clinical Infectious Diseases*, *54*, S396-S404.

Borucki, M. K., Gay, C. C., Reynolds, J., McElwain, K. L., Kim, S. H., Call, D. R. & Knowles, D. P. (2005). Genetic diversity of *Listeria monocytogenes* strains from a high-prevalence dairy farm. *Applied and Environmental Microbiology*, *71*, 5893-5899.

Burt, S. (2004). Essential oils: their antibacterial properties and potential applications in foods—a review. *International Journal of Food Microbiology*, *94*, 223-253.

Cagri-Mehmetoglu, A., Yaldirak, G., Bodur, T., Simsek, M., Bozkir, H. & Eren, N. M. (2011). Incidence of *Listeria monocytogenes* and *Escherichia coli* O157: H7 in two Kasar Cheese processing environments. *Food Control*, *22*, 762-766.

Carpentier, B. & Cerf, O. (2011). Review-Persistence of *Listeria monocytogenes* in food industry equipment and premises. *International Journal of Food Microbiology*,*145*, 1-8.

Carson, C. F. & Hammer, K. A. (2011). Chemistry and bioactivity of essential oils.In H Thormar, ed, Lipids and Essential Oils as Antimicrobial Agents. John Wiley & Sons, Chichester, UK.

Cartwright, E .J., Jackson, K. A., Johnson, S. D., Graves, L. M., Silk, B. J. & Mahon, B. E. (2013). Listeriosis outbreaks and associated food vehicles United States, 1998-2008. *Emerging Infectious Diseases*, *19*, 1-9.

Centres for Disease Control (CDC) and Prevention, (2012). "Multistate outbreak of listeriosis linked to imported Frescolina Marte Brand Ricotta Salata Cheese (Final Update)". Retrieved 30th Septmeber, 2013, from http://www.cdc.gov/listeria/outbreaks/cheese-09-12/index.html.

Centres for Disease Control (CDC) and Prevention, (2013). "Multistate uutbreak of listeriosis linked to Crave Brothers Farmstead Cheeses (Final Update)". Retirved 30th September, 2013 from http://www.cdc.gov/listeria/outbreaks/cheese-07-13/.

Chorianopoulos, N. G., Giaouris, E. D., Skandamis, P. N., Haroutounian, S. A. & Nychas, G. J. E. (2008). Disinfectant test against monoculture and mixed-culture biofilms composed of technological, spoilage and pathogenic bacteria: Bactericidal effect of essential oil and hydrosol of *Satureja thymbra* and comparison with standard acid-base sanitizers. *Journal of Applied Microbiology*, *104*, 1586-1596.

Dass, S. C., Cummins, E. J. & Abu-Ghannam, N. (2011). Prevalence and typing of *Listeria monocytogenes* strains in retail vacuum-packed cold smoked salmon in the republic of Ireland. *Journal of Food Safety*. *31*, 21-27.

Dawson, S. J., Evans, M. R., Wilby, D., Bardwell, J., Chamberlain, N. & Lewis, D. A. (2006). *Listeria* outbreak associated with sandwich consumption from a hospital retail shop, United Kingdom. *Eurosurveillance*, *11*, 89-91.

de Azeredo, G. A., De Figueiredo, R. C. B. Q., De Souza, E. L. & Stamford, T. L. M. (2012). Changes in *Listeria monocytogenes* induced by *Oreganum vulgare* L. and *Rosamarinus officinalis* L. essential oils alone and combined at subinhibitory amounts. *Journal of Food Safety*, *32*, 226-235.

de Castro, V., Escudero, J. M., Rodriguez, J. L., Muniozguren, N., Uribarri, J., Saez, D. & Vazquez, J. (2012). Listeriosis outbreak caused by Latin-style fresh cheese, Bizkaia, Spain, August 2012. *Eurosurveillance*, *17*, 1-3.

de Oliveira, M. M. M. d., Brugnera, D. F., Cardoso, M. D. G., Alves, E. & Piccoli, R. H. (2010). Disinfectant action of *Cymbopogon* sp. essential oils in different phases of biofilm formation by *Listeria monocytogenes* on stainless steel surface. *Food control, 21*, 549-553.

de Oliveira, M. M. M., Brugnera, D. F., do Nascimento, J., Batista, N. & Piccoli, R. (2012). Cinnamon essential oil and cinnamaldehyde in the control of bacterial biofilms formed on stainless steel surfaces. *European Food Research and Technology., 234*, 821-832.

de Oliveira, M. M. M., Brugnera, D. F., do Nascimento, J. A. & Piccoli, R. H. (2012b). Control of planktonic and sessile bacterial cells by essential oils. *Food and Bioproducts Processing, 90*, 809-818.

Delaquis, P. J. & Stanich, K. (2011). Antilisterial properties of cilantro essential oil. *Journal of Essential Oil Research., 16*, 409-414.

Denny, J. & McLauchlin, J. (2008). Human *Listeria* monocytogenes infections in Europe- an opportunity for improved European surveillance. *Eurosurveillance, 13*, 1-5.

Desai, M. A., Soni, K. A., Nannapaneni, R., Schilling, M. W. & Silva, J. L. (2012). Reduction of *Listeria monocytogenes* biofilms on stainless steel and polystyrene surfaces by essential oils. *Journal of Food Protection, 75*, 1332-1337.

Dimitrijević, S. I., Mihajlovski, K. R., Antonović, D. G., Milanović-Stevanović, M. R. & Mijin, D. Ž. (2007). A study of the synergistic antilisterial effects of a sub-lethal dose of lactic acid and essential oils from *Thymus vulgaris* L., *Rosmarinus officinalis* L. and *Origanum vulgare* L. *Food Chemistry, 104*, 774-782.

Djabou, N., Lorenzi, V., Guinoiseau, E., Andreani, S., Giuliani, M.-C., Desjobert, J.-M. & Muselli, A. (2013). Phytochemical composition of Corsican Teucrium essential oils and antibacterial activity against foodborne or toxi-infectious pathogens. *Food Control, 30*, 354-363.

Doordyun, Y., de Jager, C. M., van der Zwaluw, W. K., Wannet, W. J., van der Ender, A., Spanjaard, L. & Duynhoven, Y. T. (2006). First results of the active surveillance of *Listeria monocytogenes* infections in the Netherlands reveal higher than expected incidence. *Eurosurveillance, 11*, 1-3.

Ehsani, A. & Mahmoudi, R. (2013). Effects of *Mentha longifolia* L. essential oil and *Lactobacillus casei* on the organoleptic properties and on the growth of *Staphylococcus aureus* and *Listeria monocytogenes* during manufacturing, ripening and storage of Iranian white-brined cheese. *International Journal of Dairy Technology, 66*, 70-76.

European Food Saftey Authority (EFSA), (2013). "*Database of flavouring substances*". Retrieved from 11th November, 2013, from http://ec.europa.eu/food/food/chemicalsafety/flavouring/database/dsp_search.cfm.

Fallah, A. A., Saei-Dehkordi, S. S. & Mahzounieh, M. (2013). Occurrence and antibiotic resistance profiles of *Listeria monocytogenes* isolated from seafood products and market and processing environments in Iran. *Food Control, 34*, 630-636.

Farber, J.M. & Peterkin, P.I. (1991). *Listeria monocytogenes*, a food-borne pathogen. *Microbiol. Mol. Biol. Rev, 55*, 476-511.

Fonnesbech Vogel, B., Huss, H.H., Ojeniyi, B., Ahrens, P. & Gram, L. (2001). Elucidation of *Listeria monocytogenes* contamination routes in cold-smoked salmon processing plants detected by DNA-based typing methods. *Applied and Environmental Microbiology, 67*, 2586-2595.

Fox, E., Hunt, K., O'Brien, M. & Jordan, K. (2011). *Listeria monocytogenes* in Irish farmhouse cheese processing environments. *International Journal of Food Microbiology, 145*, S39-S45.

Fretz, R., Pichler, J., Sagel, U., Much, P., Ruppitsch, W., Pietzka, A. T., Stöger, A., Huhulescu, S., Heuberger, S., Appl, G., Werber, D., Stark, K., Prager, R., Flieger, R., Karpíškova, R., Pfaff, G. & Allerberger, F. (2010). Update: Multinational listeriosis outbreak due to 'Quargel', a sour milk curd cheese, caused by two different *L. monocytogenes* serotype 1/2a strains, 2009-2010. *Eurosurveiilance*, *15*, 1-2.

Gandhi, M. & Chikindas, M. L. (2007). *Listeria:* A foodborne pathogen that knows how to survive. *International Journal of Food Microbiology*, *113*, 1-15.

Garrido, V., Torroba, L., García-Jalón, I. & Vitas, A. (2008). Surveillance and outbreak reports. *Eurosurveillance,13*, 1.

Garrido, V., Vitas, A. & García-Jalón, I. (2010). The problem of listeriosis and ready-to-eat products: prevalence and persistence. *Current Research technology and Education Topics in Applied Microbiology and Microbial Biotechnology*, 1182-1189.

Gaulin, C., Ramsay, D. & Bekal, S. (2012). Widespread listeriosis outbreak attributable to pasteurized cheese, which led to extensive cross-contamination affecting cheese retailers, Quebec, Canada, 2008. *Journal of Food Protection*, *75*, 71-78.

Godshall, C. E., Suh, G. & Lorber, B. (2013). Cutaneous Listeriosis. *Journal of Clinical Microbiology*, *51*, 3591-3596.

Goulet, V., Hedberg, C., Le Monnier, A. & de Valk, H. (2008). Increasing incidence of listeriosis in France and other european countries. *Emerging infectious diseases*, *14*, 734-740.

Guerrieri, E., de Niederhäusern, S., Messi, P., Sabia, C., Iseppi, R., Anacarso, I. & Bondi, M. (2009). Use of lactic acid bacteria (LAB) biofilms for the control of *Listeria monocytogenes* in a small-scale model. *Food control*, *20*, 861-865.

Gutierrez, J., Barry-Ryan, C. & Bourke, P. (2008). The antimicrobial efficacy of plant essential oil combinations and interactions with food ingredients. *International Journal of Food Microbiology*, *124*, 91-97.

Howell, G. V. & Miller, R. (2010). Maple Leaf Foods: Crisis and containment case study. *Public Communication Review*, *1*, 47-56.

Ibusquiza, P. S., Herrera, J. J. R. & Cabo, M. L. (2011). Resistance to benzalkonium chloride, peracetic acid and nisin during formation of mature biofilms by *Listeria monocytogenes*. *Food Microbiology*, *28*, 418-425.

Iseppi, R., Pilati, F., Marini, M., Toselli, M., de Niederhäusern, S., Guerrieri, E. & Bondi, M. (2008). Anti-listerial activity of a polymeric film coated with hybrid coatings doped with Enterocin 416K1 for use as bioactive food packaging. *International Journal of Food Microbiology*, *123*, 281-287.

Jadhav, S., Bhave, M.& Palombo, E. A. (2012). Methods used for the detection and subtyping of *Listeria monocytogenes*. *Journal of microbiological methods*, *88*, 327-341.

Jadhav, S., Shah, R., Bhave, M. & Palombo, E. A. (2013). Inhibitory activity of yarrow essential oil on *Listeria* planktonic cells and biofilms. *Food Control*, *29*, 125-130.

Jordán, M. J., Lax, V., Rota, M. C., Lorán, S. & Sotomayor, J. A. (2013). Effect of bioclimatic area on the essential oil composition and antibacterial activity of *Rosmarinus officinalis* L. *Food Control*, *30*, 463-468.

Kabuki, D. Y., Kuaye, A. Y., Wiedmann, M. & Boor, K. J. (2004). Molecular subtyping and tracking of *Listeria monocytogenes* in Latin-style fresh-cheese processing plants. *Journal of Dairy Science*, *87*, 2803-2812.

Kadam, S. R., den Besten, H. M. W., van der Veen, S., Zwietering, M. H., Moezelaar, R., Abee, T. (2013). Diversity assessment of *Listeria monocytogenes* biofilm formation, Impact of growth condition, serotype and strain origin. *International Journal of Food Microbiology*, *165*, 259-264.

Kerouanton, A., Roche, S., Marault, M., Velge, P., Pourcher, A., Brisabois, A., Federighi, M. & Garrec, N. (2010). Characterization of isolates of *Listeria monocytogenes* from sludge using pulsed field gel electrophoresis and virulence assays. *Journal of applied microbiology*, *108*, 1380-1388.

Koch, J. & Stark, K. (2006). Significant increase of listeriosis in Germany. *Eurosurveillance*, *11*, 1-11.

Koch, J., Dworak, R., Prager, R., Becker, B., Brockmann, S., Wicke, A., Wichmann-Schauer, H., Hof, H., Werber, D. & Stark, K. (2010). Large listeriosis outbreak linked to cheese made from pasteurized milk, Germany, 2006–2007. *Foodborne pathogens and disease*, *7*, 1581-1584.

Kostaki, M., Chorianopoulos, N., Braxou, E., Nychas, G. J. & Giaouris, E. (2012). Differential biofilm formation and chemical disinfection resistance of sessile cells of *Listeria monocytogenes* strains under monospecies and dual-Species (with *Salmonella enterica*) conditions. *Applied and Environmental Microbiology*, *78*, 2586-2595.

Kovačević, J., Bohaychuk, V. M., Barrios, P. R., Gensler, G. E., Rolheiser, D. L. & Mcmullen, L. M. (2009). Evaluation of Environmental sampling methods and rapid detection assays for recovery and identification of *Listeria* spp. from meat processing facilities. *Journal of Food Protection*, *72*, 696-701.

Kovačević, J., McIntyre, L. F., Henderson, S. B. & Kosatsky, T. (2012). Occurrence and distribution of *Listeria* species in facilities producing ready-to-eat foods in British Columbia, Canada. *Journal of Food Protection*, *75*, 216-224.

Kushwaha, K. & Muriana, P. (2009). Adherance characteristics of *Listeria* strains isolated from three ready-to-eat meat processing plants. *Journal of Food Protection*, *72*, 2125-2131.

Latorre, A. A., Van Kessel, J. A. S., Karns, J. S., Zurakowski, M. J., Pradhan, A. K., Zadoks, R. N., Boor, K. J. & Schukken, Y. H. (2009). Molecular Ecology of *Listeria monocytogenes:* Evidence for a reservoir in milking equipment on a dairy farm. *Applied and Environmental Microbiology*, *75*, 1315-1323.

Lebert, I., Leroy, S. & Talon, R. (2007). Effect of industrial and natural biocides on spoilage, pathogenic and technological strains grown in biofilm. *Food Microbiology*, *24*, 281-287.

Leonard, C. M., Virijevic, S., Regnier, T., Combrinck, S. (2010). Bioactivity of selected essential oils and some components on *Listeria monocytogenes* biofilms. *South African Journal of Botany*, *76*, 676-680.

Lim, E., Lopez, L., Borman, A., Cressey, P. & Pirie, R. (2012). "Annual report concenrning foodborne disease in New Zealand 2011" Ministry for Primary Industry, New Zeland. Retrieved 10th Novermber, 2013, from http://www.foodsafety.govt.nz/elibrary/industry/FBI-report-2011. pdf.

Linnan, M. J., Mascola, L., Lou, X. D., Goulet, V., May, S., Salminen, C., Hird, D. W., Yonekura, M. L., Hayes, P. & Weaver, R. (1988). Epidemic listeriosis associated with Mexican-style cheese. *New England Journal of Medicine*, *319*, 823-828.

Lomonaco, S., Verghese, B., Gerner-Smidt, P., Tarr, C., Gladney, L., Joseph, L., Katz, L., Turnsek, M., Frace, M., Chen, Y., Brown, E., Meinersmann, R., Berrang, M. & Knabel,

S. (2013). Novel epidemic clones of *Listeria monocytogenes*, United States, 2011. *Emerging Infectious Diseases*, *19*, 147-150.

Luz, I. d. S., Neto, N. J. G., Tavares, A. G., Magnani, M. & de Souza, E. L. (2012). Exposure of *Listeria monocytogenes* to sublethal amounts of Origanum vulgare L. essential oil or carvacrol in a food-based medium does not induce direct or cross protection. *Food Research International*, 48, 667-672.

Mataragas, M., Zwietering, M. H., Skandamis, P. N. & Drosinos, E. H. (2010). Quantitative microbiological risk assessment as a tool to obtain useful information for risk managers — Specific application to *Listeria monocytogenes* and ready-to-eat meat products. *International Journal of Food Microbiology*, *141*, S170-S179.

McLauchlin, J. & Low, J. (1994). Primary cutaneous listeriosis in adults: an occupational disease of veterinarians and farmers. *Veterinary record*, *135*, 615-617.

Mook, P., O'Brien, S. & Gillespie, I. (2011). Concurrent conditions and human listeriosis, England, 1999-2009. *Emerging infectious disease*, *17*, 38-43.

Nguyen, V. T., Gidley, M. J. & Dykes, G. A. (2008). Potential of a nisin-containing bacterial cellulose film to inhibit *Listeria monocytogenes* on processed meats. *Food Microbiology*, *25*, 471-478.

Nilsson, R. E., Ross, T. & Bowman, J. P. (2011). Variability in biofilm production by *Listeria monocytogenes* correlated to strain origin and growth conditions. *International Journal of Food Microbiology*,*150*, 14-24.

Norwood, D. E. & Gilmour, A. (2000). The growth and resistance to sodium hypochlorite of *Listeria monocytogenes* in a steady-state multispecies biofilm. *Journal of Applied Microbiology*, *88*, 512-520.

Nostro, A., Scaffaro, R., D'Arrigo, M., Botta, L., Filocamo, A., Marino, A. & Bisignano, G. (2012). Study on carvacrol and cinnamaldehyde polymeric films: mechanical properties, release kinetics and antibacterial and antibiofilm activities. *Applied Microbiology and Biotechnology*, 96, 1029-1038.

Owen, R. J. & Palombo, E. A. (2007). Anti-listerial activity of ethanolic extracts of medicinal plants, *Eremophila alternifolia* and *Eremophila duttonii*, in food homogenates and milk. *Food Control*. *18*, 387-390.

OzFoodNet. (2012). Monitoring the incidence and causes of diseases potentially transmitted by food in Australia: Annual report of the OzFoodNet network, 2010. *Communicable diseases intelligence journal*, *36*, E213-E241.

Pan, Y., Breidt, Jr F. & Kathariou, S. (2006). Resistance of *Listeria monocytogenes* biofilms to sanitizing agents in a simulated food processing environment. *Applied and Environmental Microbiology*, *72*, 7711-7717.

Pérez-Conesa, D., Cao, J., Chen, L., McLandsborough, L. & Weiss, J. (2011). Inactivation of *Listeria monocytogenes* and *Escherichia coli* O157: H7 biofilms by micelle-encapsulated eugenol and carvacrol. *Journal of Food Protection*, *74*, 55-62.

Pérez-Rodríguez, F., Todd, E. C. D., Valero, A., Carrasco, E., García, R. M. & Zurera, G. (2006). Linking quantitative exposure assessment and risk management using the food safety objective concept: An example with *Listeria monocytogenes* in different cross-contamination scenarios. *Journal of Food Protection*, *69*, 2384-2394.

Rahnama, M., Najimi, M. & Ali, S. (2012). Antibacterial effects of *Myristica fragrans, Zataria multiflora* Boiss, *Syzygium aromaticum*, and *Zingiber* officinale Rosci essential oils, alone and in combination with nisin on *Listeria monocytogenes*. *Comparative Clinical Pathology*, *21*, 1313-1316.

Romanova, N. A., Gawande, P. V., Brovko, L. Y. & Griffiths, M. W. (2007). Rapid methods to assess sanitizing efficacy of benzalkonium chloride to *Listeria monocytogenes* biofilms. *Journal of microbiological methods*, *71*, 231-237.

Rørvik, L. M. (2000). Listeria monocytogenes in the smoked salmon industry. International Journal of Food Microbiology, 62, 183-190.

Rotariu, O., Thomas, D. J. I., Goodburn, K. E., Hutchison, M. L. & Strachan, N. J. C. (2014). Smoked salmon industry practices and their association with *Listeria monocytogenes*. *Food Control*, *35*, 284-292.

Sandasi, M., Leonard, C. & Viljoen, A. (2008). The effect of five common essential oil components on *Listeria monocytogenes* biofilms. *Food control*. *19*, 1070-1075.

Soni, K. A. & Nannapaneni, R. (2010). Removal of *Listeria monocytogenes* biofilms with bacteriophage P100. *Journal of Food Protection*, *73*, 1519-1524.

Swaminathan, B. & Gerner-Smidt, P. (2007). The epidemiology of human listeriosis. *Microbes and Infection*. *9*, 1236-1243.

Takahashi, H., Suda, T., Tanaka, Y. & Kimura, B. (2010). Cellular hydrophobicity of *Listeria monocytogenes* involves initial attachment and biofilm formation on the surface of polyvinyl chloride. *Letters in Applied Microbiology*, *50*, 618-625.

Todd, E. C. D. & Notermans, S. (2011). Surveillance of listeriosis and its causative pathogen, *Listeria monocytogenes*. *Food control*, *22*, 1484-1490.

U. S. Food and Drug Administration, (2013). "*Listeria monocytogenes* risk assessment: I. Introduction". Retrieved on 11th November, 2013, from http://www.fda.gov/food/ foodscienceresearch/risksafetyassessment/ucm184052. htm.

U. S. Food and Drug Administration, (2013b). "*CFR-Code of Federal Regulations*" Title 21. Retrieved on 11th November, 2013, from http://www.accessdata.fda.gov/SCRIPTs/cdrh/ cfdocs/cfcfr/CFRSearch. cfm?CFRPart&showFR

Vaid, R., Linton, R. H. & Morgan, M. T. (2010). Comparison of inactivation of *Listeria monocytogenes* within a biofilm matrix using chlorine dioxide gas, aqueous chlorine dioxide and sodium hypochlorite treatments. *Food Microbiology*, *27*, 979-984.

Valderrama, W. B. & Cutter, C. N. (2012). An ecological perspective of *Listeria monocytogenes* biofilms in food processing facilities. *Critical Reviews in Food Science and Nutrition*, *53*, 801-817.

Viswanath, P., Murugesan, L., Knabel, S. J., Verghese, B., Chikthimmah, N. & LaBorde, L. F. (2013). Incidence of *Listeria monocytogenes* and *Listeria* spp. in a small-scale mushroom production facility. *Journal of Food Protection*, *76*, 608-615.

Vogel, B. F., Hansen, L. T., Mordhorst, H. & Gram, L. (2010). The survival of *Listeria monocytogenes* during long term desiccation is facilitated by sodium chloride and organic material. *International Journal of Food Microbiology*. *140*, 192-200.

Walls, I., Buchanan, R. L. (2005). Use of food safety objectives as a tool for reducing foodborne listeriosis. *Food control*. *16*, 795-799.

Warriner, K. & Namvar, A. (2009). What is the hysteria with *Listeria*? *Trends in Food Science and Technology*. 20, 245-254.

Williams, S. K., Roof, S., Boyle, E. A., Burson, D., Thipareddi, H., Geornaras, I., Sofos, J. N., Weidmann, M. & Nightingale, K. (2011). Molecular ecology of *Listeria monocytogenes* and other *Listeria* species in small and very small ready-to-eat meat processing plants. *Journal of Food Protection. 74*, 63-77.

Xu, H., Lee, H. Y. & Ahn, J. (2011). Characteristics of biofilm formation by selected foodborne pathogens. *Journal of Food Safety. 31*, 91-97.

Yan, H., Neogi, S. B., Mo, Z., Guan, W., Shen, Z., Zhang, S., Li, L., Yamasaki, S., Shi, L. & Zhong, N. (2010). Prevalence and characterization of antimicrobial resistance of foodborne *Listeria monocytogenes* isolates in Hebei province of Northern China, 2005–2007. *International Journal of Food Microbiology. 144*, 310-316.

In: *Listeria monocytogenes*
Editor: Edmund C. Hambrick
ISBN: 978-1-63117-054-6

Chapter 10

PREVALENCE, ANTIMICROBIAL RESISTANCE AND GROWTH KINETICS OF *LISTERIA MONOCYTOGENES* IN READY-TO-EAT (RTE) FOODS

Yumiko Okada**, *Hiroshi Asakura and Shizunobu Igimi
Division of Biomedical Food Research,
National Institute of Health Sciences,
Setagaya-ku, Tokyo, Japan

ABSTRACT

Listeria monocytogenes is a facultative intracellular bacterium that causes life-threatening illnesses including meningitis, stillbirth, and abortion in humans. This pathogen is ubiquitously distributed in nature, and in the most cases, transmission of this pathogen to humans are achieved via the intake of contaminated foods. In order to trace the sources and routes of infection, it is a first-line to understand the prevalence of this pathogen in a series of foods. Here we review the prevalence of this pathogen in a number of ready-to-eat (RTE) foods that have been concerned as one of the most high risk foods for listerial infection. In addition, we also summarize the characteristics of *L. monocytogenes* isolates from different sources, especially in terms of their antimicrobial resistance (for clinical aspects) and growth kinetics in food environments (supporting to construct microbial risk modeling). Finally, we discuss future perspective of *L. monocytogenes* study from a viewpoint of food hygiene.

Keywords: *Listeria monocytogenes*, Ready-To-Eat (RTE) foods, epidemiology

* Corresponding author: Yumiko Okada, Ph.D. and Hiroshi Asakura, Ph.D. Address: Division of Biomedical Food Research, National Institute of Health Sciences, Kamiyoga 1-18-1, Setagaya-ku, Tokyo 158-8501, Japan. Tel: +81-3-3700-1141 Fax: +81-3-3700-9246; E-mail: yokada@nihs.go.jp (YO) and hasakura@nihs.go.jp (HA).

1. INTRODUCTION

Listeria monocytogenes is distributed in a variety of foods such as meat, dairy, seafood, and deli salad products. Infection with this pathogen often causes no symptom in healthy adults, but in some cases develops severe illnesses (i.e., abortion, stillbirth, and meningitis). Epidemiological studies have mounted evidence that this pathogen can be transmitted from animals and/or contaminated foods. Considering the public health threat of *L. monocytogenes* infection, the International Codex aliment Arius Commission determined microbiological criteria for *L. monocytogenes* in ready-to-eat (RTE) foods in 2009 [9], which defined safety levels of RTE foods as follows: (1) *L. monocytogenes* should not be detected from a total of 25 g of sample (n=>5); (2) The contamination levels should not be > 100 colony-forming units (CFU)/g; (3) Nevertheless the commission also allowed that competent authorities may choose other validated limits of acceptable listerial levels. To adopt the last idea, however, the levels of contamination and the growth kinetics of *L. monocytogenes* in target foods during their shelf-lives must be determined, respectively.

In this chapter, we summarize current status of food contamination with *L. monocytogenes*, as well as the growth kinetics of this pathogen in some types of food matrixes, and the occurrence of ampicillin (ABPC) resistance resistance in this pathogen as which is used as first-line therapeutic drug. These concepts and information would give some helpful idea to reduce the risks of health hazard relating to this pathogen.

2. PREVALENCE OF *LISTERIA MONOCYTOGENES* IN READY-TO EAT (RTE) FOODS

L. monocytogenes is a ubiquitous bacterium, which could be often isolated from soil, aquatic environments, feces of animals, and foods. To date, this pathogen is detected from various types of ready-to-eat (RTE) food, such as meat/dairy products and deli salads mainly. Corresponding to a recent increase of consumption of the RTE foods, food-borne listeriosis have often occurred mainly in western countries. Understanding the level and frequency of contamination by *L. monocytogenes* in RTE foods is thus contributable to assess the microbial risk in such foods. Followings are representatives to summarize the prevalence of this pathogen in 4 types of RTE foods which include dairy, meat, seafood, and vegetable products, respectively.

(i) Meat Products

Non-cooked meat products and deli meats are recognized as one of the major causative agents of outbreak of listeriosis. For example, a large outbreak was caused by consumption of meat products in Canada in 2009, which resulted in at least 57 illnesses and 23 deaths [19]. In European countries, *L. monocytogenes* was isolated from 20% (10/50) of meat products [44], 20.5% (23/112) of vacuum-packaged sliced salami [13] and 45.7% (48/140) of salami [54] in Italy, 1.1% (4/639) of cooked meat products in Belgium [71], 2.2% (29/1321) of cooked meat, 0.24% (1/411) of pate, and 1.9% (6/316) of fermented meat, and 0% (0/143) of smoked sausage in UK [43], 1.2% (6/507) of heat-treated meat products in Sweden [68]. These studies also included quantitative analyses, among which 24 of 140 salami samples in Italy showed the level of contamination above the Codex criterion (100CFU/g) [54]. 7.8% (6/77) of non-cooked meat products in Japan were positive for this bacterium at <100 CFU/g [49]. By contrast, as heat-processed meats, grilled chicken products (70 samples) in Mexico [12] and deli meats (40 samples) in Canada [31] were absent of this bacterium, while 3.31% (10/302) of cooked meats were still contaminated with *L. monocytogenes* in China [75].

(ii) Seafoods

Although no large outbreak occurred in recent years caused by, it is likely that seafood products are contaminated with high numbers of this pathogen. 12% (6/50) of smoked fish [39] and 34.1% (45/132) of smoked salmon [13] in Italy were contaminated. 27.8% (25/90) in Belgium [71], 4.08% (6/147) of crustacean, 6.74% (12/178) of smoked fish, and 2% (1/50) of sushi in UK were also positive to this bacterium [43]. In Sweden, listerial contaminations were found in 12% (66/558) of smoked and gravid fish [68]. Of smoked fish, one sample in Italy [44], 4 samples in Begium [71] and 3 samples in Sweden [68] were contaminated with > 100CFU/g of *L. monocytogenes*. In Canada, 5% (2/40) of fish products were positive [31]. The prevalence rates of 12.6% (12/95) and 12.2% (64/531) were reported in Japan [22, 45]. In China, 1.17% (4/343) of seafood was contaminated [75]. These findings thus suggest that the seafood products might show high frequency and high level of contamination with this pathogen.

(iii) Fruits and Vegetables

Fruits and vegetables were paid attention as causative agents for many types of foodborne disease such as *E.coli* O157:H7 and hepatitis A, recently. In USA, cantaloupe and chopped celery caused large outbreaks [18, 42]. The level of contamination in fresh vegetables were relatively low; 2% (1/50) in Italy [44], 0% (0/11 of leafy vegetables, 0/6 of root vegetables) in Croatia [31], 0.62% (2/323) in China [75]. On the other hand, higher prevalence rates were found in salad and processed vegetables; 4.76 % (2/42) of cooked marinated products in Italy [44], 7% (80/1187) in Belgium [71] and 27% (31/115) in Italy [13], 10% (3/30) of mayonnaise based salad in Iran [26], 0.9% (3/335) of green salad in UK [43], 0.5% (1/19) of cabbage salad in Croatia [31], 25.4% (88/347) of frozen vegetables and 10.2% (22/216) of freshly supermarkets prepared, cooked or raw ready-to-eat vegetable salad in Chile [11], 3.1% (16/512) of RTE vegetables in Brazil [61]. These suggest that many of salad should be contaminated when they were processed.

(iv) Daily Products

A recent outbreak diffused in three countries in Europe had occurred by the intake of contaminated sour cream cheese, resulting 34 illnesses and 8 deaths [17]. Several reports demonstrated the prevalence of this pathogen in cheese products. For example, *L.*

monocytogenes was isolated from 9% (18/200) of fresh cheese in Mexico [69], 18.2% (10/55) of cheese in Brazil [6], 11.1% (39/350) of cheese in Jordan [52], 0% (0/70) of imported natural cheese in Japan [50], 0% (0/88) of dairy products in Iran [26], 9.2% (13/142) of homemade white cheese in Turkey [3], 1.9% (2/108) of cream cheeses and 0% (0/186) of mozzarella in Italy [13], 0% (0/419) of butter and 0% (0/473) of soft cheese in UK [43], 0.4% (2/525) of soft and semi-soft cheese in Sweden [68], and 0% (0/51) of cheese in Spain [2]. From the reports by Manfreda in Italy, 2.1% (13/1489) of gorgonzola was contaminated by *L. monocytogenes* after package, whereas, the contamination rate was increased up to 4.8% (8/167) at the end of shelf life [39]. Similar results can find in the report of *L. monocytogenes* baseline survey by EFSA [14].

The prevalence rates and levels of contamination at sampling time point were compared with those at end-of shelf-life in smoked or gravad fish, meat products, and cheese samples [14]: the prevalence rates were not changed remarkably after storage, however, the number of samples that contained > 100 CFU/g of this pathogen raised by longitudinal storage of samples [14], proposing the importance of determination of accurate shelf-lives and choice of storage temperature not to increase the numbers of this pathogen in those foods.

3. GROWTH OF *L. MONOCYTOGENES* IN RTE FOODS

It is needless to say that refrigeration is an effective method to inhibit bacterial growth in foods. Many food-borne pathogens cannot grow at temperatures below 4°C, although they can survive or reduce their numbers gradually. Different from major foodborne bacteria, *L. monocytogenes* has an ability to grow even at 0°C [64]. This unique characteristic makes difficult to control the growth of this bacteria in refrigerated foods. Long-term storage of foods at refrigerating temperature in processing plants, stores, and consumer's houses has thus provided potential risks to growth of this pathogen in foods, thereby causing human infection. To determine the essential shelf-lives in various types of foods, the growth kinetics of *L. monocytogenes* in stored foods should be therefore monitored. To date, the growth of this pathogen in some types of foods such as ham [30] and smoked salmon [25, 46] has been well-monitored. In smoked salmon, the CFU count of one strain of *L. monocytogenes* increased rapidly at 30°C and exhibited a very slow growth at 4°C to a level of 10^8 CFU/g. However, in the traditional foods peculiar to limited region, the growth model of *L. monocytogenes* is unpredictable. The growth tests were performed in the traditional foods, for example, pork scrapple [1] and vacuum-packaged souse in USA [27], and oriental-style rice cakes in Korea [35]. In pork scrapple, a type of meat product, *L. monocytogenes* can grow at 4, 10 and 21°C, although 33, 7 and 2 days were needed for 6 log-fold of growth, respectively [1]. On the other hand, in another traditional meat product souse, *L. monocytogenes* survived at 5 and 10°C for 32 days with reduced CFU [27]. The pH of souse was a major factor for growth inhibition. In Korean rice cake, 6 log-fold of increase was observed during 24h at room temperature [35]. In Japan, the growth of *L. monocytogenes* in some types of traditional seafood products was examined. *L. monocytogenes* did not grow in cod roe (tarako) and spicy cod roe (mentaiko) at 4°C and 30°C [46]. The levels of contamination with *L. monocytogenes* in these types of products are high [24]. The growth of *L. monocytogenes* might be restricted in some types of fermented seafood products [74]. *L. monocytogenes* did not grow in fish source made from raw squid (shiokara), which had an Aw value of approximately 0.93, whereas "nishin-kirikomi" (another type of fermented fish product made from herring) with a similar Aw value supported the growth of *L. monocytogenes*. Trimethylamin oxide, which is found naturally in squid, was supposed as a growth inhibitor.

In other study, the growth of *L. monocytogenes* in broiled squid which shelf life was 7 days at 10°C began at day 4 after inoculation. Thus, if this sample was consumed at half of its shelf life, the risk of listeriosis would be relatively low [50]. In dried baby sardine with Aw0.957, *L. monocytogenes* increased 2.5 log-fold from the initial CFU at 10°C, whereas only 1 log-fold growth was shown at 4°C. However, in a similar product with a low Aw value (0.898), *L. monocytogenes* did not grow in this sample for 2 weeks in both temperatures [50].

From the results shown above, the kinetics of *L. monocytogens* had changed drastically according to the types of food, even though they were made from same ingredients. The difference of characteristics such as Aw, pH, biological flora, natural compounds and the presence/absence of food additives may give impact to the growth of listerial cells in foods. Sometimes consumers, particularly elderly consumers, may have RTE foods that have exceeded their shelf lives. These results suggest that there is a higher risk of listeriosis when foods are stored at 10°C than at 4°C even in their shelf lives. Many countries have their own traditional foods and the characteristics of these foods are quite diverse. To keep the contamination levels of *L. monocytogenes* in RTE foods to below an acceptable level, the initial level of contamination should be determined, and the combination of refrigerating temperature and shelf life should be considered carefully.

4. ANTIBIOTICS SUSCEPTIBILITIES OF *L. MONOCYTOGENES* ISOLATED FROM FOODS AND PATIENTS

The most essential treatment for listeriosis is antibiotic therapy, and the emergence of antibiotic resistant isolates of *L. monocytogenes* can be therefore a great problem from a viewpoint of public health. To know the situation of the resistance to antibiotic chemicals in *L. monocytogenes*, the continuous survey in the world is necessary. Especially, ampicillin (ABPC) is the first-choice of drug for treatment of invasive listeriosis, and diffusion of ABPC-resistant isolates may increase the mortality cases of listeriosis. Table 1 briefly summarizes the reports for the prevalence of ABPC-resistant *L. monocytogenes* in humans and foods.

Clinical and Laboratory Standard Institute (CLSI) has changed its description about ABPC-resistance in *L. monocytogenes* in the latest guideline M31-A3 published in 2008 [8] from the former version of the guideline M31-A2 published by National Committee for Chemical Laboratory Standard (NCCLS) in 2003 [47]. CLSI has specified that the MIC of ≦ 2 μg/ml should be considered as susceptible in *Listeria* spp., while the MIC of ≧4 μg/ml should be considered as "non-susceptible". According to the guideline M31-A2 by NCCLS, the MIC breakpoints of <=2μg/ml and >=4μg/ml were considered as ABPC-susceptible or -resistant, respectively. Considering the current trends that many studies adopt the cut-off MIC value of 4 μg/ml as "resistance" to ABPC, here we use a term of "resistant" instead of "non-susceptible" thoroughly.

Table 1. Prevalence of ABPC-resistant *L. monocytogenes* in humans and foods

Continent	Country	Published	% (No. resistant / No. tested isolates)		Reference
		year	Clinical	Food & environment	
Europe					
	France	1978	0 (0/50)	-	[15]
	Germany	1981	0 (0/44)	-	[41]
	Sweden	1985	0 (0/175)	-	[34]
	Denmark	1988	0 (0/156)	-	[55]
	UK	1990	0 (0/103)	-	[38]
	UK	1990	0 (0/617)	-	[38]
	Italy	1991	-	0 (0/98)	[16]
	Italy	1992	-	0 (0/64)	[4]
	Swiss	1992	0 (0/74)	-	[58]
	Spain	1996	-	8.7 (2/23)	[60]
	Austria, France, Norway, Switzerland	1997	0 (0/66)	-	[23]
	Ireland	2001	-	0 (0/351)	[72]
	Northern Ireland	2001	-	0 (0/45)	[21]
	Spain	2001	-	0 (0/28)	[40]
	Poland	2004	0 (0/14)	0 (0/59)	[53]
	Denmark	2005	0 (0/106)	-	[20]
	Turkey	2005	-	66 (5/9)	[76]
	Hungary	2006	0 (0/17)		[29]
	Turkey	2008	-	7.7 (1/13)	[3]
	Turkey	2008	-	73.9 (17/23)	[5]
	Italy	2009	-	1.6 (2/120)	[10]
America					
	USA	1984	100 (1/1)	-	[57]
	USA	2003	9.2 (6/65)	-	[62]
	USA	2005	-	92 (35/38)	[65]
	Costa Rica	2006	35 (7/20)	5 (1/20)	[77]
	Mexico	2006	-	23.7 (16/68)	[59]
	Canada	2007	0 (0/84)	-	[36]
	USA	2011	-	0 (0/20)	[73]
	Canada	2012	-	0 (0/2)	[31]
Africa					
	Kenya	2004	-	100 (9/9)	[48]
Asia					
	Japan	1990	0 (0/3)	-	[33]
	Japan	1991	-	0 (0/9)	[66]
	Japan	1998	50 (1/2)	-	[70]
	Korea	2001	-	0 (0/8)	[7]
	Japan	2002	-	0 (0/13)	[28]
	Japan	2008	20 (1/5)	-	[63]
	Japan	2008	16 (4/35)	-	[67]
	Japan	2009	0 (0/100)	0 (0/101)	[51]
	China	2010	-	0 (0/44)	[75]

-, not reported.

At the moment, the limited information is available for the prevalence of ABPC-resistant *L. monocytogenes* in human clinical patients, especially in geographies of Africa and south-east Asia; most of which were reported from north America [57, 62], Latin America [59, 77], Turkey [3, 5, 76], Italy [10], Spain [60], Kenya [48], and Japan [63, 67, 70].

In Europe, the ABPC-resistant isolates were detected from turkey meat [5, 76], fish [10], and dairy products [3], while were not detected from patients. In US and Latin America, the ABPC-resistant *L. monocytogenes* were detected from either food, food processing environment [59, 65, 77], and patients [57, 62]. The limited number of reports can be seen in Africa, however, 100% (9/9) of chicken samples in Kenya were positive for ABPC-resistant *L. monocytogenes* [48]. In Japan, 3 studies have demonstrated the prevalence of ABPC-resistant *L. monocytogenes* from patients: Sakai et al., reported five cases of listeriosis in cancer patients from 1995 to 2007, in which one isolate from these cases showed ABPC-resistance [63]. A materno-infantile case isolate also exhibited resistance to ABPC (MIC = 8 μg/ml) [70]. Tago et al., reviewed 38 cases of listeriosis in Japanese children from 1983 to 2007 [67], among them, ABPC was apparently used for the treatment in 25 cases. In the 4 of 25 cases (16%) performed in 1990s, ABPC did not respond to *L. monocytogenes* infection, suggesting the causative bacteria in these cases showed ABPC-resistance. By contrast, a more recent antimicrobial resistance study in Japan showed that the MIC of *L. monocytogenes* isolates (100 patient isolates and 101 food/environmental isolates) to ABPC was below 1μg/ml [51]. Thus, it might be considered that ABPC-resistant *L. monocytogenes* isolates to ABPC have been distributed in this country for at least the last two decades, while they appeared spontaneously. It remains unclear about the possible geographic differences of the ABPC-resistant isolates. However, it may possible that the differences were related with the amounts and/or ways of the ABPC usage at medical fields and animal husbandries in each country.

CONCLUSION

Listeriosis remains a public health threat worldwide. The high tolerance of this pathogen to a series of stressors makes very difficult for us to remove the bacterium from foods and the related environments. To reduce the number of cases of listeriosis, it must be effective and necessary to control temperature throughout food chains with preventing cross contamination in addition to the prevention of food contamination with this pathogen. Efficient education programs for consumers would be also required to improve awareness of salinity control, which might contribute to create novel strategies to prevent infections.

REFERENCES

[1] Adekunle, A.O., Porto-Fett, A.C.S., Call, J.E., Shoyer, B., Gartner, K., Tufft, L., & Luchansky, J.B. (2009) Effect of Storage and Subsequent Reheating on Viability of *Listeria monocytogenes* on Pork Scrapple. *J. Food Protect.* 72, 2530-2537.

[2] Arrese, E., & Arroyo-Izaga, M. (2012) Preavalence of *Listeria monocytogenes* in Idiazabal cheese. *Nutr. Hosp.* 27, 2139-2141.

[3] Arslan, S., & Özdemir, F. (2008) Prevalence and antimicrobial resistance of *Listeria* spp. in homemade white cheese. Food Cont. 19, 360-363.

[4] Barbuti, S., Maggi, A., Casoli, C. (1992) Antibiotic resistance in strains of *Listeria* spp. in meat products. *Letters in Applied Microbiology* 15, 56-58.

[5] Bilir Ormanci, F.S., Erol, I., Ayaz, N.D., Iseri, O., & Sariguzel, D. (2008). Immunomagnetic separation and PCR detection of *Listeria monocytogenes* in turkey meat and antibiotic resistance of the isolates. *Br. Poult. Sci.* 49, 560-565.

[6] Brito, J.R.F., Santos, E.M.P., Arcuri, E.F., Lange, C.C., Brito, M.A.V.P., Souza, G.N., Cerqueira, M.M.P.O., Beltran, J.M.S., Call, J.E., Liu, Y., Porto-Fett, A.C.S., & Luchansky, J.B. (2008) Retail survey of Brazilian milk and minas frescal cheese and a contaminated dairy plant to establish prevalence, relatedness, and sources of *Listeria monocytogenes* isolates. *Appl. Environ. Microbiol.* 74, 4954-4961.

[7] Choi, Y-H., Cho, S-Y., Park, B-K., Chung, D-H., Oh, D-H. (2001) Incidence and characterization of *Listeria* spp. from foods available in Korea. *J. Food Prot.* 64, 554-558.

[8] Clinical and Laboratory Standards Institute. (2008) Performance Standards for Antimicrobial Disk and Diliution Susceptibility Tests for Bacteria Isolated from Animals; Approved Standard- Third Edition. M31-A3, vol. 28, No. 8.

[9] Codex alimentarius commission. (2009) Proposed draft annex II: microbiological criteria for *Listeria monocytogenes* in ready-to-eat foods. –Alinorm 09/32/13 – Appendix II –Step 5/8. *In*: Report of the fortieth session of the codex committee of food hygiene. http://www.codexalimentarius.net/web/archives.jsp?year=09

[10] Conter, M., Paludi, D., Aanardi, E., Ghidini, S., Vergara, A., & Ianieri A. (2009) Characterization of antimicrobial resistance of foodborne ***Listeria monocytogenes. Int. J. Food Micribiol. 128, 497-500.***

[11] Cordano, A.M., & Jacquet, C. (2009). *Listeria monocytogenes* isolated from vegetable salads sold at supermarkets in Santiago, Chile: Prevalence and strain characterization. 132, 176-179.

[12] Díaz-López, A., Cantú-Ramírez, R.C., Garza-González, E., Ruiz-Tolentino, L., Tellez-Luis, S.J., Rivera, G., & Bocanegra-García, V. (2011) Prevalence of Foodborne Pathogens in Grilled Chicken from Street Vendors and Retail Outlets in Reynosa, Tamaulipas, Mexico. *J. Food Protect.* 74, 1320-1323.

[13] Di Pinto, A., Novello, L., Montemurro, F., Bonerva, E., & Tantillo, G. (2010) Occurrence of *Listeria monocytogenes* in ready-to-eat foods from pupermarkets in Southern Italy. *New Microbiologica* 33, 249-252.

[14] EFSA. (2013) *Listeria monocytogenes* baseline survey report part A. *EFSA J.* 11, 3241.

[15] Espaze, E.P., Rouveix, Y.G., Le Berre, J.Y., Courtieu, A.L. (1978) *In vitro* susceptibility of *Listeria monocytogenes* to some antibiotics and their combinations. Zentralblatt für Bakteriologie, Parasitenkunde, Infektionskrankheiten und Hygiene. Erste Abteilung Originale. *Reihe A: Medizinische Mikrobiologie und Parasitologie* 240, 76-85.

[16] Facinelli, B., Giovanetti, E., Varaldo, P.E., Casolari, P., Fabio, U. (1991) Antibiotic resistance in foodborne *Listeria*. *Lancet.* 338, 1272.

[17] Fretz, R., Pichler, J., Sagel, U., Much, P., Ruppitsch, W., Pietzka, A.T., Stöger, A., Huhulescu, S., Heuberger, S., Apple, G., Werber, D., Stark, K., Prager, R., Flieger, A., Karpíšková, R., Pfaff, G., & Allergerger, F. (2010) Update: Multinational listeriosis

outbreak due to 'quargel', a sour milk curd cheese, caused by two different *L. monocytogenes* serotype 1/2a strain, 2009-2010. *Euro Surveill.* 15, pii19543.

[18] Gaul, L.K., Farag, N.H., Shim, T., Kingsley, M.A., Silk, B.J., & Hyytia-Trees, E. (2013) Hospital-acquired listeriosis outbreak caused by contaminated diced celery--Texas, 2010. *Clin. Infect.* Dis. 56, 20-26.

[19] Government of Canada. (2009) Listeriosis Investigative Review. Final report. Report of the independent investigator into the 2008 listeriosis outbreak. http://www.listeriosis-listeriose.investigation-enquete.gc.ca/lirs_rpt_e.pdf

[20] Hansen, J.M., Gerner-Smidt, P., Bruun, B. (2005) Antibiotic susceptibility of *Listeria monocytogenes* in Denmark 1958-2001. *APMIS.* 113, 31-36.

[21] Harvey, J., Gilmour, A. (2001) Characterization of recurrent and sporadic *Listeria monocytogenes* isolates from raw milk and nondairy foods by pulsed-field gel electrophoresis, monocin typing, plasmid profiling, and cadmium and antibiotic resistance determination. *Appl. Environ. Microbiol.* 67, 840-847.

[22] Handa-Miya, S., Kimura, B., Takahashi, H., Sato, M., Ishikawa, T., Igarashi, K., & Fujii, T. (2007) Nonsense-mutated *inlA* and *prfA* not widely distributed in *Listeria monocytogenes* isolates from ready-to-eat seafood products in Japan. *Int. J. Food Microbiol.* 117, 312-318.

[23] Heger, W., Dierich, M.P., Allerberger, F. (1997) *In vitro* susceptibility of *Listeria monocytogenes*: comparison of the E test with the agar dilution test. *Chemotherapy.* 43, 303-310.

[24] Hiwaki, H., Baba, A., Ebuchi, S., Uryu, K., Miyazaki, E., & Miyamoto, T. (2006). Behavior of *Listeria monocytogenes* in the manufacturing process of karashi-mentaiko (red-pepper seasoned cod roe). *Jpn. J. Food Micribiol.* 23,85-92.

[25] Hwang, C.A., & Sheen, S. (2009). *Modeling the growth characteristics of Listeria monocytogenes* and native microflora in smoked salmon. *J. Food Sci.* 74, 125-130.

[26] Jalali, M., & Abedi, D. (2008). Prevalence of *Listeria* species in food products in Isfahan, Iran. *Int. J. Food Microbiol.* 122, 336-340.

[27] Kim, M.K., Bang, W., Drake, M.A., Hanson, D.J., & Jaykus, L.A. (2009). Impact of Storage Temperature and Product pH on the Survival of *Listeria monocytogenes* in Vacuum-Packaged Souse. *J. Food Prot.* 72, 637-643.

[28] Kitazume, H., Suzuki, M., Suzuki, M., Matsumoto, Y., Yamada, M., Mutoh, T., Fujii, K., (2002) Detection of *Listeria monocytogenes* from non-heated foods. *Annu. Rep. Yokohama Inst. Health.* 41, 91-93. Japanese.

[29] Kiss, R., Tirczka, T., Szita, G., Bernath, S., Csiko, G. (2006) *Listeria monocytogenes* food monitoring data and incidence of human listeriosis in Hungary, 2004. *Int. J. Food Microbiol.* 112, 71-74.

[30] Koseki, S., Mizuno, Y., & Yamamoto, K. (2007) Predictive modelling of the recovery of *Listeria monocytogenes* on sliced cooked ham after high pressure processing. *Int. J. Food Microbiol.* 119, 300-307.

[31] Kovačević, J., Mesak, L.R., & Allen, K.J. (2012) Occurrence and characterization of ***Listeria*** spp. in ready-to-eat retail foods from Vancouver, British Columbia. *Food Microbiol.* 30, 372-378.

[32] Kovačević, M., Burazin, J., Pavlović, H., Kopjar, M., & Piližota, V. (2013) Prevalence and level of *Listeria monocytogenes* and other *Listeria* sp. in ready-to-eat minimally processed and refrigerated vegetables. *World J. Microbiol. Biotechnol.* 29, 707-712.

[33] Kuroki, H., Sugimoto, K., Kohri, Y., Toba, T. (1990) Three cases of meningitis due to *Listeria monocytogenes* type 1b. *J. Japan. Assoc. Infect. Dis*. 64, 249-256.

[34] Larsson, S., Walder, M.H., Cronberg, S.N., Forsgren, A.B., Moestrup, T. (1985) Antimicrobial susceptibilities of *Listeria monocytogenes* strains isolated from 1958 to 1982 in Sweden. *Antimicrob. Agents Chemother*. 28, 12-14.

[35] Lee, S-Y., Gwon, S-Y., Kim, S-J., & Moon, B.K. (2009) Inhibitory Effect of Commercial Green Tea and Rosemary Leaf Powders on the Growth of Foodborne Pathogens in Laboratory Media and Oriental-Style Rice Cakes. *J. Food Prot*. 72, 1107-1111

[36] Lyautey, E., Hartmann, A., Pagotto, F., Tyler, K., Lapen, D.R., Wilkes, G., Piveteau, P., Rieu, A., Robertson, W.J., Medeiros, D.T., Edge, T.A., Gannon, V., Topp, E. (2007) Characteristics and frequency of detection of fecal *Listeria monocytogenes* shed by livestock, wildlife, and humans. *Can. J. Microbiol*. 53, 1158-1167.

[37] MacGowan, A.P., Holt, H.A., Bywater, M.J., Reeves, D.S. (1990) *In vitro* antimicrobial susceptibility of *Listeria monocytogenes* isolated in the UK and other *Listeria* species. *Eur. J. Clin. Microbiol. Infect Dis*. 9, 767-770.

[38] MacGowan, A.P., Reeves, D. S. (1990) Antibiotic resistance of *Listeria monocytogenes*. *Lancet*. 336, 513-514.

[39] Manfreda, G., De Cesare, A., Stella, S., Cozzi, M., & Cantoni, C. (2005) Occurrence and ribotypes of *Listeria monocytogenes* in Gorgonzola cheeses. *Int. J. Food Microbiol*. 102, 287-93.

[40] Margolles, A., Mayo, B., de los Reyes-Gavilán, C.G. (2001) Susceptibility of *Listeria monocytogenes* and *Listeria innocua* strains isolated from short-ripened cheeses to some antibiotics and heavy metal salts. Food Microbiol. 18, 67-73.

[41] Marklein, G. & Hafelein, J. (1981) *In vitro* susceptibility of *Listeria monocytogenes* with special reference to newer beta lactam antibiotics. Zentralblatt für Bakteriologie, Mikrobiologie und Hygiene. 1. Abt. Originale A, Medizinische Mikrobiologie, *Infektionskrankheiten und Parasitologie* 251, 40-53.

[42] McCollum, J.T., Cronquist, A.B., Silk, B.J., Jackson, K.A., O'Connor, K.A., Cosgrove, S., Gossack, J.P., Parachini, S.S., Jain, N.S., Ettestad, P., Ibraheem, M., Cantu, V., Joshi, M., DuVernoy, T., Fogg, N.W. Jr., Gorny, J.R., Mogen, K.M., Spires, C., Teitell, P., Joseph, L.A., Tarr, C.L., Imanishi, M., Neil, K.P., Tauxe, R.V., & Mahon, B.E. (2013) Multistate outbreak of listeriosis associated with cantaloupe. *N. Engl. J. Med*. 369, 944-53.

[43] Meldrum, R.J., Ellis, P.W., Mannion, P.T., Halstead, D., & Garside, J. (2010) The Welsh Food Microbiological Forum. Prevalence of *Listeria monocytogenes* in Ready-to-Eat Foods Sampled from the Point of Sale in Wales, United Kingdom. *J. Food Prot*. 73, 1515-1518.

[44] Meloni, D., Galluzzo, P., Mureddu, A., Piras, F., Griffiths, M., & Mazzette R. (2009) *Listeria monocytogenes* in RTE foods marketed in Italy: Prevalence and automated *Eco*RI ribotyping of the isolates. *Int. J. Food Microbiol*. 129, 166-173.

[45] Nakamura, H., Hatanaka, M., Ochi, K., Nagao, M., Ogasawara, J., Hase, A., Kitase, T., Haruki, K., & Nishikawa, Y. (2004) *Listeria monocytogenes* isolated from cold-smoked fish products in Osaka City, Japan. *Int. J. Food Microbiol*. 94, 323-328.

[46] Nakamura, H., Ogasawara, J., Hase, A., Kitase, T., & Ishii, E. (2006) Fate of *Listeria monocytogenes* in seafood products. Ann. Rep. Osaka City Inst. *Public Health Environ. Sci.* 68, 1-11.

[47] National Committee for Chemical Laboratory Standard, (2003) Performance Standards for Antimicrobial Disk and Dilution Susceptibility Tests for Bacteria Isolated from Animals; Approved Standard-Second Edition. M31-A2, vol.22 No. 6.

[48] Njagi, L.W., Mbuthia, P.G., Bebora, L.C., Nyaga, P.N., Minga, U., & Olsen, J.E. (2004) Sensitivity of *Listeria* species, recovered from indigenous chickens to antibiotics and disinfectants. *East African Med. J.* 81, 534-537.

[49] Okada, Y., Monden, S., Igimi, S., & Yamamoto, S. (2012) The occurrence of *Listeria monocytogenes* in imported ready-to-eat foods in Japan. *J. Vet. Med. Sci.* 74, 733-735.

[50] Okada, Y., Ohnuki, I., Suzuki, H., & Igimi, S. (2013) Growth of *Listeria monocytogenes* in refrigerated ready-to-eat foods in Japan. Food Addit. Contam. Part A, *Chem. Anal. Control Expo. Risk Assess.* 30, 1446-1449.

[51] Okada, Y., Okutani, A., Suzuki, H., Asakura, H., Monden, S., Nakama, A., Maruyama, T., & Igimi, S. (2011) Antimicrobial susceptibilities of *Listeria monocytogenes* isolated in Japan. *J. Vet. Med. Sci.* 72, 1681-1684.

[52] Osaili, T.M., Al-Nabulsi, A.A., Taha, M.H., Al-Holy, M.A., Alaboudi, A.R., Al-ROusan, W.M., & Shaker, R.R. (2012) Occurrence and antimicrobial susceptibility of *Listeria monocytogenes* isolated from brined white cheese in Jordan. *J. Food Sci.* 77, M528-532.

[53] Paciorek, J., 2004. Antibiotic susceptibility of *Listeria monocytogenes* strains isolated from 2000 to 2002 in Poland. *Pol. J. Microbiol.* 53, 279-281.

[54] Petruzzelli, A., Blasi, G., Masani, L., Calza, L., Duranti, A., Santarelli, S., Fisichella, S., Pezzotti, G., Aquilanti, L., Osimani, A., & Tonucci, F. (2010) Occurrence of *Listeria monocytogenes* in salami manufactured in the march region (Central Italy). *J. Vet. Med. Sci.* 72, 499-502.

[55] Poulsen, P.N., Carvajal, A., Lester, A., Andreasen J. (1988) *In vitro* susceptibility of *Listeria monocytogenes* isolated from human blood and cerebrospinal fluid. *APMIS.* 96, 223-228.

[56] Public Health Agency of Canada. (2010) Update to 2008 *Listeria monocytogenes* cases number. http://www.phac-aspc.gc.ca/alert-alerte/listeria/listeria 20100413-eng.php.

[57] Rapp, M.F., Pershadsingh, H.A., Long, J.W., & Pickens, J.M. (1984) Ampicillin-resistant *Listeria monocytogenes* meningitis in a previously healthy 14-year-old athlete. *Archiv. Neurol.* 41, 1304.

[58] Rohner, P., Arrigoni, E., Auckenthalert, R. (1992) *In vitro* antimicrobial activity on *Listeria monocytogenes* during a human epidemic. *Médicine et Maladies Infectieuses* 22, 220-225.

[59] Rodas-Suárez, O.R., Flores-Pedroche, J.F., Betancourt-Rule, J.M., Quiñones-Ramírez, E.I., & Vázquez-Salinas, C. (2006) Occurrence and antibiotic sensitivity of *Listeria monocytogenes* isolates isolated from oysters, fish, and estuarine water. *Appl. Environ. Microbiol.* 72, 7410-7412.

[60] Rota, C., Yangüela, J., Blanco, D., Carramiñana, J. J., Ariño, A. & Herrera, A. (1996) High prevalence of multiple resistance to antibiotics in 144 *Listeria* isolates from Spanish dairy and meat products. *J. Food Protect.* 59, 938-943.

[61] Sant'Ana, A.S., Igarashi, M.C., Landgraf, M., Destro, M.T., & Franco, B.D.G.M. (2012) Prevalence, populations and pheno- and genotypic characteristics of *Listeria monocytogenes* isolated from ready-to-eat vegetables marketed in São Paulo, *Brazil Int. J. Food Microbiol.* 155, 1-9.

[62] Safdar, A., & Armstrong, D. (2003) Antimicrobial activities against 84 *Listeria monocytogenes* isolates from patients with systemic listeriosis at a comprehensive cancer center. *J. Clin. Microbiol.* 41, 483-485.

[63] Sakai, C., Kumagai, K., Tsujimura, H., Ise, M., Mimura, N. & Satomura, H. (2008) Case reports: Listerial bacteremia in 5 cancer patients. *Rinsho to Biseibutsu.* 35, 103-107. Japanese.

[64] Seeliger, H.P.R., & Jones, D. (1986) Bergey's manual of systematic bacteriology, vol.2, 3rd ed. Baltimore (MD): The Williams and Wilkins. pp.1235-1245.

[65] Srinivasan, V., Nam, H.M., Nguyen, L.T., Tamilselvam, B. & Murinda, S. E. (2005) Prevalence of antimicrobial resistance genes in *Listeria monocytogenes* isolated from dairy farms. *Foodborne Path. Dis.* 2, 201-211.

[66] Sunaga, S., Nakauchi, K., Nakajima, H., Kataoka, Y. and Morozumi, T. (1991) Isolation and some properties of *Listeria* spp. from pig carcasses in an Abattoir. *Journal of Japanese Veterinary Medical Association*, 44, 1051-1055.

[67] Tago, K., Tomono, N., Matsushima, T. & Shiro, H. (2008) Three cases of purulent meningitis caused by *Listeria monocytogenes. J. Pediatric. Infect. Dis. Immun.* 20, 8-14.

[68] Thisted Lambertz, S., Nilsson, C., Brådenmark, A., Sylvén, S., Johansson, A., Jansson, L.M., & Lindblad, M. (2012) Prevalence and level of *Listeria monocytogenes* in ready-to-eat foods in Sweden 2010. *Int. J. Food Microbiol.* 160, 24-31.

[69] Torres-Vitela, M.R., Mendoza-Bernardo, M., Castro-Rosas, J., Gomez-Aldapa, C.A., Garay-Martinez, L.E., Navarro-Hidalgo, V., & Villarruel-López, A. (2012) Incidence of *Salmonella, Listeria monocytogenes, Escherichia coli* O157:H7, and Staphylococcal Enterotoxin in Two Types of Mexican Fresh Cheeses. *J. Food Prot.* 75, 79-84.

[70] Tsubakio, Y., Iwamura, C., Yoshimura, A., Asada, M. & Furukawa, K. (1998) Case report of Listeria monocytogenes infection in a maternal patient. *Jap. J. Pediatric.* 51, 1597-1600. Japanese.

[71] Uyttendaele, M., Busschaert, P., Valero, A., Geeraerd, A.H., Vermeulen, A., Jacxsens, L., Goh, K.K., De Loy, A., Van Impe, J.F., & Devlieghere, F. (2009) Prevalence and challenge tests of *Listeria monocytogenes* in Belgian produced and retailed mayonnaise-based deli-salads, cooked meat products and smoked fish between 2005 and 2007. *Int. J. Food Microbiol.* 133, 94-104.

[72] Walsh, D., Duffy, G., Sheridan, J.J., Blair, I.S., McDowell, D.A. (2001) Antibiotic resistance among *Listeria*, including *Listeria monocytogenes* in retail foods. *J. Appl. Microbiol.* 90, 517-522.

[73] Wang, F., Jiang, L., Yang, Q., Han, F., Chen, S., Pu, S., Vance, A., & Ge, B. (2011) Prevalence and Antimicrobial Susceptibility of Major Foodborne Pathogens in Imported Seafood. *J. Food Prot.* 74, 1451-1461.

[74] Yamamoto, T., Tateyama-Nishimura, T., Yamazaki, K., Kawai, Y., & Inoue, N. (2004) Fate of *Listeria monocytogenes* in Japanese fishery product. *Jpn. J. Food Micribiol.* 21, 254-259.

[75] Yan, H., Neogi, S.B., Mo, Z., Guan. W., Shen, Z., Zhang, S., Li, L., Yamasaki, S., Shi, L., & Zhong, N. (2010) Prevalence and characterization of antimicrobial resistance of foodborne *Listeria monocytogenes* isolates in Hebei province of Northern China, 2005-2007. *Int. J. Food Microbiol.* 144, 310-316.

[76] Yücel, N., Çitak, S., & Önder, M. (2005) Prevalence and antibiotic resistance of *Listeria* species in meat products in Ankara, Turkey. *Food Microbiol.* 22, 241-245.

[77] Zamora, J.M., Chaves, C., & Arias, M.L. (2006) Comparison of the antibiotics sensibility pattern of *Listeria monocytogenes* and *Salmonella* spp. isolates isolated from food with clinical origin samples. *Archivos Latinoamericanos de Nutrición*, 56, 171-174.

In: *Listeria monocytogenes*
Editor: Edmund C. Hambrick
ISBN: 978-1-63117-054-6

Chapter 11

NOVEL TECHNOLOGIES FOR CONTROLLING *LISTERIA MONOCYTOGENES* IN READY-TO-EAT (RTE) PRODUCTS

Rosa Raybaudi-Massilia** *and Jonathan Mosqueda-Melgar
Institute of Food Science and Technology, Central University of Venezuela,
Colinas de Bello Monte, Caracas, Venezuela

ABSTRACT

Listeria monocytogenes is an emerging Gram-positive pathogenic microorganism widely distributed in the environment, food processing areas and in foods. This microorganism has been isolated from several ready-to-eat (RTE) products from meat, poultry, seafood, dairy, fruits and vegetables origin, thus representing, high risk foods for transmitting listeriosis to susceptible individuals. In the last decades, this food-borne disease has increased due mainly to the changes in food habits of consumers, and new practices of handling, preparation and commercialization of retails and food industries. Novel technologies for the RTE products preservation have recently been developed, among them: the use of natural antimicrobials, edible films and coatings, irradiation, UV-C light, pulsed light, pulsed electric fields, high-pressure processing and modified atmosphere packaging for ensuring the safety and quality of these products. However, hurdle technologies are preferred to control *L. monocytogenes* populations and to decrease the impact on the sensory attributes in these kinds of products. This chapter provides an overview of novel technologies applied to RTE products to controlling *L. monocytogenes*.

1. INTRODUCTION

Ready-to-eat (RTE) foods include any meal that is normally consumed in its raw state, or any food handled, processed, mixed, cooked or otherwise prepared into a form in which it is

* PO Box 47097. Phone: +58-0212-7534403, Fax: +58-0212-7533871. E-mail: rosa.raybaudi@ciens.ucv.ve.

normally consumed without further processing [1]. Therefore, RTE foods may be contaminated with pathogenic and spoilage microorganisms if bad manufacture practices are carried out during handling, processing, packaging, distribution and commercialization [2, 3]. These foods could be considered high-risk vehicles for transmitting listeriosis (by *Listeria monocytogenes*) to susceptible populations, because they may support the growth of the pathogen during refrigerated storage. RTE foods are usually preserved under refrigeration and offer an appropriate environment for the multiplication of *L. monocytogenes* during manufacture, aging, transportation, and storage. The foods in this category include unpasteurized milk and products prepared from unpasteurized milk, soft unfermented cheeses, unreheated frankfurters, certain delicatessen meats and poultry products, and some seafoods [2, 4, 5].

Listeria monocytogenes is an emerging Gram-positive microorganism widely dispersed in the environment and foods [6]. Its ability to grow at refrigerated temperature in different environments makes its eradication very difficult in foods [2]. Lethality treatments such as cooking generally eliminate *L. monocytogenes* populations; however, RTE products may be re-contaminated by exposure to the environment after the lethality treatment during peeling, slicing, re-packaging, and other processing steps. By controlling sanitation in the post-lethality processing environment or implementing interventions in their products, the establishments can ensure that their RTE products do not become contaminated with *L. monocytogenes* [7, 8].

The application of new technologies such as the use of natural antimicrobials, edible films and coatings, irradiation, UV-C light, pulsed light, pulsed electric fields, high-pressure processing and modified atmosphere packaging have demonstrated be good interventions alternatives for controlling *L. monocytogenes* in RTE foods. For these reasons, this chapter has as objective show an overview about the effectiveness of these novel technologies on *L. monocytogenes* populations in different RTE foods.

2. Ready-to-eat Products Associated with *L. monocytogenes*

Recent outbreaks of the bacterial pathogen *L. monocytogenes* have drawn attention by the severity of listeriosis in humans, and by the risk of *L. monocytogenes* contamination in all foods. Most cases of human listeriosis occur sporadically; however, much of what is known about the epidemiology of the disease has been derived from outbreak-associated cases [8].

Foods that have been reported to be associated with outbreaks or sporadic cases of listeriosis have been foods that can support the growth of *L. monocytogenes* and that are ready-to-eat (including coleslaw, fresh soft cheese made with unpasteurized milk, frankfurters, deli meats, butter, among others) [6]. In general, the levels of *L. monocytogenes* in the implicated food have exceeded 10^3 CFU/g, but there have been instances where the observed level of *L. monocytogenes* in the implicated food has been substantially lower. However, there is a great deal of uncertainty concerning these estimates because the actual level of the pathogen in the serving of food consumed by an infected individual could have varied considerably from that observed in other portions of the food during a subsequent investigation [1].

Table 1. Food-borne disease outbreaks produced by *L. monocytogenes* in RTE foods in the United States

Year	Implicated RTE Food	N° Cases	N° Death	Reference
1994	Pasteurized chocolate milk	69	0	[10]
1998	Processed meat	101	15	[10]
1998	Frankfurters	108	14	[11]
1999	Frankfurters	4	-	[12]
1999	Deli meat	5	1	[12]
1999	Pate	11	-	[12]
1999	Deli meat	2	1	[12]
2000	Mexican-style cheese	13	0	[13]
2000	Delicatessen turkey	30	4	[10]
2001	Deli meat	28	0	[14]
2002	Deli meat	54	8	[15]
2003	Mexican-style cheese	12	1	[12]
2005	Grilled chicken	3	0	[12]
2005	Deli meat	13	1	[12]
2005	Mexican-style cheese	12	0	[12]
2006	Nacho salad	2	0	[12]
2006	Cheese	3	1	[12]
2007	Milk	5	3	[16]
2008	Tuna salad	5	3	[17]
2008	Sprout	20	0	[12]
2008	Mexican-style cheese	8	0	[18]
2009	Mexican-style cheese	18	0	[19]
2010	Sushi Rolls	2	0	[19]
2010	Pre-cut celery	10	0	[19]
2011	Cantaloupe	147	-	[19]
2011	Aged, blue-veined cheese	15	-	[19]

Listeriosis includes a spectrum of clinical illnesses ranging from febrile gastroenteritis to potentially fatal bacteremia and meningitis. Healthy individuals are not usually affected when they consume foods containing this organism; however, illness can be very severe in immune-compromised individuals, children and older persons and, in pregnant women [9].

Outbreaks of listeriosis are often associated with a processing or production failure. The risk is highest in refrigerated and ready-to-eat foods because *L. monocytogenes* is one of the few foodborne pathogens capable of multiplying at refrigerated temperatures.

In Table 1 is shown the outbreaks number associated to *L. monocytogenes* on a wide range of RTE foods.

3. Novel Technologies for Controlling *L. monocytogenes* in RTE Products

The food industry and research's institutions have shown a great interest in the development of novel technologies that offers the advantages of using low processing temperatures, low energy consumption, and retention of nutritional and sensory attributes [20], while inactivating pathogenic microorganisms to levels that do not cause a public health risk, among them:

3.1. Natural Antimicrobials

Food antimicrobials are chemical compounds or substances that may delay microbial growth or cause microbial death when finding their way into a food matrix [21]. Traditional antimicrobials have been extensively used for many years. However, consumers are currently demanding wholesome, fresh-like, and safe foods without addition of chemically synthesized preservatives [22]. The application of novel natural antimicrobials to assure the safety of foods while preventing quality loss is a promising alternative and feasible. Natural antimicrobials from animal (lactoperoxidase, lysozyme, and chitosan), plant (essential oils, aldehydes, esters, herbs, and spices), and microbial (nisin) origin can be used to effectively reduce pathogenic microorganisms, among them *L. monocytogenes*. Nevertheless, the use of these compounds at a commercial level is still limited due to several factors such as impact on sensory attributes or, in some cases, regulatory issues concerning their use. Therefore, further studies on the effects of each antimicrobial on food sensory characteristics are still needed.

Generally, antimicrobials have different concentration thresholds for inhibition or inactivation. These thresholds will depend on the specific targets of the antimicrobial substance, including cell wall, cell membrane, metabolic enzymes, protein synthesis, and genetic systems [23]. The efficiency of a certain antimicrobial will also depend on the type, genus, species, and strain of the target microorganism. Likewise, it will depend on environmental factors such as pH, water activity (aw), temperature, atmosphere composition, initial microbial load, and acidity of the food substrate [24].

3.1.1. Plants Origin

Essential oils (EOs), also called volatile or ethereal oils, are aromatic oily liquids obtained from plant materials (flowers, buds, seeds, leaves, twig bark, herbs, wood, fruits, and roots) [25], that can be used as food flavorings [26] and antimicrobials [22]. Although the antimicrobial properties of EOs and their components have been reviewed in the past, their mechanisms of action have not been studied in great detail. Considering the large number of different groups of chemical compounds present in EOs, it is most likely that their antibacterial activity is not attributable to one specific mechanism but to action over several specific targets in the cell [25]. Nychas et al., [26] and Burt [25] have reported the location and mechanisms of action in the bacterial cell of EOs, for instance: degradation of the cell wall, damage to cytoplasmic membrane and membrane proteins, leakage contents out of the cell, coagulation of cytoplasm, and depletion of the proton motive force. Nychas et al., [26] indicated that the mode of action of EOs is concentration dependent, indicating that low concentrations inhibit enzymes associated with energy production, while higher amounts may precipitate proteins.

Different researches have demonstrated that EOs can be used to control *L. monocytogenes* in different RTE foods. Control of *L. monocytogenes* in RTE meat products using natural antimicrobials have been reported by Hao et al., [27], who found that eugenol and pimento extract significantly inhibited the growth of *L. monocytogenes* inoculated in cooked beef slices. Ahn et al., [28] reported that *L. monocytogenes* declined 2 logs after 9 days of refrigerated storage in cooked beef treated with 1% Pycnogenol or grape seed extract (ActiVin). Sivarooban et al., [29] observed undetectable levels of *L. monocytogenes* in turkey frankfurters treated with a combination of grape seed extract (1%) plus nisin (6,400 IU/mL) after 21 days of storage at 4ºC and 10ºC. Likewise, Larson et al., [30] reported that hop extracts (1 mg/g) can be used to control *L. monocytogenes* in minimally processed food with low fat content.

Recently, studies have also demonstrated that natural antimicrobials of plant origin can also be used to control *L. monocytogenes* in whole and fresh-cut fruits and vegetables as well as their juices. Xu et al., [31] reported more than 2.5 log reductions of *L. monocytogenes* population in cucumber and lettuce treated with grape seed extract (GSE) at 0.1% during storage refrigerated (10 and 4 ◦C). Moon et al., [32] did not report recuperation of viable cells of *L. monocytogenes* in apple juice supplemented with vainillin 40 mM after 3 days of storage at 4ºC and 15ºC. Ferrante et al., [33] found reductions of 1.8 and 3.5 log in orange juice treated with 1000 and 1500 ppm of vainillin and high-power ultrasound after 15 min of exposure at treatment. Belletti et al., [34] evaluated the effect of pure citron essential oil on *L. monocytogenes* Scott A in fruit-based salad (fresh-cut apple, pineapple, melon, kiwifruit, orange and grapes). They reduced the growth of *L. monocytogenes* Scott A by more than 2 logs after 1 day of storage at the abuse temperature of 9ºC for a period of 9 days using 300 or 600 ppm citron essential oil.

3.1.2. Animal Origin

Lysozyme is a protein present in milk and eggs that catalyzes the hydrolysis of the β-1,4 linkages between N-acetylmuramic acid and N-acetylglucosamine in the peptidoglycan layer of the bacterial cell wall [22].Lysozyme is generally active against most Gram-positive bacteria, including *L. monocytogenes.*

Mangalassary et al., [35] evaluated the effect of lysozyme and/or nisin in combination with in-package pasteurization of a ready-to-eat low-fat turkey bologna on the inactivation of *L. monocytogenes*. The authors indicated that sterile bologna samples were initially treated with solutions of nisin (2 mg/mL = 5,000 AU/mL = 31.25 AU/cm^2), lysozyme (10 mg/mL = 80 AU/mL = 0.5 AU/cm^2), and a mixture of nisin and lysozyme (2 mg/mL nisin + 10 mg/mL lysozyme = 31.75 AU/cm^2). Bologna surfaces were uniformly inoculated with a *L. monocytogenes* suspension resulting in a population of approximately 0.5 log CFU/cm^2. Then samples were vacuum packaged and subjected to heat treatment (60, 62.5, or 65 degrees C). Those authors reported that the nisin-lysozyme combination and nisin treatments were effective in reducing the time required for 4-log reductions at 62.5 and 65 ºC in comparison with the control, but not at 60 ºC. Thus, 62.5 ºC, nisin-lysozyme-treated samples required 23% less time than did the control sample to achieve a 4-log reduction and 31% less time at 65 ºC. Whereas, lysozyme alone treatment did not enhance antilisterial activity with heat.

Ko et al., [36] reported that the combination of ovotransferrin with EDTA have a higher antibacterial effect over *L. monocytogenes* in brain heart infusion broth culture than Ovotransferrin plus lysozyme, Ovotransferrin in 100 mM NaHCO3 solution or Ovotransferrin in 100 mM NaHCO3 solution plus lysozyme. However, ovotransferrin plus either lysozyme or EDTA, or both, did not show any antibacterial effect in inoculated commercial hams during storage at 10ºC.

3.1.3. Microbial Origin

Lactic acid bacteria or bacteriocins produced by them can also be used to inhibit the growth of *L. monocytogenes.* Bacteriocins are ribosomally synthesized polypeptides produced by bacteria with an ability to kill or inhibit the growth of similar bacterial strains. Nisin is the most commercially important bacteriocin due to its relatively long history of safe use [37]. It is currently recognized as a safe food preservative in approximately 50 countries [38]. It has been shown that nisin is more effective against Gram-positive microorganisms such as *L. monocytogenes* because can permeabilizes the cytoplasmic membrane, thereby dissipating the membrane potential, inhibiting transport of amino acids, and causing release of accumulated [39].

Schillinger et al., [40] reported that the addition of *Lactobacillus sakei* Lb 706 prevented the growth of *L. monocytogenes* in pasteurized minced meat and comminuted cured raw pork during the 1st few days after production. Likewise, Dixon et al., [41] carried out three batches of Toroi (is a mixture of boiled green vegetable and uncooked mussels) which were contaminated with 10^3 CFU/g of *L. monocytogenes* L70; a strain derived from NZ mussels, and stored at 4-6 °C. When nisin was added to Toroi at 1000 IU/g (final concentration) *L. monocytogenes* numbers declined (c. 0.5 logs) but had returned to the initial level by 14 days. In contrast, addition of *L. sakei* Lb706 (10^7 CFU/g Toroi) resulted in a rapid decline in *L. monocytogenes* to a very low level (< 10 CFU/g) with no increase over 2 months. However, *L. sakei* Lb706 was not effective for Toroi that contained 10^5 CFU/g *L. monocytogenes.* Bredholt et al., [42] indicated that the growth of *L. monocytogenes* was inhibited when a *Lactobacillus sakei* strain was applied to cooked products at a concentration of 10^5 to 10^6 (CFU)/g immediately before slicing and vacuum-packaging and stored at both 8ºC y 4ºC. Amezquita and Brashears [43] found that the addition of a combined culture of lactic acid bacteria *Pediococcus acidilactici*, *Lactobacillus casei*, and *Lactobacillus paracasei* to frankfurters and cooked ham inoculated with *L. monocytogenes* showed bacteriostatic activity on *L. monocytogenes* in cooked ham and bactericidal activity in frankfurters. Jacobsen et al., [44] reported that live cells of bacteriocin-producing *Leuconostoc carnosum* inhibited the growth of *L. monocytogenes* in cooked, sliced, and gas-packed meat products stored at 5 °C and 10 °C for 4 weeks.

3.2. Edible Films and Coatings

The terms edible film and coating are used in most cases interchangeably to indicate that the surface of a food is covered by a thin layer of material of certain composition. However, a film is occasionally differentiated from a coating by the notion that it is a stand-alone wrapping material; whereas, a coating is applied and formed directly on food surface itself [45].

Table 2. Application of edible films and coatings to control *L. monocytogenes* in RTE foods

Type of film or coating matrix		RTE Food	Antimicrobial	Results	Reference
Coating	Film				
	Gelatin	Turkey bologna	Nisaplin (0.5%) Guardian CS1-50 (1%)	4 log reductions of *L. monocytogenes* 3 log reductions of *L. monocytogenes*	[48]
	Alginate, Zein or Polyvinyl alcohol	Sliced cooked ham	Enterocins A and B (200 or 2000 AU/cm^2)	Inhibited the growth of *L. monocytogenes* by more 60 days	[49]
	Alginate	Slices cooked ham	Enterocins A and B (200 or 2000 AU/cm^2) + HPP	Inhibited the growth of *L. monocytogenes* by more 60 days	[50]
	Alginate	Bologna and ham slices	Essential oils of: Spanish oregano (O), Chinese cinnamon (C) or Winter savory (S), all at 1%	Alginate films with C were more effective in reducing (more than 2 logs) *L. monocytogenes* in bologna and ham after 5 days of refrigerated storage than those films containing O or S.	[51]
Alginate		Smoked Salmon	Oyster Lysozyme or hen egg white lysozyme (160 μg/mL) and nisin (1000 IU/g)	Reduced by more than 2.2 to 2.8 log CFU/g the population of *L. monocytogenes* (without films) after 35 days of refrigerated storage	[52]
	Chitosan	Smoked Salmon	Nisin (500 IU/cm^2), sodium lactate (SL) (9 mg/cm^2), sodium diacetate (SD) (0.5mg/cm^2), potassium sorbate (PS) (0.6mg/cm^2) and sodium benzoate (SB) (0.2 mg/cm^2)	SL was more effective (completely inhibited *L. monocytogenes* growth during 10 days of storage at 20°C). Chitosan-coated plastic films with 4.5 mg/cm^2 SL, 4.5 mg/cm^2 SL–0.6 mg/cm^2 PS, and 2.3 mg/cm^2 SL–500 IU/cm^2 nisin were also effective to inhibit the *L. monocytogenes* growth during 6 weeks at refrigeration temperature (4°C)	[53]
Chitosan		Roast beef	Lactic acid, acetic acid	Inhibited the growth of *L. monocytogenes* being more effective the coating containing acetic acid	[54]
	Chitosan	Ham steaks	Nisin 500 IU/cm^2 , sodium lactate (SL) 0.01 g/cm^2, sodium diacetate 0.0025 g/cm^2, potassium sorbate (PB) 0.003 g/cm^2, sodium benzoate (SB) or 0.001 g/cm^2	Inhibited the growth of *L. monocytogenes* during 10 days of storage at 20°C. SL was more effective.	[55]
Chitosan	Chitosan	Mozzarella Cheese	Lysozyme	Reduced from 0.32 to 1.35 log reductions of *L. monocytogenes* population stored at 10°C.	[56]

Table 2. (Continued)

Type of film or coating matrix		RTE Food	Antimicrobial	Results	Reference
Coating	Film				
Chitosan		Cold-smoked salmon	Sodium lactate (1.2%) and sodium diacetate (0.25%), or 2.4% sodium lactate.	Reductions by more than 1.3 logs of *L. monocytogenes* were achieved during 30 days of storage at 4° C.	[57]
	Zein	Chicken	Nisin (1000 IU/g) or Calcium propionate (1%)	Treatments with nisin (N) or nisin with calcium propionate (NCP) significantly suppressed the bacterial populations of *L. monocytogenes* by 1 to 3 log cycles from day 0 to 8	[58]
Zein		Turkey Frankfurters	Nisin (N), sodium lactate (L), and sodium diacetate (D)	During 28 days, treatment with N or D alone reduced *L. monocytogenes* counts on turkey frankfurters by 6.6 or 6.3 log CFU/g, respectively.	[59]
Whey protein isolate		Turkey Frankfurters	Grape seed extract (0.5%), nisin (6000 IU/g) and malic acid (1.0%).	The *L. monocytogenes* population were reduced by more than 3 logs after 28 d at 4 ◦C.	[60]
Soja protein		Turkey Frankfurters	Nisin (10000 IU/mL) + grape seed extract (1%) or nisin (10000 IU) + green tea (1%)	*L. monocytogenes* population were decreased by more than 2 log cycles after 28 d at 4 °C and 10 °C, in the samples containing combined antimicrobials.	[6]
Soja protein		Turkey Frankfurters	Nisin (10,000 IU/mL), grape seed extract (1%), green tea (1%) or their combinations	The *L. monocytogenes* population (7.1 CFU/g) was decreased by more than 2 log cycle after 28 d at 4 °C and 10 °C, in the samples containing nisin (10000 IU) combined with either GSE (1%) or GTE (1%).	[29]
	MC / HPMC	Hot Dog Sausage	Nisin (100 and 75 mg/ml)	Significantly decreased the populations of *L. monocytogenes* on the surface of hot dogs by more than 2 logs CFU/g throughout the 60 days of storage.	[61]
	Cellulose	Frankfurter sausages	Nisin (25 mg/ml)	Populations of *L. monocytogenes* were significantly reduced by approx. 2 logs CFU/g after 14 days of refrigerated storage.	[62]
Pectin		Roasted Turkey	Sodium lactate (1.2%) and sodium diacetate (0.25%)	Reductions by more than 1.1 logs of *L. monocytogenes* were achieved during 8 weeks of storage at 4° C.	[63]

Edible films and coatings are applied on many products to control moisture transfer, gas exchange, oxidation processes and microbial growth. One major advantage of using edible films and coatings is that several active ingredients can be incorporated into the polymer matrix and to be consumed with the food, thus enhancing safety, quality or even nutritional and sensory attributes [46-47].

Several studies have demonstrated that the application of edible films or coatings with antimicrobials added to RTE foods have been effective for controlling pathogenic microorganisms such as *L. monocytogenes* (Table 2). In this context, different researchers have indicated that the incorporation of natural antimicrobials such as organic acid, salts, essential oils, enzymes or bacteriocins into polysaccharide and protein based films or coatings have been effective to inactivate or delay the growth of *L. monocytogenes* and other pathogenic microorganisms.

3.3. Irradiation

Ionizing radiation is a process in which products are exposed to radiant energy. Ionizing radiation includes gamma rays, electron beams, and x-rays. Gamma irradiation uses high-energy gamma rays from cobalt 60 or cesium 137, which has a long half-life (5.27 y and 30.1 y, respectively) and high penetration power and thus can treat bulk foods on shipping pallets. Electron beam (E-beam) irradiation uses a stream of high-energy electrons, known as beta rays, which can penetrate only about 5 cm. X-irradiation, which has intermediate properties of the 2 previously discussed irradiation methods, penetrates foods more shallowly than gamma irradiation but much more deeply than electron beams [64].

Clardy et al., [65] studied the effect of gamma irradiation against *L. monocytogenes* in inoculated ready-to-eat sandwiches. Ham and cheese sandwiches were contaminated with *L. monocytogenes* (10^7 CFU/g), frozen at -40 °C, and exposed to gamma irradiation. Counts of *L. monocytogenes* in sandwiches treated with 3.9 kGy decreased by 5 log units during the first days of a storage period of 39 days. Chen et al., [66] reported that a combination of 6,000 AU of pediocin with irradiation at 2.3 kGy or more was effective for inhibition of *L. monocytogenes* in frankfurters for 12 weeks at 4ºC or 10ºC. Knight et al., [67] evaluated the effect of a combination of potassium lactate/sodium diacetate with irradiation to control *L. monocytogenes* in frankfurters. Frankfurters formulated with 0% or 3% added potassium lactate/sodium diacetate solution and inoculated with *L. monocytogenes* before or after treatment with irradiation (0, 1.8, or 2.6 kGy), stored aerobically or vacuum packaged and refrigerated were evaluated.These authors found that the incorporation of lactate/diacetate with or without irradiation had a strong listeriostatic effect for aerobically stored frankfurters. Moreover, outgrowth was suppressed and counts were not different from initial counts (5.2 log CFU/frankfurters compared with 5.0 log CFU/frankfurters); however, those without the additive increased steadily (5.4 to 9.3 log CFU/frankfurters). Irradiation treatments alone had higher *L. monocytogenes* counts after 3 weeks. For vacuum-packaged frankfurters, both the addition of lactate/diacetate and irradiation were effective at controlling growth after 8 weeks.

Zhu et al., [64] determined the effectiveness of antimicrobials and irradiation combination treatments to control inoculated *L. monocytogenes* (10^6 CFU/cm^2) in slices of RTE turkey breast rolls. Six antimicrobial additive treatments that include no preservatives (control), 0.1% potassium benzoate (PB), 2% sodium lactate (SL), 0.1% potassium benzoate plus 2% sodium lactate (PB + SL), 2% sodium lactate plus 0.1% sodium diacetate (SL + SDA), and 0.1% potassium benzoate, 2% sodium lactate, and 0.1% sodium diacetate (PB + SL + SDA) were evaluated. Irradiation of turkey rolls, which were previously treated with PB + SL or SL + SDA, at 1.0 kGy resulted effective in suppressing the growth of *L. monocytogenes* for about 6 weeks when stored at 4°C. Whereas, no growth of *L. monocytogenes* after irradiation occurred during 42 d of storage for 2.0 kGy irradiated breast rolls formulated with 0.1% PB + 2% SL, 2% SL + 0.1% SDA or 0.1% PB + 2% SL + 0.1% SDA, and 1.0 kGy irradiated turkey breast with 0.1% PB + 2% SL + 0.1% SDA. Kudra et al., [68] investigated the efficacy of a combination of irradiation and modified atmosphere packaging (MAP) containing a high concentration of CO_2 for controlling *L. monocytogenes* on frankfurters and cooked pork chops. Frankfurters and cooked pork chops were inoculated with a cocktail of *L. monocytogenes,* packaged in vacuum or high-CO_2 MAP and Irradiated at 0, 0.5, 1.0, or 1.5 kGy. Authors indicated that no significant packaging effect was found for the radiation sensitivity of *L. monocytogenes*. However, high-CO_2 MAP was more effective than vacuum packaging for controlling the growth of survivors during refrigerated storage.

3.4. UV-C Light

UV light is used in the food industry for disinfecting surfaces. Applications include decontamination of surfaces of equipment in bakeries, cheese and meat plants, as an adjunct to usual cleaning and sanitizing practices, and for decontamination of conveyor surfaces and packaging containers such as boxes, caps, bottles, cartons, tubes, films and foils. Despite the efficacy of UV light to disinfect smooth surfaces, there are relatively few applications of this technology in the food processing industry. The restricted range of commercially available equipment for disinfecting solids may contribute to its limited use [69].

Some researchers have demonstrated that UV-C light application to RTE meat products is a good alternative to reduce *L. monocytogenes* populations. In this way, Sommers et al., [70] studied the ability of UV-C light to inactivate inoculated *L. monocytogenes* population on frankfurters that contained potassium lactate (PL) and sodium diacetate (SDA), either before or after packaging. Reductions of 1.31, 1.49, and 1.93 log of the inoculated *L. monocytogenes* population at doses of 1, 2, and 4 KJ/m^2, respectively were reached in the frankfurters. Chun et al., [71] evaluated the applicability of UV-C light on the inactivation of foodborne pathogenic bacteria in RTE sliced ham. For that, RTE sliced hams were inoculated with the pathogens *L. monocytogenes*, *S.* Typhimurium, and *C. jejuni* and submitted to UV-C light of 1, 2, 4, 6, and 8 KJ/m^2. Results indicated that foodborne pathogen populations significantly ($p < 0.05$) decreased with increasing UV-C irradiation. In particular, UV-C irradiation at 8 KJ/m^2 reduced the populations of *L. monocytogenes* in the ham by 2.74 log CFU/g.

Applications of UV-C light have also been realized in produce. In this sense, Chun et al., [72] applied different doses of UV-C light to inactivate distinct pathogens in RTE salad and found that a dose of 8 kJ/m^2 reduced the population of *L. monocytogenes* in 2.57 log CFU/g. Raybaudi-Massilia et al., [47] evaluate the effect of different doses of UV-C light (0, 0.96, 2.88, 5.76 and 8.64 kJ/m^2) and treatments with malic acid (0, 0.5, 1.0 and 1.5 %, w/v) applied into dipping solutions containing calcium lactate (1 %, w/v) and ascorbic acid (0.5 %, w/v) on the inactivation of *L. monocytogenes* inoculated onto fresh-cut ''Maradol'' papaya. A reduction of 3.15 Log10 CFU/g of *L. monocytogenes* population was reached using a combination of malic acid (1.5 %) and UV-C light (8.64 kJ/m^2).

3.5. Pulsed Light

Pulsed light is a non-thermal method for food preservation that involves the use of intense, short duration pulses of a broad spectrum to ensure microbial decontamination on the surface of either foods or packaging materials [73]. This method was approved for food surface decontamination by the Food and Drug Administration in 1999 [74]. The inactivation efficacy of pulsed light depends on the intensity (measured in J/cm^2) and number of pulses delivered.

The lethal effects of pulsed light can be attributed to its rich broad-spectrum UV content, its short duration, high peak power, and the ability to regulate both the pulse duration and frequency output of flash lamps, which play a major role in microbial destruction [75]. Various mechanisms of action of pulsed light have been proposed to explain their lethal effects, all of them related to the UV part of the spectrum and its photo-chemical and/or photo-thermal effect.

The primary cell target of pulsed light is nucleic acids because DNA is a target molecule for these UV wavelengths [73]. The germicidal effect of UV light has been attributed primarily to a photo-chemical transformation of pyrimidine bases in the DNA of bacteria, viruses, and other pathogens to form dimmers. The formation of those bonds prevents DNA unzipping for replication and the organism becomes incapable of reproduction. Without sufficient repair mechanisms, such damage results in mutations, impaired replication and gene transcription, and ultimately the death of the organism [73, 76].

The lethal action of pulsed light also can be due to a photothermal effect. Wekhof [77] proposed that with an energy exceeding 0.5 Jcm–2 the disinfection is achieved through bacterial disruption during their temporary overheating resulting from the absorption of all UV light from a flash lamp. This overheating can be attributed to a difference in UV light absorption by bacteria and that of a surrounding medium. The water content of bacteria is vaporized, generating a small steam flow that induces membrane disruption [75].

Several studies have been carried out to evaluate the effect of the pulsed light over *L. monocytogenes* in culture media. However, few studies have been realized to investigate the effect of this technology on this microorganism in RTE foods.

The inactivation of *L. monocytogenes* inoculated on raw salmon fillets by pulsed UV light was studied by Ozer and Demirci [78], using a broad spectrum lamp that generated 5.6 J/cm^2 per pulse at the lamp surface and three pulses per second. Two different surfaces were treated, the skin side and the muscle side. For *L. monocytogenes*, the maximum log reduction was 0.74 logs on the muscle side and 1.02 logs on the skin side. Some fish fillets were overheated at shorter distances and longer treatment times. The higher inactivation on the skin side was attributed to a higher temperature in comparison with the muscle side.

On the other hand, the impact of high power pulsed light (HPPL) on natural microbial contamination and inoculated *L. monocytogenes* population (1 x 10^7/ CFU/ml), health promoting components and shelf-life of strawberries were evaluated by Luksiene et al., [74]. For that, strawberries were treated with 0–1000 pulses of light (UV light dose 0-3.9 J/cm^2) and afterwards analyzed in terms of microbial contamination, shelf-life expansion, antioxidant capacity, firmness, total phenolics, total anthocyanins, ascorbic acid content and color. Results indicated that the decontamination of strawberries by HPPL was significant compared to control. Naturally distributed mesophilic bacteria on the surface of strawberries were inactivated by 2.2 logs, whereas, inoculated *L. monocytogenes* was inactivated by 1.1 logs.

This technology has a high potential like novel technology for food decontamination, and to assure the food quality and safety; however, technological problems to find ways to control food heating and to treat homogeneously foods, nutritional effects and photosensitization to foods need to be researched in deep major. In addition, the efficiency of pulsed light is depending on microbial exposure and food composition (high-protein or oily foods).

3.6. Pulsed Electric Fields

This non-thermal technology is able to inactivate spoilage and pathogenic microorganisms and to decrease the activity of enzymes in order to extend the shelf life and safety of foods without significant loss of flavor, color, and nutrients [79-80]. Thus, this processing method may offer to the consumer safety, fresh-like, and nutritious food products. PEF treatment involves the application of short pulses (1 to 10 μs) with high intensity electric field (typically 20 to 80 kV/cm) to fluid foods placed between two electrodes in batch or continuous flow treatments [79, 80]. Nevertheless, PEF application is restricted to fluid foods that can tolerate high electric fields strength, have low electrical conductivity, and do not contain or form bubbles [81]. The particle size of the food is also a limitation [82]. However, other factors such as frequency, treatment time, type and shape of the pulse, in addition to the self-microorganism characteristics could influence the efficacy of PEF processing [83].

The microbial inactivation mechanism of PEF was elucidated by Coster and Zimmermann (1975) through electroporation's Theory. This theory explains the formation of pores in the cellular membrane, which is able to generate cellular lysis with subsequent leakage out of intracellular compounds due to the induced electric field. However, this phenomenon has demonstrated to be reversible or irreversible, depending on the level of electric fields intensity [84] and/or the membrane organizational changes [85]. Mild electric fields intensity conditions form reversible pores in the cellular membrane, whereas drastic electric field intensity conditions lead to the irreversibility of this phenomenon, which results in cellular death [84, 86] Mosqueda-Melgar et al., [87] reported 3.56 and 3.41 log reductions of *L. monocytogenes* in melon and watermelon juices, respectively, treated with an electric field of 35 kV/cm, 4 μs pulse duration in bipolar mode and square shape during 1440 - 1727 μs of treatment time. These same authors, but in other study [88], and under the same conditions of electric field, pulse duration and treatment time, combined this technology with natural antimicrobials such as cinnamon bark oil (0.2%) or citric acid (1.5 - 2.0%) in melon and watermelon juices, and decreased more than 5 logs the *L. monocytogenes* populations. In this study, although PEF combined with citric acid or cinnamon bark oil ensured the microbiological quality and safety of fruit juices in the same way than thermal treatment, caused changes on some sensory properties (aroma and taste) of melon and watermelon juices.

Table 3. Application of high-pressure processing (HPP) to control *L. monocytogenes* in RTE foods

RTE Food	Treatment	Results	Reference
Apple and plum jam baby food	HPP at 100, 200, 300 and 400 MPa were used by 5, 10 and 15 min	The results showed that there not were significant differences between pressures of 100 and 200 MPa at 5, 10 and 15 min. However, at 300 MPa, there were significant reductions of *L. monocytogenes* at 15 min. When the pressure treatment was 400 MPa, significant differences were observed at pressure times of 5, 10 and 15 min.	[91]
Frankfurters	HPP at 300, 500, and 700 MPa for 9 min.	*L. monocytogenes* populations decreased by more than 1, 3, and 5 logs, respectively.	[92]
Cold-smoked salmon	HPP at 150, 200 and 250 MPa by 20 min	HPP with 250 MPa did not inactivate *L. monocytogenes* but enhanced the lag phase.	[93]
Sliced cooked ham	Nisin, Potasium lactate (2-2.5%), and HPP at 400 MPa y 10 min	A synergistic effect to inhibit a cocktail of *L. monocytogenes*, was observed between potassium lactate, high hydrostatic pressure, and storage for 84 days at 1ºC.	[94]
Spanish dry-cured hams, Iberian and Serrano	HPP at 450 MPa by 10 min	Reduction of *L. monocytogenes* populations by 1.50 and 1.16 log CFU/g in Iberian and Serrano hams, respectively.	[95]
Cheeses	HPP at 300, 400 or 500 MPa by 10 min	Counts of *L. monocytogenes* (log cfu/g) at 300, 400 and 500 MPa were 8.5, 5.4 and 2.5, respectively.	[96]
Slices cooked ham	HPP + edible film (with enterocins)	Reduction of *L. monocytogenes* at undetectable levels without recovery during 60 days of storage was observed	[50]
Slices cooked ham	HPP + bacteriocins (Enterocins A and B, Sakacin K and nisin + potassium lactate	Reductions about 4 log CFU/g of *L. monocytogenes* population when HPP was combined with bacteriocins and potassium lactate	[97]
Milk	HPP at 550 MPa by 5 min + mild-heat treatments (37ºC for 240 min or 50ºC for 10 min)	Did not recovery *L. monocytogenes* in milk during 70-days of storage at 25 ºC in samples that received mild-heat treatments following by HPP.	[98]
Slices cooked ham	HPP at 600 MPa + nisin + potassium lactate	4 logs CFU/g reductions of *L. monocytogenes* population	[99]

Table 3. (Continued)

RTE Food	Treatment	Results	Reference
Fuet (a low acid fermented sausage)	HPP at 400 MPa alone and combined with enterocin AS 48 (148 AU/g)	HHP alone had no anti-*Listeria* effect, whereas HHP + enterocin AS 48 reduced the level of *L. monocytogenes* by 2 log CFU/g	[100]
Dry-cured ham	HPP at 347-852 MPa by 2.3 to 15.75 min	Pressure and time were the most important factors determining the inactivation of *L. monocytogenes*. The result also indicated that little effect was observed below 450 MPa, whereas holding time longer than 10 min did not result in a meaningful reduction of *L. monocytogenes* counts	[101]
Two types of sliced dry-cured ham Aw of 0.92 and 14.25% fat. Aw of 0.88 and 33.26% fat.	HPP at 600 MPa by 5min alone and combined with nisin (200 AU/cm^2)	Immediate inactivation of *L. monocytogenes* by HPP ranged from 1.82 to 3.85 logs. The lower the water activity in the product, the lower inactivation induced by HPP, both immediately and during storage at 8°C for 60 days. HPP was more effective than nisin alone, however the both hurdles combined resulted better.	[102]
Smoked dry-cured ham with NaCl (standard) or KCl + potassium lactate	HPP at 600 MPa by 5 min	Elimination of *L. monocytogenes* after 14 days in ham with NaCl. Whereas in ham with KCl + potassium lactate the microorganism survived until 56 days, indicating that the NaCl-free processed dry-cured ham had lower stability than standard smoked dry-cured ham	[103]
Sliced ham and turkey breast formulated with sodium nitrite (0 or 200 ppm) and sodium chloride (1.8% or 2.4%)	HPP at 600 MPa by 3 min	3.85-4.35 log CFU/g reductions of *L. monocytogenes* population	[104]

Results obtained of these studies, demonstrated that the application of PEF treatment in combination with natural antimicrobials may be a good alternative to the thermal pasteurization for preserving the microbiological quality and safety in fruit juices, and reducing the risk of food-borne illness caused by unpasteurized juices consumption. However, further studies are needed to reduce the negatives effects on sensory attributes, when natural antimicrobials are added into fruit juices at concentrations that ensuring its safety, since PEF alone did not significantly affected the sensory properties of the juices. In addition, further research is needed to carry out experiments on RTE foods.

3.7. High-pressure Processing (HPP)

High-pressure processing (HPP), also called high hydrostatic pressure (HHP) or ultra high pressure (UHP) processing, is a novel non-thermal method that subjects to liquid and solid foods, with or without packaging, to pressures between 100 and 800 MPa [89]. Thus, HPP can inactivate microorganisms without significant changes in texture, color, or nutritional value of foods [64]. The decontamination efficacy of HPP depends on several factors such as level of pressure, treatment temperature, exposure time, pH, water activity, and food composition [90].

The mechanisms of action of HPP can be grouped into cell envelope-related effects, pressure-induced cellular changes, biochemical aspects, and effects on genetic mechanisms. It has been established that cellular morphology is altered by pressure, and that cell division slows with the application of increasing pressures. High pressures of 100 to 300 MPa can induce spore germination and resultant vegetative cells are more sensitive to environmental conditions [89]. Pressurized membranes normally show altered permeability; therefore the primary site of damage is the cell membrane of microorganisms. Pressure-induced membrane cause inhibition of amino acid uptake probably due to membrane protein denaturation [89]. Numerous studies have shown loss of intracellular constituents from microorganisms after pressure treatment. Leakages of these components from the cells indicate damage to the cellular membrane, and the higher the amount lost from cells correlates with a greater degree of death and injury.

Several reports on the effect of HHP technology applied to different RTE foods for controlling populations of *L. monocytogenes* are shown in Table 3.

The application of HPP to RTE foods has demonstrated to be effective to inhibit or reduce inoculated populations of *L. monocytogenes.*

3.8. Modified Atmosphere Packaging

A modified atmosphere can be defined as one that is created by altering the normal composition of air (78% nitrogen, 21% oxygen, 0.03% carbon dioxide and traces of noble gases) to provide an optimum atmosphere for increasing the storage length and quality of food/produce [105]. This can be achieved by using controlled atmosphere storage (CAS) and/or active or passive modified atmosphere packaging (MAP). Under controlled atmospheric conditions, the atmosphere is modified from that of the ambient atmosphere, and these conditions are maintained throughout storage. MAP uses the same principles as CAS;

however, it is used on smaller quantities of produce and the atmosphere is only initially modified. Active modification occurs by the displacement of gases in the package, which are then replaced by a desired mixture of gases, while passive modification occurs when the product is packaged using a selected film type, and a desired atmosphere develops naturally as a consequence of the products' respiration and the diffusion of gases through the film [105].

The use of modified atmosphere packaging can result a good alternative to extend the shelf life of produce, since retard the spoilage microorganisms growth. However, the presence of pathogens could be encouraged. Also, pathogens populations may further increase in numbers due to the extension of shelf life. Currently, there is concern with the psychrotrophic foodborne pathogens such as *L. monocytogenes*, *Yersinia enterocolitica*, *Aeromonas hydrophila*, as well as with nonproteolytic C. *botulinum*, *Salmonella* spp., *E. coli* O157:H7 and *Shigella* spp..

Different researches have been realized to evaluate the effectiveness of MAP to control *L. monocytogenes* growth in several can of produce. Berrang et al., [106] found that *L. monocytogenes* inoculated onto broccoli, asparagus and cauliflower was unaffected by a modified atmosphere of 3% CO_2, 18% O_2 and 79% N_2 for 10 days at 10 °C. Beuchat & Brackett [107] clearly demonstrated that *L. monocytogenes* increased significantly in number on lettuce stored in a modified atmosphere of 3% O_2 and 97% N_2. Farber et al., [108] demonstrated the effectiveness of the refrigeration applied to fresh-cut processed and packaged vegetables. They found that *L. monocytogenes* levels remained constant in all kind of vegetables processed and packaged stored at 4ºC with the exception of butternut squash and carrots on which the levels increased and decreased, respectively. Also they observed that at 10 °C, the growth of *L. monocytogenes* was supported on all vegetables tested with the exception of chopped carrots, where the population decreased by 2 log units over 9 days. Jacxsens et al., [109] reported a decline in *L. monocytogenes* on both Brussels sprouts and carrots packaged under a modified atmosphere (2 to 3% O_2). 2 to 3% CO_2, and 94 to 96% N_2) and stored at 7 °C. Under severe temperature abuse conditions (25 °C, for 1-2days) followed by storage at 4ºC or 10 °C, inoculated Caesar salad and coleslaw mix supported the growth of *L. monocytogenes*, although larger increases were noted for the coleslaw mix. Beuchat & Brackett [107] found that *L. monocytogenes* could grow on lettuce after procedures such as chlorine treatment, modified atmosphere and shredding. Jacxsens et al., [109] investigated the behavior of *L. monocytogenes* on minimally processed vegetables packaged under either a modified atmosphere (that is, 2-3% O_2, 2-3% CO_2, 94-96% N_2) or air, and found that the organoleptic quality of the produce was obviously decreased before pathogen levels increased significantly, and that the growth of the psychrotrophic pathogens was influenced more by the type of vegetable than the atmosphere used. Similar results were observed by Carlin and Nguyen [110] who inoculated *L .monocytogenes* on minimally processed green salads and found that the growth of this microorganism was faster on butterhead lettuce than on green endive and was not observed on lamb's lettuce. Francis & O'Beirne [111] evaluated the effects of storage atmosphere on *L. monocytogenes* and the competing microflora (*Pseudomonas fluorescens*, *P. aeruginosa*, *Enterobacter cloacae*, *Enterobacter agglomerans* and *Leuconostoc citreum*). They found that at increased CO_2 levels (20%), the growth of the lactic acid bacteria increased, inhibiting the growth of the pathogen, possibly due to the production of well-known antilisterial agents. Conway et al., [112] reported *L. monocytogenes* growth on sliced apples packed under controlled atmosphere.

Studies evaluating the effect of MAP over *L .monocytogenes* growth in meat products have also been realized. Mechaelsen et al., [113] determined the inhibitory effect of (MAP), both alone and in combination with potassium lactate and sodium diacetate (PLSD), on inoculated *L. monocytogenes* in slices ham. Untreated and injected with PLSD ham slices were inoculated with *L. monocytogenes*, subjected to vacuum packaging (VP) or MAP with high concentrations (99.5 to 100%) of CO_2 and stored at 4°C or 10°C. *L. monocytogenes* growth on ham slices at 4°C was inhibited for up to 28 days by all of the treatments, but after 28 days, the VP-PLSD treatment shown a greater inhibitory effect than did the other treatments. However, the results obtained with MAP plus PLSD at 10 °C were not different from those obtained with either MAP or PLSD alone.

CONCLUSION

Most novel non-thermal technologies are still developing and testing, although some emerging non-thermal processes have now been implemented in industrial-scale systems for commercial and research applications. Effective combinations of two or more preservation hurdles should be chosen to avoid a high impact on sensory attributes, and once the modes of action and cellular targets of each treatment are known. The small number of scientific studies summarized in this chapter show that combining non-thermal treatments has great potential for improving the safety of RTE foods, although many technological and regulatory barriers still need to be overcome before the food supply can receive these benefits.

REFERENCES

[1] WHO/FAO [World Health Organization/Food and Agriculture Organization of the United Nations]. Risk assessment of *Listeria monocytogenes* in ready-to-eat foods. Microbiological risk assessment series 4. 2004 [2013-08-14]. Available from: http://www.who.int/foodsafety/publications/micro/en/mra4.pdf

[2] Zhu, M., Du, M., Cordray, J., & Ahn, D. U. (2005). Control de *Listeria monocytogenes* contamination in ready-to-eat meat products. *Comprehensive Reviews in Food Science and Food Safety*. 4, 34-42.

[3] Raybaudi-Massilia R. M, & Mosqueda-Melgar, J. (2009). Factors affecting the incidence of pathogenic microorganisms of fresh-cut produce. *Stewart Postharvest Rev*iew. 4, 1-8.

[4] Swaminathan, B., & Gerner-Smidt, P. (2007). The epidemiology of human listeriosis. *Microbes Infections*. 9, 1236-1243.Ahn, J., Grün, I. U., & Mustapha, A. (2007). Effects of plant extracts on microbial growth, color change, and lipid oxidation in cooked beef. *Food Microbiology*. 24, 7–14.

[5] MAF [Ministry of Agriculture and Forestry of New Zealand]. Guidance for the control of *Listeria monocytogenes* in ready-to-eat foods. Part 1: *Listeria* Management. 2011 [2013-08-14]. Available from: http://www.foodsafety.govt.nz/elibrary/

[6] Theivendran, S., Hettiarachchy, N. S., & Johnson, M. G. (2006). Inhibition of *Listeria monocytogenes* by nisin combined with grape seed extract or green tea extract in soy protein film coated on turkey frankfurters. *Journal of Food Science.* 71(2), M39–M44.

[7] FSIS [Food Safety and Inspection Service]. FSIS Compliance Guideline: Controlling *Listeria monocytogenes* in Post-lethality Exposed Ready-to-Eat Meat and Poultry Products. 2006 [2013-06-12]. Available from: http://www.fsis.usda.gov/OPPDE/rdad/FRPubs/97-013F/LM_Rule_Compliance_Guidelines_May_2006.pdf

[8] FDA [U.S. Food and Drug Administration]. Guidance for industry: control de *Listeria monocytogenes* in refrigerated or frozen ready-to-eat foods; draft guidance. 2008 [2013-10-10]. Available from: http://www.fda.gov/Food/GuidanceRegulation/ Guidance-DocumentsRegulatoryInformation/FoodProcessingHACCP/ucm073110.htm.

[9] Suslow, T., & Harris, L. Guidelines for controlling Listeria monocytogenes in small- to medium-scale packing and fresh-cut operations. Publication 8015. 2000 [08-14-2013]. Available from: http://anrcatalog.ucdavis.edu/pdf/8015.pdf.

[10] Olsen, S. J., Patrick, M., Hunter, S. B., Reddy, V., Kornstein, L., & MacKenzie, W. R. (2005). Multistate outbreak of *Listeria monocytogenes* infection linked to delicatessen turkey meat. *Clinical Infectious Disease*. 40, 962–967.

[11] Mead, P. S., Dunne, E. F., Graves, L., Wiedmann, M., Patrick, M., Hunter, S. (2006). Nationwide outbreak of listeriosis due to contaminated meat. *Epidemiology Infectious*. 134, 744–751.

[12] Cartwright, E. J., Jackson, K. A., Johnson, S. D., Graves, L. M., Silk, B. J., & Mahon, B. E. (2013). Listeriosis Outbreaks and Associated Food Vehicles, United States, 1998–2008. *Emerging Infectious Diseases*. 19(1), 1-9.

[13] MacDonald, P. D., Whitwam, R. E., Boggs, J. D., MacCormack, J. N., Anderson, K. L., & Reardon, J. W. (2005). Outbreak of listeriosis among Mexican immigrants as a result of consumption of illicitly produced Mexican-style cheese. Clinical Infectious Disease. 40:677–82

[14] Frye, D. M., Zweig, R., Sturgeon, J., Tormey, M., LeCavalier, M., & Lee, I. (2002). An outbreak of febrile gastroenteritis associated with delicatessen meat contaminated with *Listeria monocytogenes. Clinical Infectious Disease*. 35, 943–949.

[15] Gottlieb, S. L., Newbern, E. C., Griffin, P. M., Graves, L. M., Hoekstra, R. M., & Baker, N. L. (2006). Multistate outbreak of listeriosis linked to turkey deli meat and subsequent changes in US regulatory policy. *Clinical Infectious Disease*. 42, 29–36.

[16] CDC [Centers for Disease Control and Prevention]. (2008). Outbreak of *Listeria monocytogenes* infections associated with pasteurized milk from a local dairy—Massachusetts, 2007. *MMWR*. 57, 1097–1100.

[17] Cokes, C., France, A. M., Reddy, V., Hanson, H., Lee, L., & Kornstein, L. (2011). Serving high-risk foods in a high-risk setting: survey of hospital food service practices after an outbreak of listeriosis in a hospital. *Infection Control and Hospital Epidemiology*. 32, 380–386.

[18] Jackson, K. A., Biggerstaff, M., Tobin-D'Angelo, M., Sweat, D., Klos, R., & Nosari, J. (2011). Multistate outbreak of *Listeria monocytogenes* associated with Mexican-style cheese made from pasteurized milk among pregnant, Hispanic women. *Journal of Food Protection*. 74, 949–953.

[19] CDC [Centers for Disease Control and Prevention]. (2013). Vital Signs: *Listeria* Illnesses, Deaths, and Outbreaks — United States, 2009–2011. *MMWR.* 62(22), 448-452.

[20] Ross, A. I. V., Grifiths, M. W., Mittal, G. S. & Deeth, H. C. (2003). Combining nonthermal technologies to control foodborne microorganisms. *International Journal of Food Microbiology*. 89, 125-138.

[21] Davidson P. M, & Zivanovic, S. (2003). The use of natural antimicrobials. In P. Zeuthen, & L. Bogh- Sorensen (Eds), *Food preservation techniques* (pp. 5-30). Boca Raton, FLA.: CRC Press LLC.

[22] Raybaudi-Massilia, R. M., Mosqueda-Melgar, J., Soliva-Fortuny, R., & Martín-Belloso, O. (2009). Control of pathogenic and spoilage microorganisms in fresh-cut fruit and fruit juices by traditional and alternative natural antimicrobials. *Comprehensive Reviews in Food Science and Food Safety.* 8, 157–180.

[23] Davidson, P. M. (2001). Chemical preservatives and natural antimicrobial compounds. In M. P. Doyle, L. R. Beuchat, & T. J. Montville (Eds), *Food microbiology: fundamentals and frontiers* (2nd edition. pp. 593-627) Washington, D.C.: ASM Press.

[24] Gould, G. W. (1989). Introduction. In G. W. Gould (Ed). *Mechanisms of action of food preservation procedures* (pp. 1-42). London, U.K.: Elsevier Applied Science.

[25] Burt, S. (2004). Essential oils: their antibacterial properties and potential applications in foods: a review. *International Journal of Food Microbiology*. 94, 223–253.

[26] Nychas, G. J. E., Skandamis, P. N., & Tassou, C. C. (2003). Antimicrobials from herbs and spices. In S. Roller (Ed). *Natural antimicrobials for the minimal processing of foods* (pp. 177-199). Washington, D.C.: CRC Press.

[27] Hao, Y. Y., Brackett, R. E., & Doyle, M. P. (1998). Inhibition of *Listeria monocytogenes* and *Aeromonas hydrophila* by plant extracts in refrigerated cooked beef. *Journal of Food Protection.* 61, 307–312.

[28] Ahn, J., Grün, I. U., & Mustapha, A. (2007). Effects of plant extracts on microbial growth, color change, and lipid oxidation in cooked beef. *Food Microbiology*. 24, 7–14.

[29] Sivarooban, T., Hettiarachchy, N. S., & Johnson, M. G. (2007). Inhibition of *Listeria* monocytogenes using nisin with grape seed extract on turkey frankfurters stored at 4 and 10ºC. *Journal of Food Science.* 70(4), 1017–1020.

[30] Larson, A. E., Yu, R. R., Lee, O. A., Price, S., Haas, G. J., & Johnson, E. A. (1996). Antimicrobial activity of hop extracts against *Listeria monocytogenes* in media and in food. *International Journal of Food Microbiology.* 33, 195–207.

[31] Xu, W., Qu, W., Huang, K., Guo, F., Yang, J., Zhao, H., & Luo, Y. (2007). Antibacterial effect of grapefruit seed extract on food-borne pathogens and its application in the preservation of minimally processed vegetables. *Postharvest Biology and Technology.* 45(1), 126–133.

[32] Moon, K. D., Delaquis, P., Toivonen, P., & Stanich, K. (2006). Effect of vanillin on the fate of *Listeria monocytogenes* and *Escherichia coli* O157:H7 in amodel apple juice medium and in apple juice. *Food Microbiology.* 23, 169-174.

[33] Ferrante, S., Guerrero, S., & Alzamora, S. M. (2007). Combined use of ultrasound and natural antimicrobials to inactivate Listeria monocytogenes in orange juice. *Journal of Food Protection.* 70, 1850–1857.

[34] Belletti, N., Martínez, B., Aymerich, T., & Garriga, M. (2001). Inhibición de *Listeria monocytogenes* y *Salmonella* mediante alta presión hidrostática en un producto listo para el consumo como el *bloc* de *foie gras*. *Eurocarne*. 197, 56-60.

[35] Mangalassary, S., Han, I., Rieck, J., Acton, J., & Dawson, P. (2008). Effect of combining nisin and/or lysozyme with in-package pasteurization for control of *Listeria monocytogenes* in ready-to-eat turkey bologna during refrigerated storage. *Food Microbiology*. 25, 866-870.

[36] Ko, K. Y., Mendoca, A. F. & Ahn, D. U. (2008). Effect of ethylenediaminetetraacetate and lysozyme on the antimicrobial activity of ovotransferrin against *Listeria monocytogenes*. *Poultry Science*, 87, 1649-1658.

[37] Chen H., & Hoover D. G. (2003). Bacteriocins and their food applications. *Comprehensive Reviews in Food Science and Food Safety*. 2, 82-100.

[38] Delves-Broughton, J., Blackburn, P., Evans, R.J., & Hugenholtz, J. (1996). Applications of the bacteriocin, nisin. *Antonie Leeuwenhoek*. 69, 193 – 202.

[39] Abee, T., Rombouts, F. M., Hugenholtz, J., Guihard, G., & Letellier, L. (1994). Mode of action of Nisin Z against *Listeria monocytogenes* Scott A grown at high and low temperatures. *Journal of Applied and Environmental Microbiology*. 60, 1962–1968.

[40] Schillinger, U., Kaya, M., & Lucke, F. K. (1991). Behaviour of *Listeria monocytogenes* in meat and its control by a bacteriocin-producing strain of *Lactobacillus sakei*. *Journal of Applied Bacteriology*. 70, 473-478.

[41] Dixon, L., Donnison, A., Ross, C., McDonald, I. R. (2008). Addition of bacteriocins to inhibit *Listeria monocytogenes* in Toroi: a traditional food of New Zealand Mãori. *Annals of Microbiology*. 58(2), 207-212.

[42] Bredholt, S., Nesbakken, T., & Holck, A. (2001). Industrial application of an antilisterial strain of *Lactobacillus sakei* as a protective culture and its effect on the sensory acceptability of cooked, sliced, vacuum-packaged meats. *International Journal of Food Microbiology*. 66, 191-196.

[43] Amezquita, A., & Brashears, M. M. (2002). Competitive inhibition of Listeria monocytogenes in ready-to-eat meat products by lactic acid bacteria. *Journal of Food Protection*. 65, 316-325.

[44] Jacobsen, T., Budde, B. B., & Koch, A. G. (2003). Application of *Leuconostoc carnosum* for biopreservation of cooked meat products. *Journal of Applied Microbiology*. 95, 242–249.

[45] Pavlath, A. E., Gossett, C., Camirand, W., Robertson, G. H. (1999). Innomeric films of alginic acid. *Journal of Food Science*. 64, 61–63.

[46] Rojas-Graü, M. A., Tapia, M. S., & Martín-Belloso, O. (2008). Using polysaccharide-based edible coatings to maintain quality of fresh-cut Fuji apples. *LWT e Food Science and Technology*. 41, 139-147.

[47] Raybaudi-Massilia, R. M. & Mosqueda-Melgar, J. (2012). Polysaccharides as carriers and protectors of additives and bioactive compounds in foods. In D. N. Karunaratn (Ed.). *The Complex World of Polysaccharides* (Chapter 16, pp. 429-454). Rijeka, Croacia: Editorial InTech.

[48] Min, B. J., Han, I. Y., Dawson, P. L. (2010). Antimicrobial gelatin films reduce Listeria monocytogenes on turkey bologna. *Poultry Science*. 89(6), 1307-1314.

[49] Marcos, B., Aymerich, T., Monfort, J., & Garriga, M. (2007). Use of antimicrobial biodegradable packing to control *Listeria monocytogenes* during storage of cooked ham. *International Journal of Food Microbiology*. 120(1-2), 152-158.

[50] Marcos, B., Aymerich, T., Monfort, J. M., & Garriga, M. (2008). High-pressure processing and antimicrobial biodegradable packaging to control *Listeria monocytogenes* during storage of cooked ham. *Food Microbiology*. 25, 177-182.

[51] Oussalah, M., Caillet, S., Salmieri, S., Saucier, L., Lacroix, M. (2007). Antimicrobial effects of alginate- based film containing essential oils o *Listeria monocytogenes* and *Salmonella thyphimurium* present in bologna and ham. *Journal of Food Protection*. 70(4), 901-908.

[52] Datta, S., Janes, M. E., Xue, Q. G., Losso, J., & La Peyre, J. F. (2008). Control of *Listeria monocytogenes* and *Salmonella anatum* on the Surface of Smoked Salmon Coated with Calcium Alginate Coating containing Oyster Lysozyme and Nisin. *Journal of Food Science*. 73(2), M67-M71.

[53] Ye, M., Neetoo, H., & Chen, H., (2008). Effectiveness of chitosan-coated plastic films incorporating antimicrobials in inhibition of Listeria monocytogenes on cold-smoked salmon. *International Journal of Food Microbiology*. 127, 235–240.

[54] Beverlya, R. L., Janes, M. E., Prinyawiwatkul, W., & No, H. K. (2008). Edible chitosan films on ready-to-eat roast beef for the control of *Listeria monocyotgenes*. *Food Microbiology*. 25, 534-537.

[55] Ye, M., Neetoo, H., & Chen, H., (2008). Control of Listeria monocytogenes on ham steaks by antimicrobials incorporated into chitosan-coated plastic films. *Food Microbiology*. 25, 260–268.

[56] Duan, J., Park, S., Daeschel, M., & Zhao, Y. (2007). Antimicrobial chitosan-lysozyme (CL) films and coatings for enhancing microbial safety of mozzarella cheese. *Journal of Food Science*. 72(9), M355-M362.

[57] Jiang, Z., Neetoo, H., & Chen, H. (2011). Control of *Listeria Monocytogenes* on Cold-Smoked Salmon Using Chitosan-Based Antimicrobial Coatings and Films. *Journal of Food Science*. 76(1), M22-M26.

[58] Janes, M. E., Kooshesh, S., & Johnson, M. G. (2002). Control of *Listeria monocytogenes* on the surface of refrigerated, ready-to-eat chicken coated with edible zein film coatings containing nisin and/or calcium propionate. *Journal of Food Science*. 67(7), 2754–2757.

[59] Lungu, B., & Johnson, M. G. (2005). Fate of *Listeria monocytogenes* inoculated onto the surface of model turkey frankfurter pieces treated with zein coatings containing nisin, sodium diacetate, and sodium lactate at 4°C. *Journal of Food Protection*. 68(4), 855–859.

[60] Gadang, V. P., Hettiarachchy, N. S., Johnson, M. G., & Owens, C. (2008). Evaluation of Antibacterial Activity of Whey Protein Isolate Coating Incorporated with Nisin, Grape Seed Extract, Malic Acid, and EDTA on a Turkey Frankfurter System. *Journal of Food Science*. 73(8), M389-M394.

[61] Franklin, N. B., Cooksey, K. D., & Getty, K. J. (2004). Inhibition of Listeria monocytogenes on the surface of individually packaged hot dogs with a packaging film coating containing nisin. *Journal of Food Protection*. 67, 480–485.

[62] Nguyen, T., Gidley, M., & Dykes, G. (2008). Potential of a nisin-containing bacterial cellulose film to inhibit *Listeria monocytogenes* on processed meats. *Food Microbiology.* 25(3), 471-478.

[63] Jiang, Z., Neetoo, H., & Chen, H. (2011). Efficacy of freezing, frozen storage and edible antimicrobial coatings used in combination for control of *Listeria monocytogenes* on roasted turkey stored at chiller temperatures. *Food Microbiology.* 28(7), 1394-1401.

[64] Zhu, M. J., Mendoca, A., Ismail, H. A., & Ahn, D. U. (2009). Fate of *Listeria monocytogenes* in ready-to-eat turkey breast rolls formulated with antimicrobials following electron-beam irradiation. *Poultry Science.* 88, 205-213.

[65] Clardy. S., Foley, D. M., Caporaso, F., Calicchia, M. L., & Prakash, A. (2002). Effect of gamma irradiation on Listeria monocytogenes in frozen, artificially contaminated sandwiches. *Journal of Food Protection.* 65, 1740–174.

[66] Chen, C. M., Sebranek, J. G., Dickson, J. S., & Mendonca, A. F. (2004). Combining pediocin with postpackaging irradiation for control of Listeria monocytogenes on frankfurters. *Journal of Food Protection.* 67(9), 1866-1875.

[67] Knight, T. D., Castillo, A., Maxim, J., Keeton, J. T., & Miller, R. K. (2007). Effectiveness of potassium lactate and sodium diacetate in combination with irradiation to control *Listeria monocytogenes* on frankfurters. *Journal of Food Science.* 72(1), M026-M030.

[68] Kudra, L. L., Sebranek, J. G., Dickson, J. S., Larson., Mendonca, A. F., Prusa, K. J., Cordray, J.C., Jackson-Davis, A., & Lu, Z. (2012). Control of *Listeria monocytogenes* on frankfurters and cooked pork chops by irradiation combined with modified atmosphere packaging. *Journal of Food Protection.* 75(6), 1063-1070.

[69] Koutchma, T. (2008). UV Light for Processing Foods. *IUVA News.* 10(4), 24-29.

[70] Sommers, C. H., Cooke, P. H., Fan, X., & Sites, J. E. (2009). Ultraviolet light (254 nm) inactivation of *Listeria monocytogenes* on frankfurters that contain potassium lactate and sodium diacetate. *Journal of food* 74(3), M114-M119.

[71] Chun, H. H., Kim, J. Y., Lee, B. D., Yu, D. J., & Song, K. B. (2010). Effect of UV-C irradiation on the inactivation of inoculated pathogen s and quality of chicken breasts during storage. *Food Control.* 21(3), 276-280.

[72] Chun, H. H., Kim, J. Y., Lee, B. D., Yu, D. J., & Song, K. B. (2010). Effect of UV-C irradiation on the inactivation of inoculated pathogen s and quality of chicken breasts during storage. *Food Control.* 21(3), 276-280.

[73] Elmnasser, N., Guillou, S., Leroi, F., Orange, A., Bakhrouf, A., & Federighi, M. (2007). Pulsed-light system as a novel food decontamination technology: Review, *Canadian Journal of Microbiology.* 53, 813–821.

[74] Luksiene, Z., Buchovec, I., & Viskelis, P. (2013). Impact of High-Power Pulsed Light on Microbial Contamination, Health Promoting Components and Shelf Life of Strawberries. *Food Technology & Biotechnology.* 51(2), 284-292.

[75] Takeshita, K., Shibato, J., Sameshima, T., Fukunaga, S., Isobe, S.,Arihara, K., & Itoh, M. (2003). Damage of yeast cells induced by pulsed light irradiation. International Journal of Food Microbiology. 85, 151–158.

[76] Wang, T., MacGregor, S. J., Anderson, J. G., & Woolsey, G. A. (2005). Pulsed ultra-violet inactivation spectrum of *Escherichia coli. Water Research.* 39, 2921–2925.

[77] Wekhof, A. (2000). Disinfection with flash lamp. *PDA Journal of Pharmaceutical Science and Technology*. 54, 264–276.

[78] Ozer, N. P., & Demirci, A. (2006). Inactivation of Escherichia coli O157:H7 and Listeria monocytogenes inoculated on raw salmon fillets by pulsed UV-light treatment. *International Journal of Food Science and Technology.* 41, 354-360.

[79] Cserhalmi, Z., Sass-Kiss, A., Tóth-Markus, M., & Lechner, N. (2006). Study of pulsed electric field treated citrus juices. *Innovative Food Science & Emerging Technologies*, 7, 49−54.

[80] Soliva-Fortuny, R., Balasa, A., Knorr, D., & Martín-Belloso, O. (2009). Effects of pulsed electric fields on bioactive compounds in foods: a review. *Trends in Food Science and Technology.* 20, 544-556.

[81] Mosqueda-Melgar, J., Elez-Martínez, P., Raybaudi-Massilia, R. M., & Martín-Belloso, O. (2008). Effects of pulsed electric fields on pathogenic microorganisms of major concern in fluid foods: a review. *Critical Review in Food Science and Nutrition.* 48, 747–759.

[82] IFT [Institute of Food Technologists]. Effect of preservation technologies and microbiological inactivation in foods. In *Comprehensive Reviews in Food Science and Food Safety*. Vol 2. 2001 (2013-10-12) Available from: http://www.ift.org.

[83] Wouters, P. C., Alvarez, I., & Raso, J. (2001). Critical factors determining inactivation kinetics by pulsed electric field food processing, *Trends in Food Science and Technology.* 12, 112–121.

[84] Barbosa-Cánovas, G. V., Góngora-Nieto, M. M., Pothakamury, U. R., & Swanson, B. G. (1999). *Preservation of foods with pulsed electric fields*, London: Academic Press.

[85] Tsong, T. Y. (1990). On electroporation of cell membranes and some related phenomena. *Bioelectrochemistry Bioenergetics.* 24, 271–295.

[86] Ho, S. Y., & Mittal, G. S. (1996). Electroporation of cell membranes: a review. *Critical Reviews in Biotechnology.* 16, 349–362.

[87] Mosqueda-Melgar, J., Raybaudi-Massilia, R. M., & Martín-Belloso, O. (2007). Influence of treatment time and pulse frequency on *Salmonella* Enteritidis, *Escherichia coli* and *Listeria monocytogenes* populations inoculated in melon and watermelon juices treated by pulsed electric fields. *International Journal of Food Microbiology.* 117, 192–200.

[88] Mosqueda-Melgar, J., Raybaudi-Massilia, R. M., & Martín-Belloso, O. (2008). Combination of high-intensity pulsed electric fields with natural antimicrobials to inactivate pathogenic microorganisms and extend the shelf-life of melon and watermelon juices. *Food Microbiology.* 25, 479–491.

[89] FDA [U.S. Food and Drug Administration]. Kinetics of Microbial Inactivation for Alternative Food Processing Technologies - High Pressure Processing. 2011 [2013-10-23]. Available from: http://www.fda.gov/Food/FoodScienceResearch/ SafePractices forFoodProcesses/ucm101456.htm

[90] Hugas, M., Garriga, M., & Monfort, J. M. (2002). New mild technologies in meat processing: High pressure as a model technology. *Meat Science*, 62, 359–371.

[91] Préstamo, G., Sanz, P. D., Fonberg-Broczek, M., & Arroyo, G. (1999). High pressure response of fruit jams contaminated with *Listeria monocytogenes. Letters in Applied Microbiology*. 28(4), 313-316.

[92] Lucore, L. A., Shellhammer, T. H., & Yousef, A. E. (2000). Inactivation of *Listeria monocytogenes* Scott A on artificially contaminated frankfurters by high-pressure processing. *Journal of Food Protection*. 63(5), 662-664.
[93] Lakshmanan, R. & Dalgaard, P. (2004). Effects of high-pressure processing on *Listeria monocytogenes*, spoilage microflora and multiple compound quality indices in chilled cold-smoked salmon. *Journal of Applied Microbiology*. 96, 398–408.
[94] Aymerich, T., Jofré, A., Garriga, M., & Hugas, M. (2005). Inhibition of Listeria monocytogenes and Salmonella by natural antimicrobials and high hydrostatic pressure in sliced cooked ham. *Journal of Food Protection*. 68(1), 173-177.
[95] Morales, P., Calzada, J., & Nuñez, M. (2006). Effect of high-pressure treatment on the survival of *Listeria monocytogenes* Scott A in sliced vacuum-packaged Iberian and Serrano cured hams. *Journal of Food Protection*. 69(10), 2539-2543.
[96] López-Piedemonte, T., Roig-Sagués, A., De Lamo, S., Hernández-Herrero, M., & Guamis, B. (2007). Reduction of counts of *Listeria monocytogenes* in cheese by means of high hydrostatic pressure. *Food Microbiology*. 24(1), 59-66.
[97] Jofré, A., Garriga, M., & Aymerich, T. (2007). Inhibition of *Salmonella* sp. *Listeria monocytogenes* and *Staphylococcus aureus* in cooked ham by combining antimicrobials, high hydrostatic pressure and refrigeration. *Meat Science*. 78, 53–59.
[98] Koseki, S., Mizuno, Y., & Yamamoto, K. (2008). Use of mild-heat treatment following high-pressure processing to prevent recovery of pressure-injured *Listeria monocytogenes* in milk. *Food Microbiology*. 25(2), 288-293.
[99] Jofré, A., Aymerich, T., & Garriga, M. (2008). Improvement of the food safety of low acid fermented sausages by enterocins A and B and high pressure. *Food Control*. 19, 634–638.
[100] Ananou, S., Garriga, M., Jofré, A., Aymerich, T., Gálvez, A., Maqueda, M., Martínez-Bueno, M., & Valdivia, E. (2010). Combined effect of enterocin AS-48 and high hydrostatic pressure to control food-borne pathogens inoculated in low acid fermented sausages. *Meat Science*. 84, 594 - 600.
[101] Bover-Cid, S., Belletti, N., Garriga, M., & Aymerich, T. (2011). Model for Listeria monocytogenes inactivation on dry-cured ham by high hydrostatic pressure processing. *Food Microbiology*. 28(4), 804-809.
[102] Hereu, A., Bover-Cid, S., Garriga, M., & Aymerich T. 2012. High hydrostatic pressure and biopreservation of dry-cured ham to meet the Food Safety Objectives for Listeria monocytogenes. *International Journal of Food Microbiology*. 154(3), 107-112.
[103] Stollewerk, K., Jofré, A., Comaposada, J., Arnau, J., & Garriga, M. (2012). The effect of NaCl- free processing and high pressure on the fate of *Listeria monocytogenes* and Salmonella on sliced smoked dry-cured ham. *Meat Science*. 90(2), 472-477.
[104] Myers, K., Montoya, D., Cannon, J., Dickson, J., & Sebranek, J. (2013). The effect of high hydrostatic pressure, sodium nitrite and salt concentration on the growth of *Listeria monocytogenes* on RTE ham and turkey. *Meat Science*. 93(2), 263-268.
[105] Farber, J. N., Harris, L. J., Parish, M. E., Beuchat, L. R., Suslow, T. V., Gorney, J. R., Garrett, E. H., & Busta, F. F. (2003). Microbiological Safety of Controlled and Modified Atmosphere Packaging of Fresh and Fresh-Cut Produce. *Comprehensive Reviews in Food Science and Food Safety*. 2, 142 – 160.

[106] Berrang, M. E., Brackett, R. E., & Beuchat, L. R. (1989). Growth of Listeria monocytogenes on fresh vegetables stored under controlled atmosphere. *Journal of Food Protection*. 52(10), 702-705.

[107] Beuchat, L. R., & Brackett, R. E. (1990). Survival and growth of Listeria monocytogenes on lettuce as influenced by shredding, chlorine treatment, modified atmosphere packaging and temperature. *Journal of Food Science*. 55(3), 755-758, 870.

[108] Farber, J. M., Wang, S. L., Cai, Y., & Zhang, S. (1998). Changes in populations of *Listeria monocytogenes* inoculated on packaged fresh-cut vegetables. *Journal of Food Protection*. 61(2), 192-195.

[109] Jacxsens, L., Devlieghere, F., Falcato, P., & DeBevere, J. (1999). Behavior of *Listeria monocytogenes* and *Aeromonas* spp. on fresh-cut produce packaged under equilibrium modified atmosphere. *Journal of Food Protection*. 62(10), 1128-1135.

[110] Carlin, F., & Nguyen-The, C. (1994). Fate of Listeria monocytogenes on four types of minimally processed green salads. *Letter in Applied Microbiology*. 18, 222-226.

[111] Francis, G. A., & O'Beirne, D. (1998). Effects of storage atmosphere on *Listeria monocytogenes* and competing microflora using a surface model system. *International Journal of Food Science & Technology*. 33, 465-476.

[112] Conway, W. S., Janisiewicz, W. J., Watada, A. E., & Sams, C. E. (1998). Survival and growth of Listeria monocytogenes on fresh-cut apple slices. *Phytopathology*. 88, S18.

[113] Michaelsen, A. R., Sebranek, J. G., Dickson, J. S. (2006). Effects of microbial inhibitors and modified atmosphere packaging on growth of Listeria monocytogenes and Salmonella enterica Typhimurium and on quality attributes of injected pork chops and sliced cured ham. *Journal of Food Protection*. 69(11), 2671-2680.

In: *Listeria monocytogenes*
Editor: Edmund C. Hambrick

ISBN: 978-1-63117-054-6

Chapter 12

THE EFFECT OF SOIL ABIOTIC AND BIOTIC FACTORS ON THE PRESERVATION AND REPRODUCTION OF *LISTERIA MONOCYTOGENES*

Marina L. Sidorenko[1,2] ***and Lyubov S. Buzoleva***[2,3]
[1]Institute of Biology and Soil Science FEB RAS, Vladivostok, Russia
[2]Research Institute of Epidemiology and Microbiology SB RAMN, Vladivostok, Russia
[3]Far Eastern Federal University, Vladivostok, Russia

ABSTRACT

Soil is one of the ecological niches of *Listeria monocytogenes*, where these bacteria are equal members of the microbial community, although the mechanisms of their adaptation and forms of existence in soil ecosystems are studied insufficiently. Up to now, scientists considered the influence of soil on the reproduction of *L. monocytogenes* without regard for the soil genesis and soil chemical and physical properties. It is obvious that the type of soil and different soil properties play a certain part in the reproduction of pathogenic microorganisms, because soils differ sharply in their humus content, concentrations of nutrients, pH, particle-size composition, and microbial associations.

In this relation, the physical and chemical properties of brown forest, brown podzolic, tidal marsh, maritime meadow and urban soils that affect the reproduction of *L. monocytogenes* were studied. The results of the experiments suggest that a high content of exchangeable bases, domination of the fraction of humic acids bound with clay minerals and the fraction of fulvic acids in humus, and high soil temperature stimulate the reproduction of the studied bacteria in soils. High humus content, a predominance of humates, and a low soil temperature have an inhibiting effect on bacteria.

Influence of mineral fertilizers on the reproduction of *L. monocytogenes* in soil ecosystems under different temperatures (4-6^{0}C and 20-22^{0}C) were studied. Application of the fertilizers favored preservation and active reproduction of *L. monocytogenes* in brown podzolic and brown forest soils. Duration of application of the mineral fertilizers positively correlated with preservation and reproduction of *L. monocytogenes* in different types of soil.

The biological activity of volatile metabolites of volatile metabolites of saprophytic soil bacteria and germinating seeds of plant against *L. monocytogenes* was studied. It was shown that volatile metabolites are transfer factors and can be the sole carbon and energy source for *L. monocytogenes*. Different character of inter specific relationships between bacteria, influencing their propagation, can be observed on the metabolic level. Volatile compounds produced by microorganisms are capable to act as both intra- and interspecific regulators of microbial communities. In this connection the propagation of *L. monocytogenes* inhabiting soil may be stimulated or inhibited by the metabolic products of soil microorganisms. Methanol is the main substance affecting their growth and reproduction.

1. *LISTERIA MONOCYTOGENES* IN THE SOILS

The existence of pathogenic bacteria in the environment as saprophytes is doubtless. They, long conserved in soil and water, are of especial interest. The vital program of such microorganisms consists in continuous transition from an environment into an organism of the warm-blooded animals and man, where they live as parasites, and returning in an environment, where they function as saprophytic (Somov, 1979; Litvin et al., 2010). Such microorganisms include *Listeria monocytogenes*. They can live in various, apparently, contrast conditions and master different under the characteristics ecological niches. One of environments of such bacteria is the soil. The question on possibility existence of pathogenic *Listeria* in various soils was studied by many researchers. V. Ya. Golovachova (1978) and other authors approved, that *L. monocytogenes* are capable of the long conservation, within 6-12 months, and multiplication in soils. The data on wide occurrence *L. monocytogenes* in soils are given by V. A. Znamenskii (1975), M. Shayegani et al., (1981) and others.

The high degree of *Listeria* multiplication in soil is proved: for instance, for 6 days at 18-20^{0}C their concentration increased 3000 times (Pomanskaya, 1963). At low positive temperatures and sufficient humidity there is an accumulation of Listeria in soil, by that the seasonal fluctuations of bacteria amount have been observed. The period of bacteria multiplication is in spring and autumn seasons. In winter, their numbers are stabilizing. In summer characterized by high temperatures and aridity, the bacteria amount decreases in soil (Gershun, 1980a). The researches, carried out in Germany, proved a role of soil as environment of *L. monocytogenes*. During these researches 154 strains have been isolated from the identical number of samples taken from the soil and only 16 strains from the excrements of animals (Weiss, Seeliger, 1975). A. Frouin (1989), H. J. Welshimer (1981) and others inform about wide *Listeria* occurrence in soil. Basing on many experiments they have proved long occurrence and multiplication of *L. monocytogenes* in soil, and as well as V. I. Gershun (1981), consider soil to be a source of *Listeria* in nature.

The bacterium well multiplies in cultivated agricultural soils and it does not multiply in sandy, brackish and sour soils with poor content of organic substances (Somov, Litvin, 1988). It is established, that the pathogenic bacterium multiplies perfectly on bentonite clay soil. The organic-mineral clay-humus soil complexes are the basic factors to activate multiplication of these bacteria on bentonite clay by L. S. Buzoleva's data (1990), they play a role to provide organic and mineral base feeding for bacteria.

The analysis of the literary data shows, that the researchers, as a rule, have considered existence of pathogenic microorganisms in soil without taking into account soil chemical and physical-chemical characteristics. In the literature there are no data on multiplication of pathogenic bacteria in various types of soils. The soil is multicomponent system, characterized by different temperatures, acidity, humidity, various humus structure and variable microbocenoses. Therefore microorganisms, living in soil, are much affected by all above mentioned factors. Hence, it is important to consider with their influence upon increase and multiplication of *L. monocytogenes*.

In the soil, there is a wide variety of micro-zone, including quite favorable on the parameters for microbial life. In addition, the pathogenic bacteria getting into the soil, there are unfavorable for them to physical and chemical conditions and the antagonistic effect of various soil biota. A wide variety and combination of components constituting soil observed even within the same limited area. This, apparently, explains the significant differences on the question of survival of pathogenic microbes in the soil (Skvortsov et al., 1966).

Pathogenic bacteria living in the environment mostly fall into the conditions in respect of limiting nutrient, both organic and mineral. Therefore, you should comprehensively study the abiotic characteristics of the different soil types in terms of their effects on pathogenic bacteria.

Soil is one of the ecological niches of *L. monocytogenes*, where these bacteria are equal members of the microbial community (Gershun, 1988a; Litvin, Pushkareva, 1994; Somov, Litvin, 1988; Shaegani et al., 1981), although the mechanisms of their adaptation and forms of existence in soil ecosystems are studied insufficiently. Up to now, scientists considered the influence of soil on the reproduction of *L. monocytogenes* without regard for the soil genesis and soil chemical and physical properties. It is obvious that the type of soil and different soil properties play a certain part in the reproduction of pathogenic microorganisms, because soils differ sharply in their humus content, concentrations of nutrients, pH, particle-size composition, and microbial associations.

In this relation, it is necessary to study the physical and chemical properties of particular soil types affecting the preservation and reproduction of *L. monocytogenes* in these soils. Investigation of this problem will give a further elucidating glimpse into the regularities of the distribution and reproduction of these bacteria in soils of different genesis.

The samples of brown forest (forest area) and brown podzolic (arable land, cabbage field) soils distributed mostly in the Sikhote-Alin foothills and in the Khanka–Razdol'naya Plain, and of urban soils (the lawns in the city of Vladivostok) taken from the upper (0–10 cm) horizon were used as the study objects. The choice of these soils was not accidental: the data of Gershun (Gershun, 1980b; 1988b) suggest that most isolation of *L. monocytogenes* fell on forest soils; arable soils used for growing cabbage and urban soils taking for comparison.

In the experiments, we used the reference strains of *L. monocytogenes* (P, A, K, 10CN, 4B, and 1/2A), which were received from the All-Russia State Control Institute of Veterinary Preparations (Moscow, Russia).

The following parameters were determined in the soil samples: the humus by Tyurin, the group composition of humus by Tyurin in modification by Ponomareva and Plotnikova, the actual acidity, the exchangeable acidity, the hydrolytic acidity, the exchangeable bases (Ca^{2+} and Mg^{2+}), the cation exchange capacity, and the base saturation (Arinushkina, 1971; Orlov, Grishina, 1991). The preparations of humic acids were obtained according to the commonly adopted method (Laboratory Manual…, 1980).

To study the growth the kinetics of *L. monocytogenes* in different media (soil samples, humin preparations, extracts of humic acids) t the temperatures of 20-22°C and 4-6°C, bacterial cultures washed out three times in physiological solution were used: the bacterial cultures were introduced o the tested media at the rate of 100 cells per 0.1 ml of the medium.

The number of bacterial cells per 1 ml of suspension was determined according to the optical standard of turbidity. The study was carried out under periodical cultivation. To build the growth curve, the dynamics of the bacteria number was determined by the way of periodic inoculations with 0.1 ml of inoculate (from a particular serial dilution) into differentiation-diagnostic media (yeast agar). The stages of the growth curves and the maximum concentrations of bacteria were determined by way of counting the number of colonies (colony-forming units, CFU) grown on Petri dishes (Labinskaya, 1978).

To carry out an experiment on determination of the number of viable cells of *L. monocytogenes* in the soil ecosystems, suspensions of soil samples were prepared that contained 1 g of soil per 100 ml of physiological solution. The suspensions were subject to steaming for sterilization 20 min a day for 3 days under the temperature of 100°C with flowing steam, then they were inoculated with daily cultures of *Listeria* in the concentration 10^3 CFU/ml (according to the turbidity standard). The initial dose in the control inoculation from the soil suspension comprised 100 CFU/ml. In all, five variants of the experiment were studied in triplicate under temperatures of 20–22°C and 4–6°C.

The reproduction of bacteria on the humic preparations from the soils was studied in the following way. Special media composed of distilled water and 0.02% humic preparation (Buzoleva, Sidorenko, 2001) were prepared preliminary and sterilized with flowing steam 20 min a day for 3 days fewer than 100°C. The prepared sterile media in the tubes were inoculated with daily cultures of *L. monocytogenes* in the amount 10^3CFU/0.1 ml. The cultures were incubated under the temperatures 4–6°C and 20–22°C. Solid nutrient media were inoculated during 15 days with daily intervals. In all, 120 variants of the experiment were studied. The study was carried out in triplicate.

The following method was used to reveal the effect of soil humus composition on the reproduction of *L. monocytogenes*. Daily cultures were introduced to the tubes with media. The media were prepared with distilled water by adding the particular extracts of humic acid or fulvic acid fractions (depending on the variant of the experiment) in concentration 0.02% (Buzoleva, Sidorenko, 2001). The readymade media were sterilized with flowing steam 20 min a day for 3 days under 100°C. Then, they were inoculated in tubes with daily cultures of *L. monocytogenes* in the amount 103CFU/0.1 ml. Then, Petri dishes with proper media were inoculated periodically from these tubes with 0.1 ml of suspension, and the number of colonies per dish was counted. In all, 420 variants of the experiment were carried out under the temperature 20–22°C and 420 variants of the experiment were carried out under the temperature 4–6°C. The study was carried out in triplicate.

L. monocytogenes were cultivated in the following media: nutrition agar (pancreatic hydrolyzate of sprat 17.9 g, agar 11.2 g, sodium chloride 7.7 g, distilled water up to 1 l, pH 7.3±0.2); yeast agar with glucose, nalidixic acid, and trypaflavine (sodium chloride 5.5 g, hydrolyzate of fodder yeast 12 g, agar 12.5 g, distilled water up to 1 l, glucose 2 g, nalidixic acid 20 mg, trypaflavine 30 mg, pH 7.2–7.4).

The dynamics of *L. monocytogenes* reproduction in different soil types was studied at the first stage of research. It was found that the preservation and reproduction of *L. monocytogenes* in soils depended on the type of soil as well as on the biological properties of the bacteria. The substrates of urban and brown podzolic soils had positive effects on the reproduction of most strains (A, P, 10CN, 4B, and 1/2A) of *L. monocytogenes*. The reproduction of the *L. monocytogenes* reached the value of 4.5 lg in urban soil and 4.2 lg in brown podzolic soil, whereas the death of bacteria was observed in brown forest soil already on the third day. The strain K of *L. monocytogenes* was an exception (Figure 1.1); it reproduced better in brown forest (5.4 lg) and urban (4.1 lg) soils as compared with the brown podzolic soil (2 lg).

It should be noted that *L. monocytogenes* reproduced actively under the temperature 20–22°C, and slow growth of a bacterial culture was observed under 4–6°C.

It was revealed that soil acidity being an abiotic environmental factor effected the reproduction of *L. monocytogenes* in the studied soils. The pH range 6.6–7.4 (the acidity of urban and brown podzolic soils is usually in this range) appeared to be optimal for the growth of these bacteria, and this fact is in agreement with the data presented in the works of V. I. Gershun (1988b), and I. A. Maksimenkova and V. Yu. Litvin (1985), because *L. monocytogenes* have growth optimums in this range of pH. For example, V. I. Gershun (1988b) experimentally determined that *Listeria* reproduce well under pH 6.7. The departure of the pH value from these values is destructive for the bacteria. Hence, *L. monocytogenes* reproduce better in soils with the acidity close to neutral.

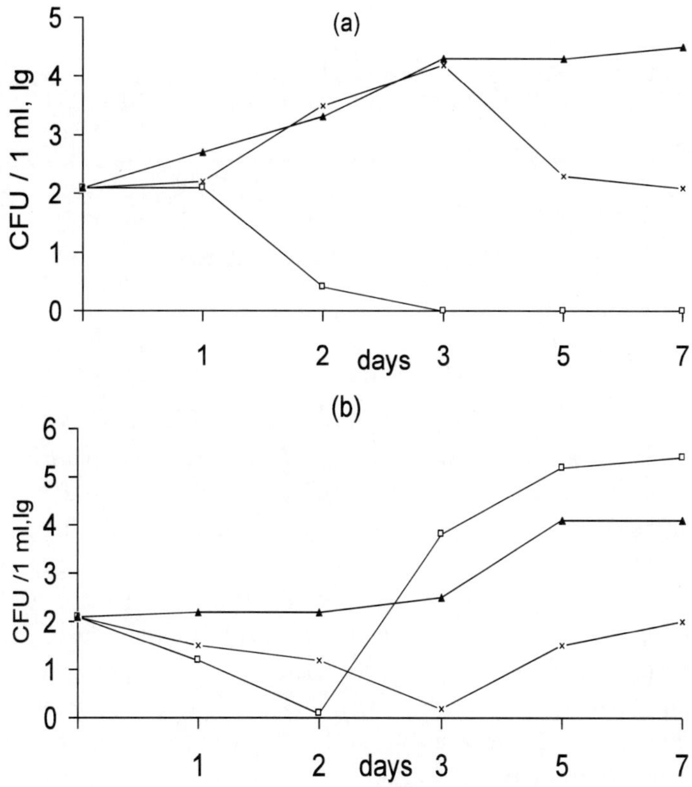

Figure 1.1. Dynamics of reproduction of *L. monocytogenes* strains P (a) and K (b) in different soils under 20–22°C: (-x-) brown podzolic soil, (-□-) brown forest soil, (-3-) urban soil.

The acidity of brown forest soil (pH 5.8) was the least suitable for these bacteria, so most strains practically did not reproduce in this soil. Additionally, the data of Table 1.1 show that the brown forest soil had higher values of the exchange and hydrolytic acidity than the urban and brown podzolic soils and this had an effect on the base saturation. The brown forest soil not only had an extremely low pH for *L. monocytogenes* reproduction but also the least value of the exchangeable bases (37.5 mg-equiv/100 g).

It was found out that the cation exchange capacity (CEC) of the brown podzolic soil exceeded those of the other studied soils. This soil contained greater amounts of calcium and magnesium ions, which are vitally important for normal growth and reproduction of *Listeria* (Rabotnova, Pozmogova, 1993). Consequently, brown podzolic soil is favorable for reproduction of *L. monocytogenes*, because it has optimal values of the pH and base saturation and contains sufficient amounts of calcium and magnesium ions.

Since the physicochemical and chemical properties did not allow explaining the reproduction of pathogenic bacteria in urban soil, its humus composition was studied. The content of total carbon reached 4.5% in the urban soil, in the brown podzolic soil it was 2.6%, and in the brown forest soil it was 1.1% (Table 1.2). It could be assumed that the composition of the organic substances in the soil has an effect on the *L. monocytogenes* in the soil ecosystems.

Table 1.1. Physicochemical and chemical properties of soils

Soil	pH		Exchange acidity	Hydrolytic acidity	Cation exchange capacity	Exchangeable bases		Base saturation, %
	H_2O	KCl				Ca^{2+}	Mg^{2+}	
			mg-equiv/ 100g of soil					
Brown forest	5,85	5,00	3,25	7,20	44,70	30,3	7,5	83,9
Brown podzolic	7,36	6,36	1,15	1,74	80,74	65,1	14,2	97,8
Urban	6,60	5,58	1,53	2,88	45,88	35,2	8,1	93,7

Urban soil had the most optimal content of humic and fulvic acids (Table 1.2). For example, L. S. Buzoleva and M. L. Sidorenko (2001) demonstrated that *Enterobacter* reached maximum reproduction in the solutions with a low concentration (0.02%) of humic acids; higher concentrations had an inhibiting effect on the bacteria reproduction. It should be noted that the sample of the urban soil contained half as much of the first aggressive fraction of fulvic acids as the other two soils. It is obvious that this fact had a positive effect on the growth and reproduction of the pathogenic bacteria in this environment, because the fulvic acids of this fraction can inhibit the growth of bacteria. This could be probably explained by the fact that the first aggressive fraction of fulvic acids is connected with mobile sesquioxides, which adversely influence the surface structures of bacteria and make it difficult for the exchange processes in the latter.

All soils had humus of fulvate-humate type (C_{ha}/C_{fa}= 1–2). However, the humus in the brown forest soil contained a greater amount of fulvic acids (C_{ha}/C_{fa} = 1) than the brown podzolic soil (C_{ha}/C_{fa} =1.7) or the urban soil (C_{ha}/C_{fa} = 1.4), and this might be unfavorable for the reproduction of *L. monocytogenes*. The total amount of fulvic acids in the brown forest soil was 1.4 times higher, and that of the second fraction of fulvic acids was 8 times higher than in the brown podzolic soil.

Table 1.2. Group composition of humus (percentage of total soil carbon)

Soil	C, %	Humic acids				Fulvic acids					$C_{ha} + C_{fa}$	C_{ha} / C_{fa}	Humin
		1	2	3	Σ	1a	1	2	3	Σ			
	1,1	12,73	13,63	15,54	41,90	5,45	1,82	24,54	8,90	40,71	82,61	1,1	17,39
Brown forest	2,6	6,11	25,95	16,84	48,90	5,72	5,35	3,05	14,84	28,96	77,86	1,7	22,14
Brown podzolic	4,5	3,10	9,73	16,37	29,20	2,21	2,29	13,50	2,88	20,88	50,08	1,4	49,92
Urban													

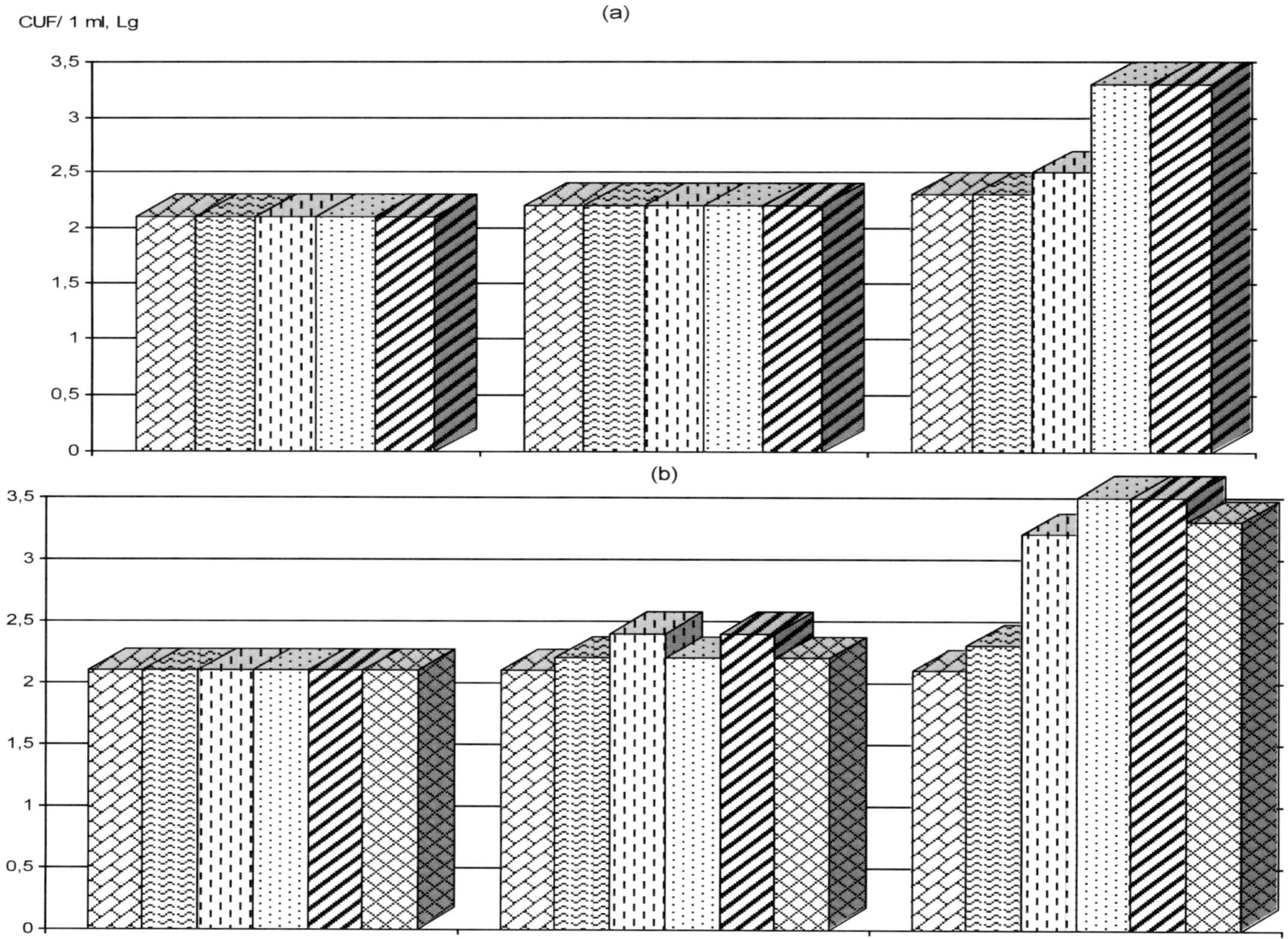
CUF/ 1 ml, Lg
(a)
3,5
3
2,5
2
1,5
1
0,5
0
(b)
3,5
3
2,5
2
1,5
1
0,5
0

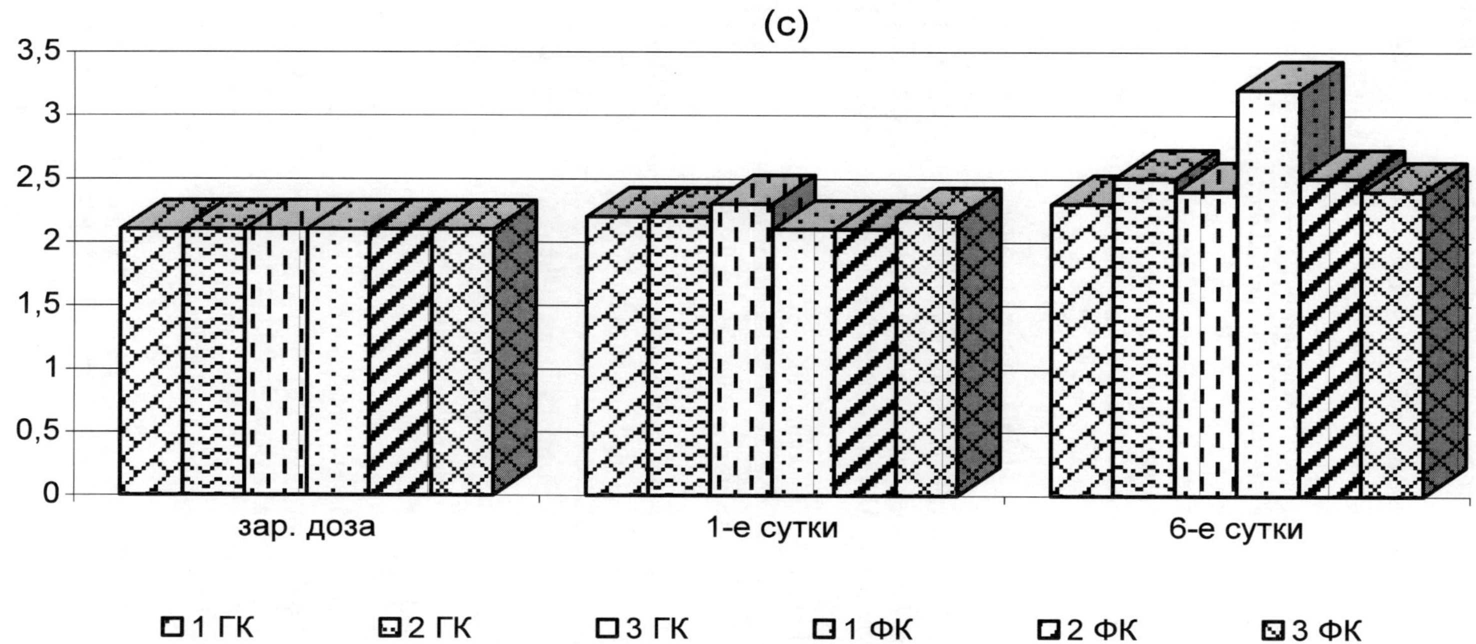

Figure 1.2. Growth of *L. monocytogenes* strain P on different fractions of humic acids under 4–6°C: (a) brown podzolic soil, (b) brown forest soil, and (c) urban soil.

There was no significant difference between the total amounts of humic acids in the brown forest and rown podzolic soils. However, it is known that the humic acids of brown forest soil are weakly consumed n comparison with those of brown podzolic soil (Buzoleva, Sidorenko, 2001). This might be the reason for the better reproduction of *L. monocytogenes* in the brown podzolic soil. Moreover, the brown podzolic soil was used for a long time in agriculture and was treated mechanically and chemically (with ameliorants). Therefore, the humic acids in this soil can be more easily available for bacteria, and the soil absorbing complex contains greater amounts of exchangeable bases, especially of calcium and magnesium ions. The preferable development of the studied bacteria in urban soil is explained by the composition of the humic acids more than by their amount. Humic acids are known to be characterized by heterogeneity and polydispersity (Orlov, 1990).

Hence, the preservation and reproduction of *L. monocytogenes* in soils of different types are affected by the organic substances of the soils and the soil physicochemical properties as well as the biological properties of studied strains. The *L. monocytogenes* reproduce the best in urban and brown podzolic soils, and this is explained by the optimal pH values in these soils and the lower concentrations of fulvic acids in comparison with brown forest soil. Additionally, brown podzolic soil has great amounts of exchangeable calcium and magnesium ions in its soil absorbing complex. Because the study of the composition of the humic and fulvic acids did not explain the preference of pathogenic bacteria for urban soil, the effects of different fractions of the humic and fulvic acids on the reproduction of *L. monocytogenes* should be studied separately.

To study the effect of the humus composition of the soil on the reproduction of different strains of *L. monocytogenes*, an experiment was carried out for which acid and alkaline extracts of humic and fulvic acids were obtained from the studied soils. The results of the experiment suggest that humic and fulvic acids possess biological activity toward the studied bacteria. It was determined that all the strains of *L. monocytogenes* used in the experiment reproduced 0.5–1.5 lg better under 20–22°C than under 4–6°C. For example, the reproduction of *L. monocytogenes* 4.5 lg under 20–22°C, and 4.5 and 3.5 lg, respectively, under 4–6°C. It was found that particular extracts of the fractions of humic and fulvic acids had different effects on the reproduction of strains of *L. monocytogenes*. These effects depended on the bacterium strain characteristics and the temperature of cultivation. For example, the strain P of L. monocytogenes reproduced equally well in all the variants of the experiment under both temperatures of cultivation, and more intense growth was observed on the extracts of the fulvic acid fractions than on the extracts of the humic acid fractions (Figure 1.2).

It is difficult to say unambiguously which fraction extracts had a stimulating effect and which had an inhibiting effect on the development of the studied bacteria. Because similar extracts of fractions of humic and fulvic acids from different soils differ in composition, the results were ambiguous. In some cases, *L. monocytogenes* grew and reproduced better in the media with fulvic acids than in the media without them; this phenomenon can be explained by the fact that fulvic acids in comparison with humic acids contain less benzenepolycarboxylic acids in their nucleus and a greater number of chains with polypeptide bonds in the peripheral part of the molecule; i.e., they are more hydrophilic, hence, more available for bacteria. Fulvic acids have a lower molecular weight than humic acids, contain less carbon and nitrogen, and the nitrogen of fulvic acids is mostly hydrolyzable; i.e., it is confined mostly to the aliphatic part of the molecule and is, therefore, more available for microbes. For example,

D. S. Orlov (1990) noted that fulvic acids are one of most available for the microbe groups of soil humus and because of this are quickly utilized by microorganisms.

The extract of the first aggressive fraction of fulvic acids engaged our attention. The data of Tables 1 and 2 suggest a similarity in composition of the soil absorbing complex in the brown forest and urban soils, and the latter soil contained half as much of the first aggressive fraction of fulvic acids as the brown forest soil. This fact was probably the cause of the inhibition of growth and reproduction of different strains of *L. monocytogenes* in the brown forest soil.

The pH values in all extracts of the fractions of humic and fulvic acids were in the range 6–8. Only the media with the extract of the first aggressive fraction of the fulvic acids, where the pH was 2.5–2.8, were the exception. It was noted already that the optimal values for the development of *L. monocytogenes* are in the range 6–8. Because of this, the acidity did not influence the preservation and reproduction of *L. monocytogenes* in the variants of the experiment without the extract of the first aggressive fraction of fulvic acids. The adverse effect of the first aggressive fraction of fulvic acids in this experiment is explained by the low pH value unfavorable for the reproduction of *L. monocytogenes*.

The dynamics of the *L. monocytogenes* reproduction was studied in the preparations of humic acids in the form of ammonium humate. The results of the experiment suggest that *L. monocytogenes* reproduced practically equally on the humates from the brown forest soil and the brown podzolic soil (Figure 1.3). All strains of *L. monocytogenes* reached maximum reproduction on the humates of brown podzolic soil (5.4 lg), and these values are comparable with those obtained for the reproduction of these bacteria on laboratory nutrient media.

Hence, it resulted from the experiments that the total humus content, its fraction composition, its acid–alkaline conditions, and the temperature regime of the soil environment, as well as the strain characteristics of the bacteria, influenced the existence of *L. monocytogenes* in the soils.

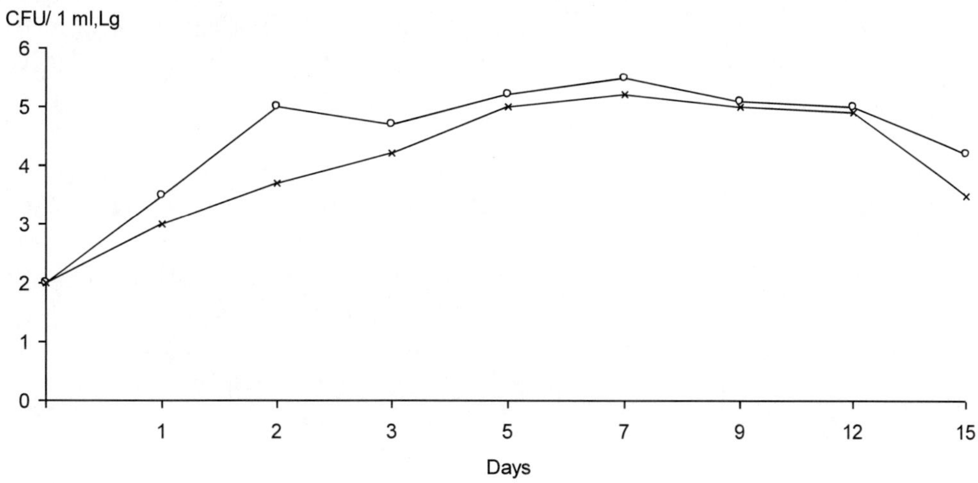

Figure 1.3. The curves of bacterial growth on humus preparations from soil *L. monocytogenes* strain 10 CN: (-o-) ammonium humates of brown podzolic soil, (-x-) ammonium humates of brown forest soil.

Conclusion

(1) It was determined that urban soils with a low content of humic acids had the optimal conditions for reproduction of *L. monocytogenes*. The reproduction of these bacteria in brown podzolic soil was determined by the presence of increased amounts of calcium and magnesium in the soil adsorption complex necessary for normal reproduction of microorganisms.

(2) The temperature affects the reproduction of *L. monocytogenes* in soil. These bacteria reproduce more intensely under the temperature of 20–22°C, whereas the intensity of their reproduction decreases under 4–6°C.

(3) Humic and fulvic acids are biologically active with respect to the reproduction of *L. monocytogenes* in all the soils studied. Particular extracts of the fractions of humic and fulvic acids from every soil type had different effects on the reproduction of these bacteria.

(4) Reproduction of *L. monocytogenes* on the fractions of the humic acids depended on the strain of the studied bacterium and the environmental temperature. All the *Listeria* strains reproduced more intensely on the fractions of fulvic acids. Fraction 3 of the humic acids was most active as relative to the growth of the studied bacteria irrespectively of the soil type.

2. *Listeria monocytogenes* in the Soils of Sea Coasts

According to generally accepted statistics, infected by pathogenic bacteria and armored fish marine creatures cause 10.5% of all food borne disease (Liston, 1990), which exceeds the proportion of meat and poultry in countries (Japan, India, Cuba, Iceland, Bangladesh, England) where the seafood in the diet of people make up 80% (Shahamat, Seaman, 1980). According to F. E. Ahmed (1992), the real incidence is much higher, since 55% of infections caused by contact with the marine environment or the use of marine organisms in raw form, are of unknown etiology.

The reservoir role of the marine environment and its inhabitants against pathogenic bacteria and viruses is recognized by many authors (Tulchinskaya, Gubanov, 1979; Ahmed, 1992; Olafsen et al., 1993). Currently, there are a significant number of reports on the detection of *Listeria* in coastal waters and sediments (Colburn et al., 1990; Ben Embarek, 1994; Papa et al., 1996; Farber, Peterkin, 1991; Lone, 2001). Inoculation of *Listeria monocytogenes* from seawater according to G. Lone (2001) is 33%. There's also a work covering issues of survival of *L. monocytogenes* in sea water (Grimes, 1991; Beleneva, 1996). Found that strains of *L. monocytogenes*, adapted to low positive temperatures, multiply in sterile sea (Terekhova et al., 2002) and river (Sidorenko et al., 2003) water at low positive temperatures and stored for a long time. In this *Listeria* remains viable in non-sterile sea water - up to 2 months in the sea ice - up to 3 months.

L. monocytogenes often isolated from a wide variety of seafood. For example, in Iceland, in assessing microbial contamination of seafood found in 56% of samples of raw fish and in 9% of the samples of shrimp (Hartemink, Georgsson, 1991). In Japan, according to A.

Nakama et al. (1992) and Y. Kokubo (1991), 1.3-1.8% of fish products contaminated with *L. monocytogenes.* In Japan, according to A. Nakama et al., (1992) and Y. Kokubo (1991), 1.3-1.8% of fish products contaminated with *L. monocytogenes*. In Russia in the analysis of fish products in the distribution network of St. Petersburg 39% of the samples were contaminated *L. monocytogenes*. (Dmitrieva et al., 2001). I. A. Beleneva and E. F. Maslennikov (2002) report on the allocation of *L. monocytogenes* from the organs of Japanese-marine bivalve molluscs.

According to some researchers *L. monocytogenes* are not autochthonous marine life and the marine environment infection caused by these bacteria removal of *L. monocytogenes* liquid and solid river flow (Watkins, Sleath, 1981; Schwartzbrod et al., 1989; Grimes, 1991).

Thus, it can be noted that the infectious agents, including *L. monocytogenes*, can long exist and multiply in seawater. But the researchers did not analyze growth of pathogenic bacteria in soils of the coast, which in its properties and formation conditions are quite different from a number of automorphic soils. There are anecdotal evidence pointing to the possibility of *Listeria* in arable soils, but an indication of the bacteria from the soil the coast is not known.

The economic and ecological significance of the sea coasts is widely recognized. The first one is connected with relatively large density of population in the seaside areas caused by the appropriate conditions to live in there. It should be noted, that about 1/3 population of the Earth is concentrated on a coastal strip of the World Ocean, 50 km in width. The biocenoses of the sea coasts are under high anthropogenic (man-made) effects. For instance, various coastal and maritime ecosystems of the Mediterranean countries are most subjected to influence of the destroying factors of environment, including tourist visiting, melioration and agriculture development and pollution by pesticides (Shlyakhov, 1996). Therefore, infection of human and animal infections and agents may be present in the soil microbocenosis coast.

In this connection there was a need to study the abiotic and biotic factors of soil seaside affecting the growth and proliferation of *L. monocytogenes* in them. This will reveal the patterns of distribution of *L. monocytogenes* in soils of different origin and environmental conditions of formation and predict their growth and reproduction in soil ecosystems coast.

Soil plains coastlines are formed mainly by accumulation (created by the activities of the Sea) forms of relief that are stacked by coastal sea or altered clastic sediments. The sea has a strong influence on the hydrological regime of the soils of coastal plains. The relatively narrow band of the adjacent soil tests also significant geochemical effects, which is a number of ways, and not in the inland areas. These soils are often, though not always, predominant in the saline sea water readily soluble salts. The chemistry of the salinity of the chloride to sulfate, the distribution of salts in the profile is very different. Among the clay minerals, the most common kaolinite and quartz is present (Shlyakhov, Kostenkov, 2000).

The structure and morphology of coastal plain soils are very diverse. Therefore, as a general can only transfer such common features as a full or partial gleying profile, which leads to the appearance of cold tones of the soil mass, ocher, rust stains and layers, lack of structure, bedding, unrelated to the processes of soil formation, traces of polycyclic pedogenesis, the presence of biogenic inclusions in the whole profile (clamshells, unmetabolized residues), the absence or weak expression of genetic horizons of mineral often with good expression of biogenic (Shlyakhov, 1997).

Microorganisms living in the soil of sea coast, largely determine the chemistry of their environment. They are responsible for processes such as sulfate reduction, oxidation of sulfur and sulphides, etc. and play methanogenesis role in the genesis of marine fauna of soil is significantly higher than most continental soils. Its feature is that it includes both coastal marine and coastal terrestrial species, which are interconnected food chains.

All spectrums of soils of sea coast on S. A. Shiyakhov (1996) refer to as "Thalassosols". Thalassosols is a group of types of soils, including underdeveloped of ground on fresh sea adjournment. The thalassosols can be determined as not of zone soil of a rather narrow flat coastal strip with grassy vegetation testing modern versatile influence of the sea, in result, which at them the specific features of morphology and properties are formed. The special place in group of thalassosols occupy of mangrove soil which is taking place behind frameworks of the present work. Other thalassosols on the moisture regimes can be broken on three parts at a level of subgroups of types: mid-flight, maritime marsh and maritime meadow. In each of subgroups till two types: in mid-flight - organogenic and mid-flight actually; in maritime marsh - organogenic and meadow-marsh; in maritime meadow - organogenic and actually.

Of mid-flight soils are formed on littoral-sea adjournment under rarefied associations of various kinds of vegetation, steady to strong salinization, test constant saturation by water or periodic flooding having flown in a combination to a close arrangement of waters, the salts in a soil structure get with sea having flown and earth waters.

Maritime soils are formed on rough fragments (sand, shingle) coastal sea adjournment under average and low grasses by meadow groupings with a grid-iron and motley grasses, test the allowable moisture regimes of spreading breeds, sometimes than bottom part of a structure (is lower 0,50м) (Kostenkova, 1987).

At realization of researches samples mid-flight and maritime meadow soils (coast of the Japanese sea), selected from top (0- 0.1 м) horizon were used. And also reference strains as *L. monocytogenes* (“П”, “A”, “K”, “10CN”, “4B”, “1/2A”), typical on their biological and biochemical properties, have been studied. The strains of Listeria are received from the All-Russia state control institute of veterinary preparations (Moscow, Russia).

In soil samples were determined: humus on Tyurin, urgent acidity, the structure of a soil absorbing complex, exchange acidity, hydrolytic acidity, exchange bases (Ca^{2+} and Mg^{2+}), capacity of a cation exchange, degree of a saturation by the bases (Arinushkina, 1971; Orlov, Grishina, 1990). The preparations of humus acids were received from mid-flight and maritime meadow soils on the standard technique (Somov, Litvin, 1988).

For study of growth kinetics of bacteria at temperatures 20-22^{0}C and 4-6^{0}C in various medium (researched samples of soils, preparations of humus, and extract of humus acids) used the culture, three times washed in a physiological solution, of bacteria, which brought in the examinees of medium in quantity 100 crates in 0.1 ml of medium. Number of bacterial cells in 1 ml of the suspension was determined by an optical turbidity standard. The researches carried out at periodic cultivation of bacteria. For construction of a curve of growth defined dynamics of number of bacteria by periodic sowing on 0.1 ml of a subculture (making necessary serial cultivations) on differential-diagnostic medium (yeast agar). The stages of a curve of growth and greatest contents of bacteria defined on number colony, evolved on cups Petri with differential-diagnostic medium containing agar (the unit forming a colony, UFC) (Frouin, 1989).

For definition of number of viable crates *L. monocytogenes* in soils prepared a suspension containing 1 g ground in 100 ml of physiological solutions. Use of natural soil was not represented possible because of a high sorbtion by soil particles of crates of bacteria

(Pomanskaya, 1963; Shaegani et al., 1981). The prepared suspension sterilized 3 times for 20 minutes with an interval 24 hours at temperature 100^0C to the fluid ferry, then they were inoculated with overnight culture of Listeria and Yersinia in an amount of 10^3 cfu / ml (under the standard of turbidity). The initial infecting doze at control sowing from a soil suspension has made 1000 / ml. Five variants of experiences in total were investigated at $20\text{-}22^0$C and five - at $4\text{-}6^0$C. Repeat three-multiple.

Multiplication of researched bacteria on humus preparations of soils studied as follows. Previously prepared special medium consisting of distilled water and a humus preparation (0.02 %) (Pomanskaya, 1963), which sterilized 3 times for 20 minutes with an interval 24 hours at temperature 100^0C to the fluid ferry. Sterile medium in test-tube subculture by daily cultures *L. monocytogenes* and *Y. pseudotuberculosis* in quantity 103 / 0.1 ml. Incubate at temperatures $4\text{-}6^0$C and $20\text{-}22^0$C. Sowing made on dense nutritious medium within 15 day with periodicity per one day. 120 variants of experiences (repeat three-multiple) in total were investigated.

The duration of experiments depend on change of a curve of growth of researched bacteria and made on the average 30 day, as it was necessary to establish the fact of multiplication of bacteria, was to receive three phases of a curve of growth (initial stationary phase, phase of a delay of multiplication, phase logarithmic) enough.

For cultivation of *L. monocytogenes* used the next medium: nutritious agar (pancreatic hidrolizate of sprat – 17.9 g, agar dry – 11.2 g, NaCl – 7.7 g, water distilled - up to 1 l, pH = 7.3±0.2), yeast agar with glucose (NaCl – 5.5g, hidrolizate of fodder yeast - 12 g, agar dry – 12.5 g, water distilled - up to 1 l, glucose - 2 g, pH = 7.2-7.4).

During our experiments was established, that all researched strains of *L. monocytogenes* best were made multiple copies in maritime soil, where their concentration was increased on 3,2 Lg. In mid-flight ground less intensive multiplication of bacteria was observed, the increase of which number did not exceed 1.6 Lg. Besides the multiplication of bacteria in soils of sea coast depend on individual features of strains. So, strain "K" *L. monocytogenes* was equally well made multiple copies both in mid-flight and in maritime soil (Figure 2.1).

Besides individual features strains of bacteria on multiplication of *L. monocytogenes* in soils of sea coast renders influence the characteristic of soils. On size of pH water extracts it is visible, that the mid-flight soil is subacid (pH = 6.1), and maritime - neutral (pH = 6.76) (Table 2.1). Subacid and neutral medium of researched soils are favorable for multiplication *L. monocytogenes*. So, on the data V. I. Gershun (1980a) optimum acidity limits for *L. monocytogenes* lay in an interval pH 7.6 – 7.8. Therefore maritime soil is most favorable for growth of *L. monocytogenes*.

The received results, namely more active multiplication of *L. monocytogenes* in maritime soil, is explained to that this soil has higher parameters of a degree of a saturation by the bases (87.35 %), capacities of a caution exchange (46.08 %), than mid-flight soil (85.99 and 6.78 accordingly) (Table 1). It specifies that in maritime soil higher concentration of ions calcium and magnesium, which are necessary for normal growth and multiplication of pathogenic bacteria (Rabotnova, Pozmogova, 1993). As it is visible from the Table 1, the maritime soil surpasses in the contents of the exchange bases mid-flight: Ca^{2+} -20 mg /100r of soil, Mg^{2+}-5.2 mg/100r of soil; $Ca2^{2+}$ - 21.9 mg/100r of soil, Mg^{2+}- 8.7 mg/100r of soil accordingly.

Besides the contents of humus is one of the limiting factors at multiplication of *L. monocytogenes* in researched soils. So, contents of humus of mid-flight soil – 2.06 %, maritime – 5.43 % (Table 2.2). Therefore, we also investigated influence of humus on multiplication of pathogenic bacteria. For what the pure preparations of humus acids as humate ammonium were received. The results of experience have shown, that *L. monocytogenes* were well made multiple copies as on humate of maritime meadow soil (achieving multiplication of bacteria up to 6 Lg strain 10CN), and on humate of mid-flight soil (up to 5 Lg strain 10CN), giving back nevertheless some preference to humats, allocated from maritime meadow soil (the difference in 1Lg) (Figure 2.2).

Table 2.1. Structure of a soil absorbing complex

Soil	pH		Exchange acidity	Hydrolytic acidity	Cation exchange capacity	Exchangeable bases		Base saturation, %
	H_2O	KCl				Ca^{2+}	Mg^{2+}	
			mg-equiv/ 100g of soil					
Mid-flight	6,01	5,20	0,10	0,95	3,74	2,09	6,78	85,99
Maritime	6,76	6,08	0,16	5,83	31,2	9,05	46,08	87,35

Table 2.2. Agrochemical parameters of soils

	Mid-flight soil	Maritime soil
Natural moisture content, % by volume	69,8	63,2
Degree of water saturation, %	92,6	85,5
Total moisture content, %	137,1	144,9
Porosity aeration, %	6,3	31,6
Total porosity, %	76,1	44,5
Amount of particles > 0.01 mm,%	96,1	55,5
Amount of particles < 0.01 mm%	3,9	49,5
Readily soluble salt layer 20 cm kg/m2	1,87	4,33
Humus, %	2,06	5,43

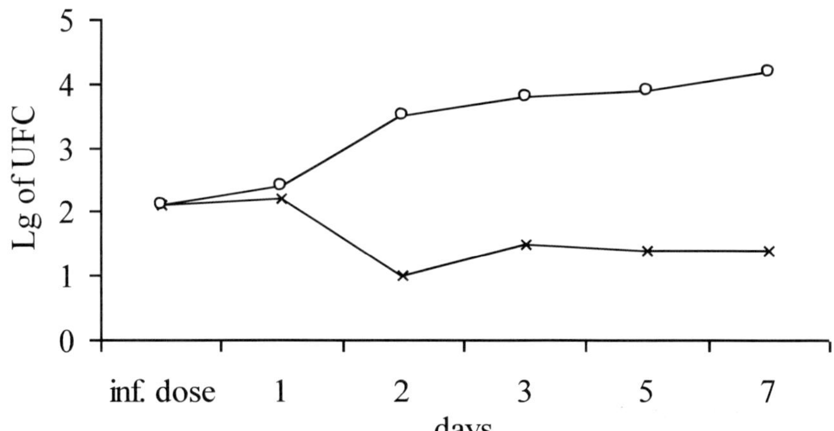

Figure 2.1. Curves of growth *L. monocytogenes* strain10CN, in soil of sea coast (-○- maritime soil, -x- mid-flight soil).

Mid-flight and maritime soils as 1.87 kg/m^2 and 4.33 kg/m^2 accordingly differ by quantity of stocks of highly soluble salts in a layer 20 sm. Aggregate fractional void volume of mid-flight soil – 76.1 %, maritime – 44.5 %. Natural humidity is 69.8 % of volume for mid-flight soil and 63.2 % of volume for maritime soil. Degree of saturation by water is 92.6 and 85.5 accordingly. Under the contents of physical sand and physical clay of mid-flight soil are estimated as is connected sandy (sum of particles > 0.01 mm (fraction of physical clay) 96.1 %), and maritime is light clay, (55.5 %) (Table 2.2).

Thus, is established, that the abiotic characteristic of soil of sea coast render direct positive influence on preservation and multiplication in them of *L. monocytogenes*. This is promoted by a degree of saturation by the bases, capacities of a cationic exchange, quantity of humus in these soils, and as property of strain of bacteria.

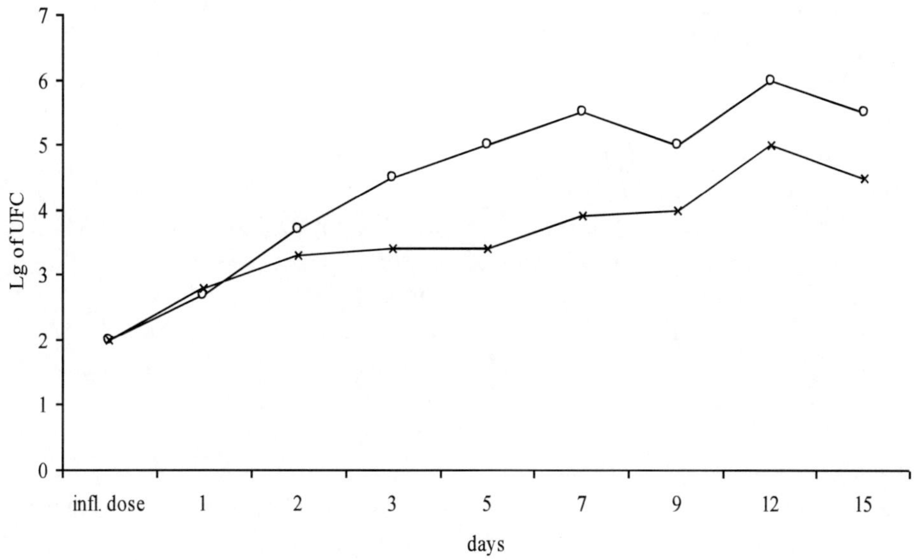

Figure 2.2. Curves of growth are conditional - *L. monocytogenes* strain 10CN on preparations of humus of soils of sea coast: -o- humate of NH_3 of maritime soil, -x- humate of NH_3 of mid-flight soil.

3. Character of Interactions of Saprophytic Soil Bacteria and *Listeria monocytogenes* via Gaseous Metabolites

Soil is a multiple-factor system with a variety of diverse species of microorganisms and relationships between them; the study of all the components of a microbial ecosystem is therefore complicated. The study of the mechanism of regulation of activity of soil microorganisms is one of the central problems of soil microbiology. The presence of microorganisms in a given natural zone is determined not only by the environmental conditions but also by the existence of control performed by other members of the biocenosis. Such a control is one of the causes of microbial associations being formed in natural ecosystems.

Soil microbiocenosis is one of the most complicated biological communities. Various interrelationships occur between bacteria in the process of their activity, including those at the metabolic level (Egorov, 1986; Dudchenko et al., 1979; Gusev, Koronelli, 1981; Zavarzin, Bonch-Osmolovskaya, 1981; Zenova et al., 1983; Kubat et al., 1979; Bull, Slater, 1982; Stotzky, Schenck, 1976). The interactions between populations via metabolites (Egorov, Landau, 1982; Korovina et al., 1974), including gaseous substances (Zavarzin, 1976; Larsson, et al., 1978), are of crucial importance for maintaining the stability of microbial communities and the control of their species composition and production capacity. Both the stimulating and inhibiting action of volatile compounds of microbial origin on bacterial growth has been noted (Dreyfuss, 1980; Rai et al., 1981).

Considering the fact that volatile compounds produced by microorganisms are able to act as intra- and interspecies regulators of microbial communities (Zavarzin, 1976; Zvyagintsev, 1979; Tambiev, Telitchenko, 1971; Berestetsky, Kravchenko, 1984), we attempted to study the character of the interaction between saprophytic soil bacteria and *L. monocytogenes* by means of gaseous metabolites.

To study the interactions between soil bacteria and *L. monocytogenes* from established microbial associations of brown forest and brown podzolic soils (the south of the Far East of Russia), we isolated saprophytic bacteria. The microorganisms were isolated and grown for the experiments on peptone agar (PA) (1% peptone (Mikrogen, FGUP NPO, Ministry of Health of the Russian Federation) and 0.5% NaCl in distilled water with 2% agar; pH 7.4). In order to compare the effect of gaseous metabolites on the growth of microorganisms, nutrient agar (NA) and plant agar (PA) (Buzoleva, 2001) were used.

A total of 20 strains of microorganisms differing in their cultural and biochemical properties were isolated. They were assigned to the genera *Agrobacterium, Acinetobacter, Aeromonas, Micrococcus (M. roseus), Pseudomonas (P. fluorescens, P. aerugonosa, P. putida), Flavobacterium,* and *Bacillus (B. cereus, B. mesentericus)* according to Bergey's Manual of Determinative Bacteriology (1997) by means of API (Analytical Profile Index) tests (BioMerieux, France).

The method proposed by Tirranen (1989), which modified by us, was used for the quantitative assessment of the action of volatile biologically active substances produced by saprophytic soil microflora. The culture whose volatile substances were studied is further referred to as the study culture; the culture on which this action was tested is referred to as the test culture. All saprophytic bacteria were tested as the study cultures and all strain *L. monocytogenes* were tested as the test cultures. In order to obtain suspensions of the study cultures, the cells were washed off solid media. Petri dishes with PA medium were inoculated with this suspension and incubated at 20°C for 72 h. Suspensions of test cultures were prepared and distributed into the wells of an immunological plate. A multichannel pipette was used to prepare replicas (0.1 ml each) on the surface of sterile media (NA or PA, depending on the experimental variant). Since the amount of medium in a Petri dish may affect the results, the medium were distributed uniformly (15 ml per dish). One half of a Petri dish with a lawn of the study culture was covered with the other half with the freshly applied replicas of the test cultures and incubated in the thermostat at 20°C. The growth of the test organisms depended on the action of the volatile metabolites produced by the study culture grown on the opposite side of the dish.

The reactions of the test cultures to the effect of volatile metabolites of the study cultures were assessed by the difference in the colony size of the test organisms in the experiment and control on the second or third day of incubation, when the growth of the test culture colonies in the control was well-pronounced. The dishes with the test cultures that were not subjected to the action of the volatile products of metabolism of the study organisms served as controls. The statistical data processing was carried out according to G. F. Lakin (1990). The mean of the colony diameter and the mean error were considered. The assessment criterion was the normalized standard deviation value to which the real value of this criterion was compared for the 95% level of significance.

The effect of the study culture on the test culture was assessed as positive (stimulating) or negative (inhibitory) when the test colony size in the experiment was increased or decreased by 20% or more, respectively, compared to the control. The effect of the study culture was assessed as zero if the colony sizes in the experiment did not differ from the control ones by more than ± 20%. The results of the comparison of the colony sizes were expressed in millimeters. A total of 400 variants of the experiment with five repeats were performed.

The analysis of the volatile metabolites of saprophytic bacteria with gas–liquid chromatography (GLC) was performed using a Shimadzu-16A gas chromatograph (Japan); helium was the carrier gas, the column temperature was 35°C, the vaporizer temperature 100°C, the flame ionization detector temperature 100°C, and the hydrogen and air pressure on the detector was 0.5 kg/cm^2. The chromatogram calculation and processing were carried out using a specialized Shimadzu-cR-4A computer (Japan). The qualitative composition of volatile organic substances was determined by comparing the chromatographic peaks obtained to the peaks acquired during the chromatographic analysis of pure volatile organic substances (references) without any impurities.

For chromatographic analysis, bacteria were grown in tightly capped vessels on the NA medium. After 24, 48, and 72 h of incubation, 1 ml of air was sampled from the vessels through the rubber caps with a syringe and introduced into the chromatograph dispenser.

The results obtained enabled us to assess the degree of influence of the volatile metabolites of saprophytic bacteria on the growth of *L. monocytogenes* during their interaction. Among them, 42% were negative (the volatile metabolites of the study cultures inhibited the growth of the test cultures), 30% were positive (the volatile metabolites of the study cultures stimulated the growth of the test cultures), and the remaining results (28%) were neutral. The zero interactions observed in the experiments may be weak positive or negative effects (less than 20% of the control), which were not determined by the method of investigation used.

The experimental results presented in the table demonstrate that among the bacteria studied an interaction exists via gaseous metabolites. Most of the study cultures released inhibitory volatile substances, which had a negative effect on the growth of the test cultures. The stimulating, i.e., positive effect of the soil bacteria occurred less frequently.

All the study cultures exerted a selective, both inhibitory and stimulating, effect on the growth of the test cultures. Bacteria probably produce a range of volatile compounds, their spectrum of action varying from broad to narrower. Of all the strains of the study cultures, the bacteria of the genera *Pseudomonas* and *Acinetobacter* revealed the highest inhibitory activity in relation to the *L. monocytogenes* (Figure 3.1). The stimulating action of these bacteria was observed in no more than 28% of cases. The volatile metabolites of *Aeromonas* exhibited the greatest stimulating activity. The inhibitory action of these bacteria was observed in no more than 8% of cases. Hence, it may be suggested that it is *Pseudomonas* and *Aeromonas* that exert the most significant effect on the growth and development of *L. monocytogenes* in the soils studied.

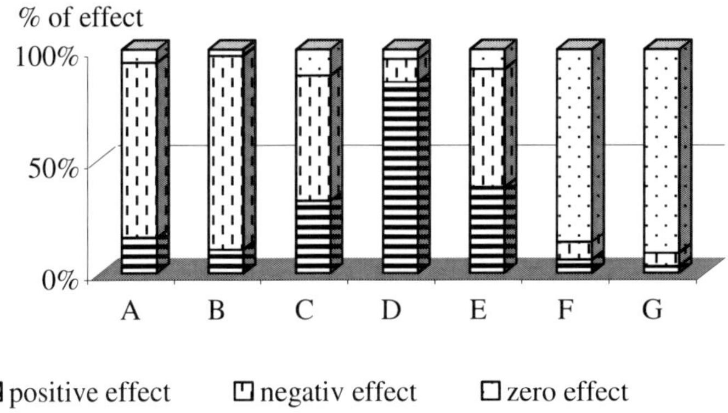

Figure 3.1. Character of interrelationships of saprophytic soil microflora of soils and *L. monocytogenes* via gaseous metabolites. A, *Pseudomonas aureus;* B, *Pseudomonas fluorescens*; C, *Micrococcus;* D, *Aeromonas;* E, *Acinetobacter;* F, *Bacillus;* G, *Flavobacterium.*

It should be noted that the volatile metabolites of the *Flavobacterium* and *Bacillus (B. cereus, B. mesentericus)* strains investigated in the experiment did not appreciably influence the growth of the *L. monocytogenes* (Table 3.1.).

The reaction of the test cultures to the volatile metabolites of the study microorganisms was diverse. Thus, most of the strains of *L. monocytogenes* (P, 4B, 1B, 1C, 2755, 4A, 5, 10527) exhibited a mainly negative reaction. An interesting feature was noted in relation to *L. monocytogenes* strain 4C, which revealed a neutral reaction to the action of volatile substances in 50% of cases. No strains showing a surge of growth in most cases were revealed.

Analysis of the experimental data revealed direct correlation (for all the strains studied, except bacilli and flavobacteria, which showed a neutral effect) between the inhibiting and stimulating effect of the culture; the absence of stimulation correlated with the presence of inhibition.

It is known that the growth rate of bacteria in different media may be different. Bacterial metabolites accumulated on different media may differ both in their qualitative and quantitative composition. We therefore studied the influence on bacterial growth of the gaseous metabolites of the same cultures grown on nutrient agar and plant agar. These comparative studies did not reveal a substantial difference between the actions of the volatile metabolites of the cultures grown on different media (the difference is statistically in significant, $\rho > 0.05$).

Thus, the stimulating and inhibiting effect of the volatile metabolites of saprophytic soil bacteria on *L. monocytogenes* was noted. Some gaseous substances released by soil bacteria may serve as nutrients for *L. monocytogenes*. For example, D. G. Zvyagintsev (1987) stated that certain components of volatile metabolites of microbial origin (acetaldehyde, ethanol) could act as nutrients for microorganisms. G. M. Larionov (1988) demonstrated that long-term cultivation of *Pseudomonas pseudomallei* in soil extracts resulted after six months in the activation of the fermentative properties and the synthesis of an inhibitor of concomitant microflora. In the opinion of L. S. Tirranen (1989), the interaction of microorganisms by means of their gaseous metabolites is a widespread phenomenon, which may play a certain ecological role in natural habitats. The qualitative composition of volatile metabolites may influence positively or negatively the process of consumption of organic compounds.

Table 3.1. Influence of the volatile metabolites of saprophytic soil bacteria on the growth of *L. monocytogenes*

Study cultures	Test cultures																			
	L. mono -cyto- genes st. A	*L. mono -cyto- genes* st. P	*L. mono- cyto- genes* st. K	*L. mono- cyto- genes* st. 10CN	*L. mono- cyto- genes* st. 1A	*L. mono- cyto- genes* st. 4B	*L. mono- cyto- genes* st. 1B	*L. mono -cyto- genes* st. 1C	*L. mono- cyto- genes* st. 4C	*L. mono- cyto- genes* st. 74-T	*L. mono- cyto- gene* st. 2755	*L. mono- cyto- genes* st. 5105	*L. mono -cyto- genes* st. 3A	*L. mono- cyto- genes* st. 4A	*L. mono -cyto- genes* st. 5	*L. mono- cyto - genes* st. 6	*L. mono -cyto- genes* st. 7	*L. mono- cyto- genes* st. 88	*L mono- cyto- genes* st. 10527	*L. mono -cyto- genes* st. 75936
P. aeruginosa st. 11	0	-	-	+	-	-	-	-	+	0	-	+	-	-	-	-	-	-	-	-
P. aeruginosa st.2	+	-	-	+	-	-	-	-	+	0	-	0	-	-	-	-	+	-	-	-
P. aeruginosa st. 12	0	-	-	+	-	-	-	-	+	-	0	+	-	-	-	-	-	-	-	-
P. fluorescens st. 7	-	-	-	-	-	0	-	+	-	-	-	-	+	-	-	-	-	-	-	+
P. fluorescens st. 13	-	-	+	-	-	-	-	+	-	-	-	-	0	-	-	-	-	+	-	-
P. fluorescens st. 14	-	-	-	-	-	-	-	+	-	-	0	-	-	-	-	-	-	0	-	+
P. fluorescens st. 6	-	-	+	-	-	-	-	+	-	-	-	-	0	-	-	-	-	0	-	+
Micrococcus st. 3	+	-	+	-	+	-	0	-	+	-	-	-	+	-	-	+	0	+	-	-
Micrococcus st. 9	+	-	+	-	+	-	0	-	-	-	-	-	+	-	-	+	0	+	-	+
Aeromonas st. 1	+	+	+	-	+	+	+	+	0	+	+	+	+	+	+	+	+	-	+	+
Aeromonas st. 10	+	+	+	+	+	+	0	+	0	+	+	+	-	+	+	+	+	+	+	+
Aeromonas st. 5	+	-	+	0	+	+	+	+	0	+	+	+	-	0	+	+	+	+	+	+
Aeromonas st. 20	+	+	+	0	+	+	+	+	0	+	+	+	-	+	+	+	+	+	+	+

Table 3.1. (Continued)

Study cultures	Test cultures																			
	L. monocytogenes st. A	*L. monocytogenes* st. P	*L. monocytogenes* st. K	*L. monocytogenes* st. 10CN	*L. monocytogenes* st. 1A	*L. monocytogenes* st. 4B	*L. monocytogenes* st. 1B	*L. monocytogenes* st. 1C	*L. monocytogenes* st. 4C	*L. monocytogenes* st. 74-T	*L. monocytogene* st. 2755	*L. monocytogenes* st. 5105	*L. monocytogenes* st. 3A	*L. monocytogenes* st. 4A	*L. monocytogenes* st. 5	*L. monocyto-genes* st. 6	*L. monocytogenes* st. 7	*L. monocytogenes* st. 88	*L. monocytogenes* st. 10527	*L. monocytogenes* st. 75936
Acinetobacter st.8	-	+	0	-	0	-	-	+	+	+	-	-	+	+	-	-	-	-	-	-
Acinetobacter st. 17	-	+	0	-	0	-	-	+	0	+	-	-	+	+	-	-	+	-	-	-
Bacillus st. 4	0	0	0	0	0	0	+	0	0	+	0	0	0	0	0	0	0	0	0	0
Flavobacterium st. 16	0	0	0	-	0	0	0	0	0	0	0	0	0	+	0	0	0	0	0	0
Flavobacterium st. 18	0	0	0	0	0	+	0	0	0	0	-	0	0	+	0	0	0	0	-	0

Note: 0, no action; +, the action is positive; –, the action is negative.

In order to determine the relationship between growth of the *Pseudomonas* and the qualitative composition of volatile metabolites of soil bacteria, we analyzed the volatile samples by gas–liquid chromatography.

Chromatographic analysis of the volatile metabolites of *Pseudomonas* and *Aeromonas* which revealed relatively high are content of acetaldehyde (1μl/ml in Pseudomonas and 0.55 μl/ml in *Aeromonas*), methanol (0.43 and 1.2 μl/ml, respectively), and ethanol (0.16 and 0.2 μl/ml, respectively) (Figure 3.2). A greater content of methanol was noted in the composition of the volatile metabolites of *Aeromonas*, compared to *Pseudomonas*. A greater content of acetaldehyde, which inhibits bacterial growth, was noted in the composition of the volatile metabolites of *Pseudomonas*.

It may be suggested that methanol is utilized by the *L. monocytogenes* as a carbon source, since some microorganisms are known to grow on C1 compounds, including methanol (Buzoleva, Somov, 1999). In our preliminary researches was shown that The *L. monocytogenes* strains can grow on methanol, formate, and dimethylamine (Buzoleva, Somov, 1999). Methanol was the most preferential substrate for bacterial growth, and dimethylamine the least preferential substrate. The growth rate on C1-compounds varied between the strains. Optimal growth was observed at concentrations of C1-compounds from 0,1 to 1% and higher concentrations (2-3%) suppressed the growth of *Listeria*. The optimal temperature for growth of the bacteria on C1-compounds was 20-25°C, whereas at 6-8°C specific growth rate of bacterial in log phase was 2-3-fold lower, and at 37°C the bacterial cells died. Similar to obligatory autotrophs, the carbon of C1-compounds may be assimilated autotrophically, i.e., after oxidation of CO2 in the ribulose diphosphate cycle although it is possible that a fraction of methanolic carbon can be assimilated at a more reduced level (formaldehyde, formate) as methyl derivatives of tetrahydrofolic acid (Loginova et al., 1981; Cox, Quagle, 1975). However, this issue requires further studies including the assay of activities of key enzymes.

The data obtained allow us to assert that, at the metabolic level, a diverse character of interspecies interrelationships is observed between the bacteria, directly influencing *L. monocytogenes* growth and preservation in soils. The volatile compounds produced by soil bacteria may act as both intra- and interspecies regulators of microbial communities. In this regard, the preservation the *L. monocytogenes* in soils may be regulated by the products of metabolism of saprophytic soil bacteria. An important role in this process is played by methanol released by saprophytic bacteria into the environment.

4. Effect of Volatile Metabolites from Germinating Seeds on the Reproduction of the *Listeria monocytogenes*

In recent decades, increasing attention has been attracted to a group of pathogens with a life program consisting of a continuous transition from the environment, where they lead a saprophytic life, into the bodies of humans and warm-blooded animals, where they exhibit parasitic properties, causing infectious processes, and a reversion to the saprophytic life when they return into the environment. The possibility of saprophytic existence of a number of pathogenic microorganisms in the environment, which was denied in the recent past, is now increasingly recognized (Somov, 1979; Golovachova, 1978; Znamensky, 1975; Litvin et al., 1998; Gershun, 1988b; Litvin, 1999).

Scientific and technological progress and the associated urbanization have violated the ecological balance and opened a way for the penetration of pathogenic microorganisms from the environment into human surroundings. In this respect, striking examples are and *L. monocytogenes* (the causative agent of listeriosis) (Timchenko et al., 2000; Tirranen, Shilenko, 2008; Markova et al., 2002; 2005; Somov, Buzoleva, 2004; Alfano, Collmer, 1996; Ausubel, 2005).

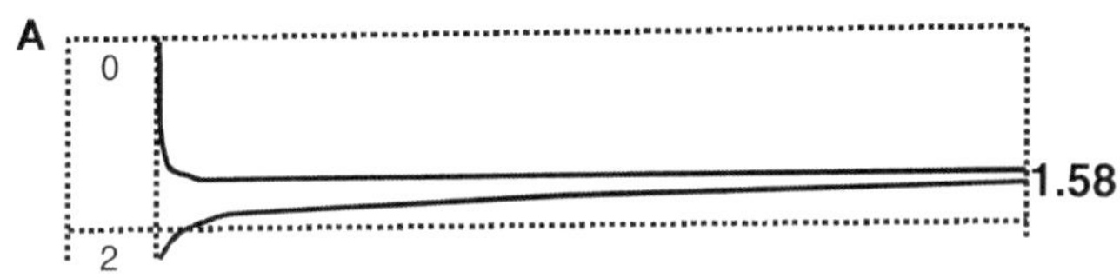

(A)

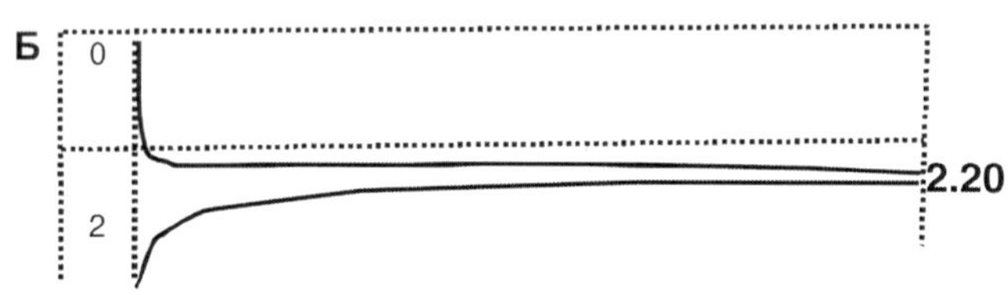

(B)

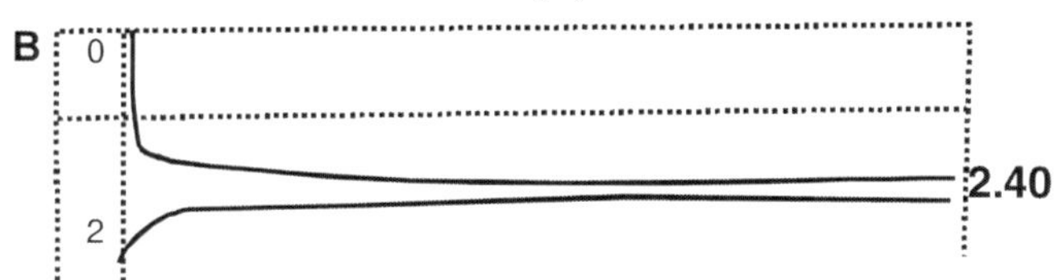

(C)

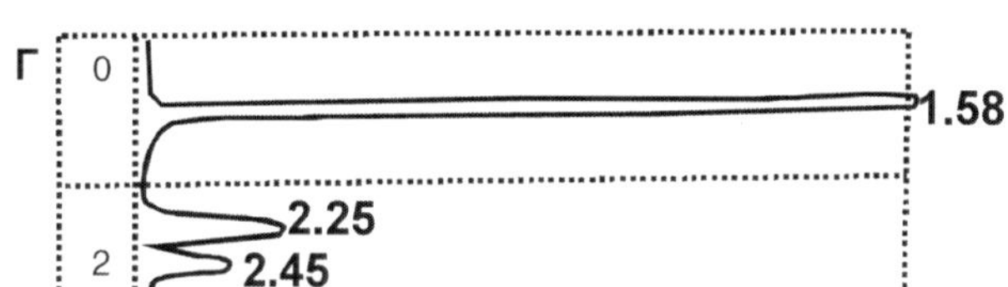

(D)

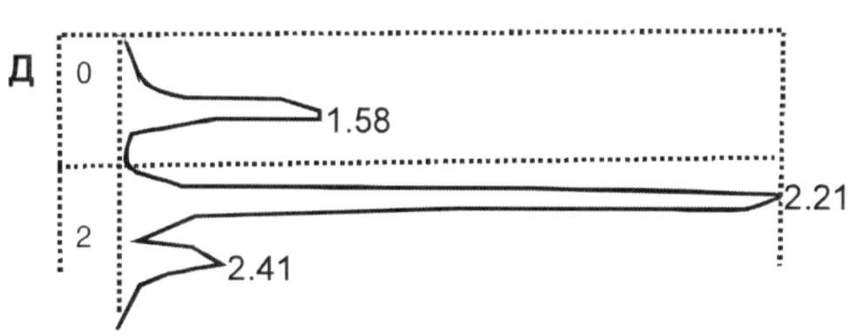

(E)

Figure 3.2. Data of the chromatographic analysis of volatile metabolites of the bacteria of the genus Pseudomonas (D) and the genus Aeromonas (E). Control: A, acetaldehyde; B, methanol; C, ethanol.

An analysis of numerous outbreaks of these diseases showed that most factors of the transmission of bacteria to humans are vegetables and root vegetables (dishes made from them) and, first of all, corn and salad for Listeria. Bacteria are actively multiplied and accumulated in large amounts in these vegetables, while maintaining a high degree of virulence (Rozhkova, 1974; Godova, et al., 2009; Taylor, 1996; Vlasov, Pavlova, 2009).

Different relationships are formed inside the complex associations between microorganisms and plants in natural and artificial systems (Morello, 2004; Mounier et al., 2006). In any community, these relationships are complex and various. Bioactive metabolites secreted by plants may play a regulatory role for the bacterial population in the plant–microorganism community. Volatile organic secrets from germinating seeds are of particular interest among them. They can be a carbon and energy source for microorganisms due to their high penetrating ability and availability for digestion. It has been proven that a number of bacteria and fungi can consume the volatile metabolites of germinating seeds as the sole carbon source (Burmistrova, 2004; Tirranen, Gitelson, 2006; Tirranen et al., 2001; Tirranen, 2001; Andreev, Talieva, 1996; Vlasov, Pavlova, 2009; Ermolaeva et al., 1999; Tartakovskii et al., 2002).

Plants are known to be the leading components of the biosphere, which determine the main form of life on Earth. Therefore, the study of the interaction between plants and bacteria is important to understand the spreading of pathogens outside warm-blooded organisms and to elucidate the adaptation mechanisms of eukaryotic organisms to changes in the existing conditions, especially under the influence of extreme factors.

We used bacterial strains from the museum of the Research Institute of Epidemiology and Microbiology (Siberian Branch, Russian Academy of Medical Sciences): Gram-positive *Listeria monocytogenes* (strain 2L, 1/2a). We also used germinating plant seeds, which are transmission factors for *Listeria* (cabbage *Brassica oleacia L.*, carrot *Daukus carota L.*, salad *Lactuca sativa L.*, corn *Zea mays L.*) (Golovachova, 1978; Timchenko et al., 2000). A mixture of plant seeds (salad, corn, cabbage, and carrot) was used as a source of volatile metabolites. In addition, *L. monocytogenes* were separately cultivated on metabolites of the cabbage and carrot mixture and the salad and corn mixture, respectively.

L. monocytogenes were cultivated on a liquid mineral medium containing NH_4NO_3 (0.3%), KH_2PO_4 (0.1%), $MgSO_4$ $7H_2O$ (0.05%), and KCl (0.025%), and pH 6.8–7.0 without a carbon source.

The effect of volatile metabolites of germinating plant seeds on the reproduction of bacteria was studied on a specially designed installation according to the Schenk and Stotsky scheme (Gershun, 1988b) in our modification (Figure 4.1). The treated seeds were placed in a vessel containing sterile soil (3). The vessel was tightly closed to avoid the entry of outside air. Sterile air, purified from CO2, was pumped through the vessel with a pump (5). The degree of carbon dioxide removal from the air was checked before the experiment with a Dukat-B portable analyzer (Delta, Russia). The air was sterilized using cotton filters (2) and purified from CO2 in a trap (1) filled with 30% KOH. The air passed through the soil and seeds in the vessel, carried volatile metabolites, and entered into a flask (4) containing a liquid mineral medium inoculated with a culture of the studied microorganism. Depending on the plant species, 20–40 seeds were placed into a vessel (3). The system was kept at room temperature (20–22°C).

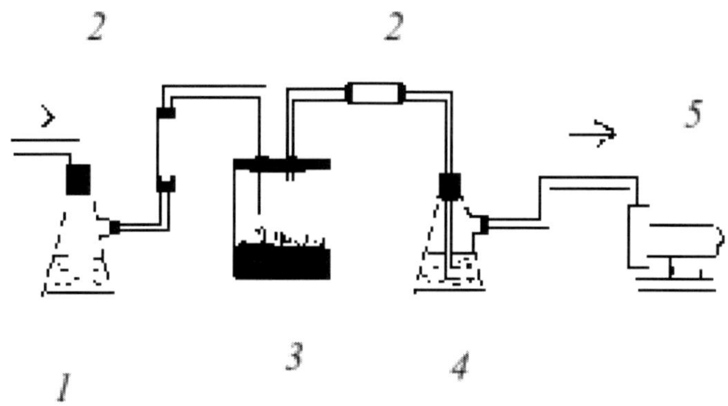

Figure 4.1. Scheme of an installation for the investigation of the effect of volatile metabolites of germinating seeds on the reproduction of pathogenic bacteria. 1, a flask with KOH; 2, cotton filters; 3, a vessel with soil and germinating seeds; 4, a flask with a liquid medium inoculated with pathogenic bacteria; 5, a peristaltic pump.

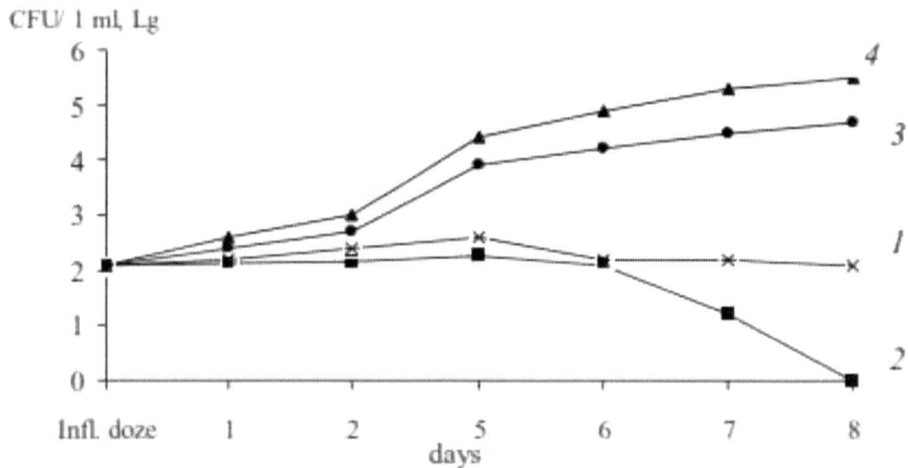

Figure 4.2. Reproduction of L. monocytogenes strain 10CN (b) on the volatile secretions of germinating seeds. 1, a seed mixture; 2, control; 3, salad; 4, corn.

The mineral medium (100 ml) was inoculated using 0.1 ml of the 1-day bacterial culture grown on an agar medium containing 102 colony-forming units (CFUs). Aliquots from the reaction mixture were taken at regular intervals for counting the number of colonies. For this, the inoculated liquid (by 0.1 ml with the necessary dilutions) was seeded on solid media. The casein–yeast agar with glucose was used to indicate *Listeria*. The number of counted colonies was expressed in common units of lg CFU/0.1 ml. A vessel with soil without seeds was used as a control.

The interior surfaces of the glass parts of the installation were carefully treated with 70% ethanol, and the rest of the details were autoclaved. The seeds were treated with a solution of potassium permanganate and washed with sterile distilled water. The soil was treated three times with flowing steam for 30 min. After the treatment, all parts of the installation, seeds, and the soil were examined for the presence of foreign micro flora using nutrient agar.

The volatile secretions of the germinating seeds were estimated by gas–liquid chromatography. For a chromatographic analysis, the seeds were germinated in sealed vessels. The air from the vessels was taken in 24, 48, and 72 h through rubber caps with a1-ml syringe and injected into the chromatography dispenser. GLC of the volatile metabolites of the germinating seeds was performed on a Shimadzu-16A gas chromatograph (Shimadzu, Japan). The chromatographic parameters were as follows: the carrier gas was helium; the stationary phase was polyethylene; the glass column was 0.32 cm in diameter and 10 m in length; the temperature of the column, evaporator, and flame ionization detector was 35, 100, and 100°C, respectively; and the hydrogen and air pressure on the detector was 0.5 kg/cm. The calculation and processing of the obtained chromatograms was performed on a special Shimadzu-cR-4A computer (Shimadzu, Japan). The estimation of the qualitative composition of the volatile organic substances was carried out by comparing the resulted chromatographic picks with those of pure volatile organic etalons.

The experiments were repeated three times. The results were processed statistically by the computer program Excel 7 using Student's t-test.

The effect of volatile metabolites of germinating plant seeds (salad, corn, carrot, and cabbage) on the reproduction of *L. monocytogenes* was studied using a specially designed installation. It was shown that these metabolites increased the reproduction of the studied bacteria by factors of 10–100 as compared to the control (Figure 4.2). An analysis of the data led to the conclusion that the *L. monocytogenes* reproduction intensity in the presence of the volatile metabolites of the plant mixture depended on the bacterial strain. The amount of strain "K" in the mineral medium increased by a factor of 2 on the seventh day of growth as compared to strain "10CN". The results are confirmed by literature data, indicating that not all bacteria are capable of consuming volatile metabolites of germinating seeds (Morello, 2004). It should be noted that the *L. monocytogenes* reproduction intensity significantly decreased on the sixth day (Figure 4.2).

The effect of volatile organic compounds is known to depend on both the bacterial species and germinating seeds (Tirranen et al., 2001). In this regard, we have studied the growth of *L. monocytogenes* on plant metabolites depending on the type of the germinating seeds. To do this, the studied bacteria were cultured on the volatile metabolites of germinating seeds of individual plant species, which are transfer factors to them.

It was found that the use of volatile plant compounds by bacteria depended on both the properties of the bacteria and the plant species. When comparing the data obtained by culturing microorganisms on volatile substances of seed mixtures or individual plants, we observed selective growth of bacteria on the metabolites of germinating seeds, depending on the plant species (Figure 4.2). *L. monocytogenes* preferably multiplied on volatile metabolites of corn, reaching a value of 260200 CFU/0.1 ml with an infecting dose of 130 CFU/0.1 ml, which is by 25759800 CFU/0.1 ml higher than the growth rate on volatile metabolites of the plant mixture. In these experiments, we registered an increase in the number of *L. monocytogenes* throughout the entire observation period (eight days). When all plant seeds were used simultaneously, no increase was observed in the number of Listeria, and the reproduction intensity of these bacteria in the control dropped to zero.

According to some authors (Morello, 2004; Tirranen et al., 2001), the qualitative composition of volatile secretions can have a positive or negative influence on the consumption of these compounds by bacteria. Some researchers (Mounier et al., 2006; Andreev, Talieva, 1995) note that the major components in the secretions of germinating seeds that serve as carbon and energy sources are ethanol and acetaldehyde. Other authors suggest that the volatile metabolites of germinating seeds contain methanol, propanol, formic and propionic aldehydes, acetone, and ethylene (Burmistrova, 2004; Tirranen, 2001).

To clarify the relationship between the increase in the number of bacteria and the qualitative composition of the volatile metabolites of germinating seeds, a gas chromatographic analysis of the volatile substances was carried out. It is seen in Figure 4.3 that volatile compounds of germinating seeds of corn, carrots, salad, and cabbage contain acetaldehyde, methanol, and ethanol in sufficiently high concentrations. The greatest amount of methanol was secreted by the corn seed lings; the seedlings of carrots, cabbage, and salad contained much smaller amounts of methanol. Unlike the experiments with salad and the seed mixture, a more intense reproduction process of L. monocytogenes was observed in the presence of germinating seeds of corn. This indicated that one of the main substances that affect the reproduction of *L. monocytogenes* was methanol.

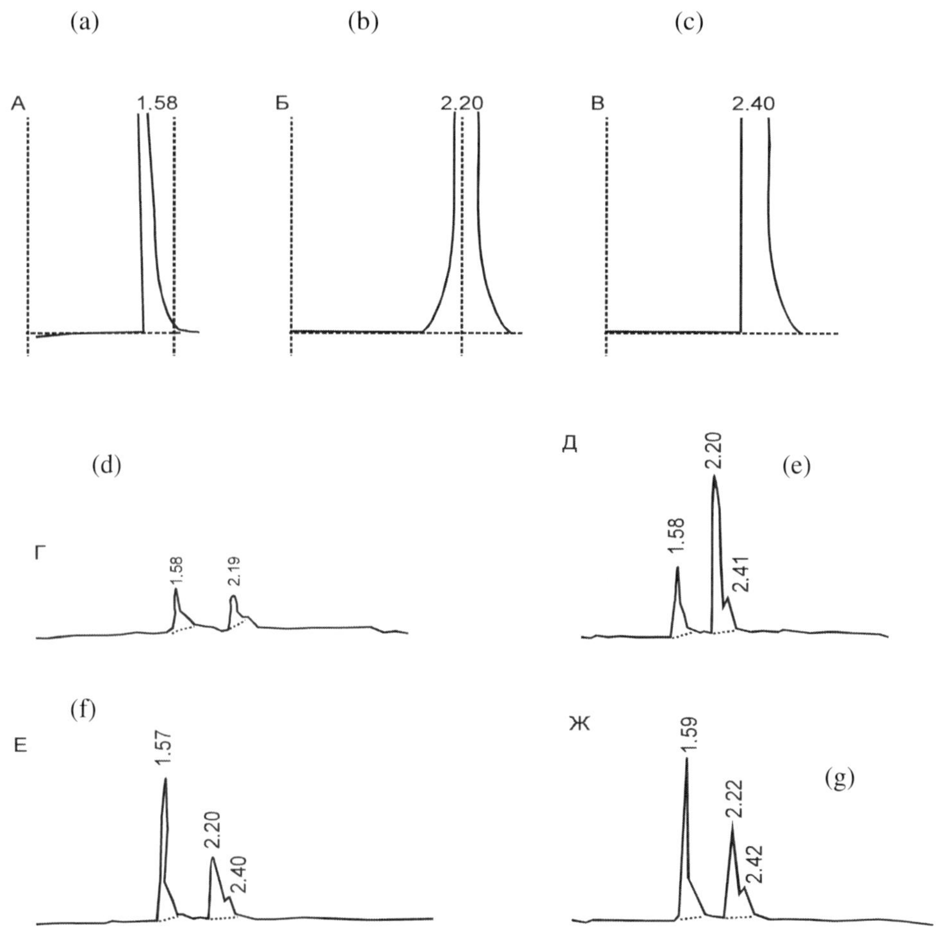

Figure 4.3. Chromatograms of volatile metabolites of germinating seeds (the most informative regions). (a) acetaldehyde (control), (b) methanol (control); (c) ethanol (control); (d) carrots; (e) corn; (f) cabbage; (g) salad. The numbers under the peaks indicate the retardation time of the compound for the determination of its maximum absorption.

According to the chromatographic studies, the volatile secrets of cabbage and carrots contained fewer methanols than acetaldehyde (Figure 4.3). Moreover, they produced fewer methanols than corn. The data suggest that Yersinia, as well as Listeria, used methanol for nutrition under the conditions of a limited carbon source. In previously published studies, we showed that methanol stimulates the reproduction of *L. monocytogenes* in a physiological solution and in a saline–phosphate buffer (Buzoleva, Somov, 1999; Sidorenko, Buzoleva, 2008).

Thus, one of the mechanisms of adaptation of *L. monocytogenes* to the environment was found: volatile metabolites of germinating seeds may be the only source of carbon and energy for these bacteria. It was shown that the main volatile substance that influences the growth and reproduction of the studied bacteria is methanol.

REFERENCES

Ahmed, F. E. Review: Assessing and Managing risk due to consumption of seafood contamination with microorganisms, parasites and natural toxins in the United States. *Int. J. Food Science Technol.,* 1992, V. 27(3), p. 243-260.

Alfano, J. R.; Collmer A. Bacterial Pathogens in Plants: Life up Against the Wall. *Plant Cell,* 1996, V. 8, pp. 1683–1698.

Andreev, L. N.; Talieva, M. N. Role of physiologically active substances in relationships between host plants and pathogenic fungi. *Russian Journal of Plant Physiology,* 1996, V. 43 (5), pp. 577-582.

Arinushkina, E. V. *A Guide to the Chemical Analysis of Soils.* Moscow: Moscow State University Press; 1971.

Ausubel, F. M. Are Innate Immune Signaling Pathways in Plants and Animals Conserved *Nature Immunology,* 2005, V. 6(10), pp. 973–979.

Beleneva, I. A. *Psychophilic Listeria monocytogenes and its ecological and epidemiological significance:* Dis. Candidate. biol. Science. Vladivostok, 1996, 29 p.

Beleneva, I. A.; Maslennikov, E. F. Opportunistic bacteria found in cultured mussels, *Zh. Microbiology,* 2002, V. 2, pp. 81-83.

Ben Embarek, P. K. Presence, detection and growth of *Listeria monocytogenes* in seafood's: a review. *Int. J. Food Microbiol.,* 1994, V. 23, pp.17-34.

Berestetsky, O. A.; Kravchenko, L. V. Volatile Products of Plant Residue Decomposition and Their Effect on Soil Microflora. *Soil. Biol. Conserv. Biosphere,* 1984, V. 1, pp. 419–425.

Bull, A. T.; Slater, J. H. *Microbial Interactions and Community Structure, Microbial Interactions and Communites.* London: Academic, 1982.

Burmistrova, A. L. *Antibiotics and Antibiotic Resistance: Problems and Solutions.* Chelyabinsk: Dom pechati, 2004.

Buzoleva, L. S. *Adaptation of Pathogenic Bacteria to Abiotic Environmental Factors.* Doctoral (Biol.) Dissertation: Vladivostok, 2001.

Buzoleva, L. S. *Biphase bentonite clay for diagnostics of intestinal infections:* Abstract of dissertation of PhD. Vladivostok, 1990, 25 p.

Buzoleva, L. S.; Sidorenko, M. L. Effect of Organic Substances from Humic Acids on the Reproduction of Enterobacteria. Zh. Mikrobiol. *Epidemiol. Immunobiol.,* 2001, V. 2, 89–91.

Buzoleva, L. S.; Somov, G. P. Autotrophic Assimilation of CO2 and C1-Compounds by Pathogenic Bacteria. *Biochemistry,* 1999, V. 64(10), pp. 1146–1149.

Colburn, K. G.; Kausner, C. A.; Abeyta, C. Listeria species in California coast estuarine environment. *Appl. Environ. Microbiol.,* 1990, V. 56, pp. 2007-2011.

Cox, R. B.; Quagle, S. R. The autotrophic growth of Micrococcus denitrificans on methanol. *Biochemical Journal,* 1975, V. 150 (3), pp. 569-571.

Dmitrieva, E. Yu.; Mukhina, L. B.; Al'Askhabi, N. I. Problems specific identification of *Listeria monocytogenes* in fish products. Neuroinfections: rabies, glubokoobraznaya encephalopathy of cattle, Creutzfeldt-Jakob disease and other prion diseases, listeriosis, Aujeszky's disease, Teschen disease. *Cover,* 2001, pp. 129-132.

Dreyfuss, M. A. Fungicidial and Bactericidial Gas from the Mycellum of a Paecilomyces Strain. *Experientia,* 1980, V. 36(4), pp. 500–501.

Dudchenko, V. G.; Ulyashova, R. M.; Slyusar', I. T.; Pota, L. G. *Effect of the Water-Air Regimeon the Microbiological Processes of Organogenic Soil,* in Sezonnaya dinamika pochvennykh protsessov. In: Seasonal Dynamics of Soil Processes. Tallinn, 1979, pp. 66–72.

Egorov, N. S. *Osnovy Ucheniya ob Antibiotikakh (Basic Principles of the Theory of Antibiotics),* Moscow: Vysshaya Shkola, 1986.

Egorov, N. S.; Landau, N. S. Biosynthesis of Biologially Active Compounds by Mixed Microbial Cultures. *Applied Biochemistry and Microbiology,* 1982, V. 18(6), pp. 835–849.

Ermolaeva, S.; Belyi, Y.; Tartakovskii, I. Characteristics of induction of virulence factor expression by activated charcoal in *Listeria monocytogenes. FEMS Microbiol. Letts.,* 1999, V. 174 (1), pp. 137-141.

Farber, J. M.; Peterkin, P. I. *Listeria monocytogenes*, Foodborn pathogen. *Microbiol. Rev.,* 1991, V. 55, pp. 476-511.

Frouin, A. Listeria, écologie et épidémiologie. *Cah. Nutr. Diet,* 1989, V. 24 (4), pp. 302-305.

Jackson, A. O.; Taylor, C. B. Plant-microbe interactions: Life and death at the interface. *Plant Cell,* 1996, V. 8 (10), pp. 1651–1668.

Hartemink, R.; Georgsson, F. Incidence of Listeria species in seafood and seafood salads. *Int. J. Food Microbiol.,* 1991, V. 12, pp. 189-195.

Gershun, V. I. Effect of Abiotic Factors on the Vital Functions of Listeria in Soil. *Sib. Vestn. S-Kh. Nauki,* 1980, V. 3, 78–81.

Gershun, V. I. Distribution of Listeria in objects of environment. News of AS of Kazakh SSR. A series biological. *Alma-Ata,* 1980, V. 6, pp. 42-44.

Gershun, V. I. Distribution and viability of Listeria in objects of external environment and influence of the temperature factor on their morphological properties. In: Questions of nidus natural of illnesses. *Alma-Ata,* 1981, Release 12, pp. 78-89.

Gershun, V. I. *Ecology of Listeria and Their Circulation Routes in a Natural Center.* In: Ecology of Sapronosis Pathogens. Moscow: Moscow State University Press; 1988, pp. 80–85.

Gershun, V. I. *Ecology of Pathogens Causing Sapronoses.* Moscow: Nauka, 1988.

Godova, G. V.; Pushkareva, V. K; Kalashnikova, E. A.; Borisova, E. Yu.; Ermolaeva, S. A.; Litvin, V. Yu. Vegetable Plants as Possible Source of Listeria. *Izv. Timiryazev. Sel'skokhoz. Akad.,* 2009, V. 4, pp. 80–89.

Golovachova, V. Ya. *About duration of a survival of pseudo-tubercular microbe in soil.* In: Far East fever similar to scarlatina (pseudo-tuberculosis of the man). Leningrad: Nauka, 1978, pp. 188-189.

Grimes, D. J. Ecology of estuarine bacteria capable of causing human disease: a review. *Estuaries,* 1991, V. 14(4), pp. 345-360.

Gusev, M. V.; Koronelli, T. V. Association of Cyanobacteria and Oil-Oxidizing Bacteria under Conditions of Petroleum Contamination. *Microbiology,* 1981, V. 50(6), pp. 1092–1097.

Kokubo, Y.; Nakama, A.; Iida, T. Listeria and safety of food products. *Mod. Media,* 1991, V. 37(10), pp. 3-10.

Korovina, V. P.; Sazonova, A. A.; Vaisman, I. Sh. Experimental Study of the Regulatory Mechanism of Population Magnitude on Bacterial Cultures. *Ekologiya,* 1974, V 6, pp. 5–9.

Kostenkova, A. F. *Mid-flight soil of Pacific Ocean coast, their salt structure and electro conductivity.* In: A soil cover of Far East, problem of its effective utilization, melioration and protection. – Vladivostok: Dal'nauka, 1987, pp. 67-73.

Kubat, J.; Kralov, M.; Novak, B. Influence of Fluctuating Temperature on Soil Microflora. *Zbl. Bakteriol., Parasitenk., Infektionskrankh. und Hyg. Abt,* 1979, V. 134 (3), pp. 229–236.

Laboratory Manual of Soil Science. Ed. by Kaurichev I. S., Moscow: Kolos, 1980.

Labinskaya, A. S. *Microbiology and Microbiological Technique.* Moscow: Meditsina, 1978.

Lakin, G. F. *Biometriya (Biometry).* Moscow: Vysshaya shkola, 1990.

Larionov, G. M. Ecologo–Biochemical Content of Sapronoses. *Zh. Mikrobiol. Epidemiol. Immunobiol.,* 1988, V. 3, pp. 36–39.

Larsson, L.; Mardh, P.; Odham, G. Analysis of Amines and Other Bacterial Products by Head-Spase Gas Chromatography. *Acta Pathol. Microbiol. Scand.,* 1978, V. B86(4), pp. 207–213.

Liston, J. Microbial hazards of seafood consumption. Toxins, bacteria and viruses are the principal causes of seafood borne diseases. *Food. Technol.,* 1990, V. 44(12), p. 58-62.

Litvin, V. Yu.; Pushkareva, I. V. A Possible Formation Mechanism of Epidemic Sapronosis Pathogens in Soil or Water. *Zh. Mikrobiol. Epidemiol. Immunobiol.,* 1994, V. 5, 89–95.

Litvin, V. Yu. Natural-Focus Infections: the Key Questions and New Vantage Points. *Zh. Mikrobiol. Epidemiol. Immunol.,* 1999, V. 5, pp. 26–33.

Litvin, V. Yu.; Gintsburg, A. L.; Pushkareva; V. K, Romanova, Yu. M.; Boev, B. V. *Epidemiological Aspects of Bacterial Ecology.* Moscow: Fargus-Print, 1998.

Litvin, V. Yu.; Somov G. L.; Pushkareva V. I. Saprozoonozy as a Natural Focal Disease. *Epidemiology and Vaccination,* 2010, no. 1, pp. 10–16.

Loginova, N. V.; Govorukhina, N. I.; Trotsenko, Yu. A. Autotrophic metabolism of methanol in Blastobacter viscosus**.** *Microbiology,* 1981, V.50, p. 591.

Lone, G. Potential hazards in cold-smoked fish: *Listeria monocytogenes, J. Food Science,* 2001, V. 66 (7), pp. 1071-1081.

Maksimenkova, I. A.; Litvin, V. Yu. *Effect of Some Soil Factors on the Dynamics of Pseudotuberculosis Microbes.* In: Problems of Microbiology, Pathogenesis, and Laboratory Diagnostics of Yersiniosis. Novosibirsk, 1985, pp. 29–33.

Markova, Yu. A.; Romanenko, A. S.; Dukhanina, A. V. Isolation of Bacteria of the Family Enterobacteriaceae from Plant Tissues. *Microbiology,* 2005, V. 74(5), pp. 575–578.

Markova, Yu. A.; Romanenko, A. S.; Igumnova, E. K.; Salyaev, R. K. Plants as Possible Source of Bacteria, Pathogenic for Humans and Animals. *Dokl. Biol. Sci.,* 2002, V. 386(2), pp. 277–279.

Morello, J. E. Negative cross-communication among wheat rhizosphere bacteria: Effect on antibiotic production by the biological control bacterium Pseudomonas aureofaciens 30-84. *Appl. Environ. Microbiol.,* 2004, vol. 70 (5), pp. 3103–3109.

Mounier, J.; Goerges, S.; Gelsomino, R.; Vancanneyt, M.; Vandemeulebroecke, K.; Hoste, B.; Brennan, N. M.; Scherer, S.; Swings, J.; Fitzgerald, G. F.; Cogan, T. M. Sources of the adventitious microflora of a smear-ripened cheese. *J. Appl. Microbiol.,* 2006, V. 101(3), pp. 668–681.

Nakama, A.; Maruyama, T.; Kokubo, Y.; Iida, T.; Umek, F. Incidence of *Listeria monocytogenes* in foods in Japan. In: Lisyeria 1992: *11th Int. Synp. Probl. Listeriosis,* Copenhagen, 11-14 May, 1992: ISOPOL XI: Booc Abstr.. - S. I., s. a., 1992, pp. 317-318.

Olafsen, J. A.; Mikkelsen, H. V.; Glaever H. M. Indigenous bacteria in hemolimh and tissues of marine bivalves at low-temperatures. *Appl. Environ. Microbiol.,* 1992, V. 59 (6), pp. 1848-1854.

Orlov, D. S. Soil Humus Acids and the General Theory of Humification. Moscow: Moscow State University Press; 1990.

Orlov, D. S.; Grishina, L. A. *Laboratory Manual of Humus Chemistry Moscow:* Moscow State University Press; 1991.

Papa, A.; Arvanitidou-Vayiona, M.; Konstantinidis, T. S. Causes of incidence of Listeria in superficial water. *Acta microbiol. Hell.,* 1996, V. 41 (16), pp. 575-579.

Pomanskaya, L. A. About multiplication of Listeria in soil. *J. Microbiol., Epidemiol. and Immunobiol,* 1963, V. 6, pp. 99-101.

Rabotnova, V. I.; Pozmogova, I. N. A feed of microorganisms. *J. Microbiol., Epidemiol. and Immunobiol,* 1993, V. 5, pp. 112-116.

Rai, B.; Srivastava, A. K.; Singh, D. B. Volatile and Non-Volatile Metabolites of Actinomycetes and the Growth of Some Litter Decomposing Fungi. *Soil Biol. Biochem.,* 1981, V. 13(1), pp. 75–76.

Rozhkova, L. P. *Far-Eastern Scarlatiniform Fever (Human Pseudotuberculosis).* Vladivostok: Dal'nauka, 1974.

Schwarzbrod, J.; Papadopoulus, O.; Burdin C. Detection et comportement des Listeria dans les bones d'epuration. *Microbiol., alim., nutr.,* 1989, V. 7 (3), pp. 225-232.

Shaegani, M.; Forge, J. De.; McGlinn, D. et al., Characteristics of Yersinia enterocolitica and Related Species Isolated from Human, Animal, and Environmental Sources. *J. Clin. Microbiol.,* 1981, V. 14 (3), pp. 304–312.

Shahamat, M.; Seaman, A. Woodbine M. Survival of *Listeria monocytogenes* in high salt concentrations. *Zbl. Bacteriol.,* 1980, Abt. 1, A246, V. 4, pp. 506-511.

Shlyakhov, S. A. *Classification of soil of sea coasts.* Vladivostok: Dal'nauka, 1996, 35 p.

Shlyakhov, S. A. *Soil flat sea coasts.* Vladivostok: Dal'nauka, 1997, p. 55.

Shlyakhov, S. A.; Kostenkov, N. M. *Soils Pacific coast of Russia, their classification, evaluation and use.* Vladivostok Dal'nauka, 2000, p. 183.

Sidorenko, M. L.; Buzoleva, L. S. Character of interactions of saprophytic soil microflora via gaseous metabolites. *Microbiology,* 2008, V. 77(2), pp. 2235–239.

Sidorenko, M. L.; Terekhova, V. E.; Buzoleva, L. S. *Features of biology agent of listeriosis in habitation in the environment.* In: Priority Russia XXI century: from the biosphere and the technosphere to the noosphere. Proceedings of the international scientific-practical conference on 21-22 January, Penza, 2003, pp. 159-161.

Skvortsov, V. V.; Kiktenko, V. S., Kucherenko, V. D. *Survival and indication of pathogenic microbes in the environment.* Moscow: Medicine, 1966, 350p.

Somov, G. P. Some preconditions to allocation of group of circumpolar infections. *Bulletin AMS USSR,* 1979, V. 6, pp. 55-60.

Somov, G. P.; Buzoleva, L. S. *Adaptation of Pathogenic Bacteria to the Abiotic Environmental Factors.* Vladivostok: Primpoligrafkombinat, 2004.

Somov, G. P.; Litvin, V. Yu. *Saprofitizm and Parasitic Pathogens Bacteria: Ecological Aspects.* Novosibirsk: Nauka, 1988.

Stotzky, G.; Schenck, S. Volatile Organic Compounds and Microorganisms, CRC Crit. *Revs Microbiol.,* 1976, V. 4(4), pp. 333–382.

Tambiev, A. Kh.; Telitchenko, M. M. *Rol' letuchikh I vodorastvorimykh biologicheski aktivnykh soedinenii biogennogo proiskhozhdeniya.* In: Role of Volatile and Water-Soluble Biologically Active Compounds of Biogenic Origin. Moscow: Moscow State University Press; 1971.

Tartakovskii, I. S.; Maleev, V. V.; Ermolaeva, S. A. *Listerii: rol' v infektsionnoi patologii cheloveka i laboratornaya diagnostika.* In: Listeria: Role in Infectious Human Diseases and Laboratory Diagnostics. Moscow: Meditsina dlya vsekh, 2002.

Terekhova, V. E.; Buzoleva, L. S.; Aizdaicher, N. A.; Yaroslavtseva, L. M.; Sergeeva, E. P.; Shakirov, R. B.; Zaitseva, E. A.; Somov, G. P. Marginal Sea north western part of the Pacific Ocean as a habitat *Listeria monocytogenes* and epidemiological significance of this phenomenon. *Epidemiology and Infectious Diseases,* 2002, V.1, pp. 43-46.

Timchenko, N. F.; Bulgakov, V. P.; Bulakh, E. V.; Yasnetskaya, E. G.; and Zhuravlev, Yu. N. The interaction of Yersinia, Listeria and Salmonella with plant cells. *Zh. Mikrobiol. Epidemiol. Immunol.,* 2000, V. 1, pp. 6–10.

Tirranen, L. S. *Rol' letuchikh metabolitov v mezhmikrobnom vzaimodeistvii.* In: Role of Volatile Metabolites in the Interaction of Microorganisms. Novosibirsk: Nauka, 1989.

Tirranen, L. S. Formation of higher plant component microbial community in closed ecological system. *Acta Astronaut.,* 2001, V. 49(1), pp. 47–52.

Tirranen, L. S.; Borodina, E. V.; Ushakova, S. A.; Rygalov, V. Ye.; Gitelson, I. I. Effect of volatile metabolites of dill, radish and garlic on growth of bacteria. *Acta Astronaut.,* 2001, V. 49(2), pp. 105–108.

Tirranen, L. S.; Gitelson, I. I. The role of volatile metabolites in microbial communities of the LSS higher plant link. *Adv. Space Res.,* 2006, V. 38 (6), pp. 1227–1232.

Tirranen, L. S.; Shilenko, M. P. Microorganisms: Indicators of the state of higher plants. *Aerospace and Environmental Medicine,* 2008, V. 42 (6/1), pp. 36–39.

Tulchinskaya, V. P.; Gubanov, V. V. Study of mussel contamination of E. coli bacteria and vibrios paragemoliticheskimi. In: Harvesting clams, mussels, and their role in ecosystems. Leningrad: *Nauka,* 1979, p. 120-122.

Vlasov, A. A.; Pavlova, I. B. Survival and Morphologyof Listeria in Plant Tissues Feed Oats. Sel'skokhoz. *Biol.,* 2009, V. 4, pp. 89–92.

Weiss, J.; Seeliger, H. Incidence of *Listeria monocytogenes* in nature. *Appl. Microbiol.,* 1975, V. 30 (1), pp. 29-32.

Welsimer, H. J. *The genus Listeria and related organisms.* In: The Prokaryotes. A Handbook on Habitats, Isolations and Identification of Bacteria. New-York: Springer-Verlang, 1981, pp. 1680-1687.

Watkins, J.; Sleath, K. P. Isolation and enumeration of *Listeria monocytogenes* from sewege, sewege sluge and river water. *J. Appl. Bacteriol.,* 1981, V. 50, pp. 655-657.

Zavarzin, G. A. Extensive Microbiology. *Izv. Akad. Nauk SSSR, Ser. Biol.,* 1976, V. 1, pp. 121–134.

Zavarzin, G. A.; Bonch-Osmolovskaya, E. A. Syntrophic Interactions in Microbial Communities. *Microbiology,* 1981, V. 50 (2), pp. 165–173.

Zenova, G. M.; Rydkina, E. B.; Kalakutskii, L. V. Growth and Antimicrobial Activity in an Association of an Actinomycete and a Green Alga. *Biologicheskie Nauki,* 1983, V.3, pp. 81–85.

Znamensky, V. A. *Far East fever similar to scarlatina (pseudo-tuberculosis).* In: An illness of naturofocal in Primorye Territory. Vladivostok, 1975, pp. 136-160.

Zvyagintsev, D. G. Gas Phase of Soil and Microorganisms, in microorganisms' role in cycle of gazes in nature. In: Role of Microorganisms in Gas Turnover in Nature. Moscow: *Nauka,* 1979, pp. 92–104.

Zvyagintsev, D. G. *Soils and Microorganisms.* Moscow: Moscow State University Press; 1987.

In: *Listeria monocytogenes*
Editor: Edmund C. Hambrick

ISBN: 978-1-63117-054-6

Chapter 13

LISTERIA PATHOGENICITY ISLAND 1: STRUCTURE AND FUNCTION

Spiros Paramithiotis**, Agni Hadjilouka and Eleftherios H. Drosinos
Laboratory of Food Quality Control and Hygiene,
Department of Food Science and Human Nutrition,
Agricultural University of Athens, Athens, Greece

ABSTRACT

Listeria pathogenicity island 1 (LIPI-1) is located between *prs* and *orfX* loci. It consists of six genes that possess central role in the infection cycle of *L. monocytogenes* and is divided into three transcriptional units; the *hly* monocistron is considered as the center, downstream of which *mpl*, *actA* and *plcB* are found and upstream of which *plcA* and *prfA* are located. LIPI-1 sequence analysis has been used to improve our understanding regarding the phylogeny and evolution of both pathogenic and nonpathogenic *Listeria* spp. Moreover, a significant amount of data has been published, especially over the last decade, regarding their regulation and the effect of environmental stimuli, such as growth conditions and food processing interventions on their transcription rate. In this chapter all information relevant to LIPI-1 structure and function is integrated and critically reviewed.

INTRODUCTION

The ubiquitous nature of the microorganisms classified to the *Listeria* species has been already established in the literature; the ecological niches from which they are isolated are very diverse and extend in nearly all agricultural and food associated environments where they survive as saprophytes. *Listeria* species consists of ten members, namely *L. innocua*, *L. monocytogenes*, *L. welshimeri*, *L. seeligeri*, *L. ivanovii*, *L. grayi*, *L. rocourtiae*, *L. marthii*, *L.*

* Corresponding author. e-mail: sdp@aua.gr; Tel.: +30 210 5294705 Fax: +30 210 5294683

weihenstephanensis and *L. fleischmannii*, with the latter four species being only recently described (Leclercq et al., 2010; Lang Halter et al., 2013; Graves et al., 2010; den Bakker et al., 2013). In Figure 1, their relative phylogenetic position based on Neighbor-Joining distance analysis of their 16S-rRNA gene sequences is shown. It seems that *L. monocytogenes*, *L. innocua*, *L. welshimeri*, *L. marthii*, *L. ivanovii* and *L. seeligeri* form a six-species cluster; *L. rocourtiae* and *L. weihenstephanensis* a two-species cluster while *L. grayi*, and *L. fleischmannii* are excluded from both. Concerning the six-species cluster, 16S-rRNA gene sequences cannot offer better resolution on their phylogenetic relationships, due to the high degree of resemblance, and sequence analysis of additional genomic regions is required.

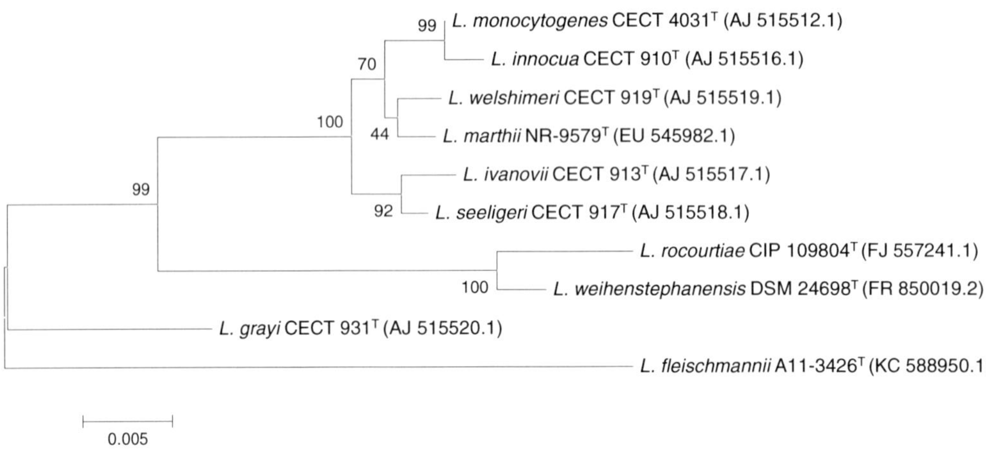

Figure 1. Phylogenetic position of the *Listeria* species based on Neighbor-Joining distance analysis of 16S-rRNA gene sequences. Bootstrap values at the node were calculated from 1000 samplings. GenBank/EMBL/DDBJ accession numbers are given in parenthesis.

Only two of the species currently assigned to the *Listeria* genus are recognized to cause listeriosis, namely *L. monocytogenes* and *L. ivanovii*. The former is associated with infection of humans as well as a wide range of animals, including mammals and birds, whereas the latter is mainly pathogenic for ruminants (Orsi et al., 2011). Both human and animal listeriosis appears to be predominantly food- and feed- borne, respectively. The key features that distinguish these two species and enable them to survive and spread within the host are the ability to enter the host cells, i.e., break the intestinal barrier, and cell-to-cell spread. By these functions, the microorganisms may evade both the extracellular environment and the humoral effectors of the immune system, and concomitantly become pathogens for the host. This intracellular parasitism requires the expression of a series of genes, denoted as virulence genes due to the effect on the host, many of which are physically gathered in a 9-kb gene cluster, named Listeria Pathogenicity Island 1 (LIPI-1).

LIPI-1 is present and functional in *L. monocytogenes* and *L. ivanovii*, a non-functional version is present in *L. seeligerii* and is totally absent from *L. innocua*, *L. welshimeri*, *L. marthii* and *L. grayi*. As far as *L. rocourtiae*, *L. weihenstephanensis* and *L. fleischmannii* were concerned, no such data are currently available. However, these species have been described as non-haemolytic and avirulent. Acquisition of LIPI-1 has been and still remains the subject of extensive speculation (Vazquez-Boland et al., 2001). Although it is still not clear whether this gene cluster has been acquired as a cassette or via a step-by-step assembly process, combination of the above data referring to LIPI-1 occurrence with the phylogenetic analysis based on 16S-rRNA gene sequences indicates the existence of 4 sublines of descent, three of which seem to have evolved from the common ancestor before acquisition of LIPI-1. The first subline corresponds to *L. fleischmannii*, the second to *L. grayi* and the third to *L. rocourtiae* and *L. weihenstephanensis*. Acquisition of LIPI-1 seems to have occurred after the latter two species diversified from the rest listeriae. Then, LIPI-1 was stabilized (*L. monocytogenes* and *L. ivanovii*), modified (*L. seeligerii*) or even lost from the genome (*L. innocua*, *L. welshimeri* and *L. marthii*).

LIPI-1 PHYSICAL ORGANIZATION

Physical organization of LIPI-1 is shown in Figure 2; it is flanked by *prs* and *orfX* and consists of *prfA*, *plcA*, *hly*, *mpl*, *actA* and *plcB*. Identical is the physical organization of the LIPI-1 present in *L. ivanovii*. On the other hand, the version of LIPI-1 that is present in *L. seeligeri* is more elaborated and, furthermore, non-functional. The later has been attributed to the insertion of *orfE* between *prfA* and *plcA* genes that lead to a disruption of the positive autoregulatory loop that is necessary for appropriate expression of *prfA*-dependent virulence genes. From an evolutionary point of view, a series of evidence supports the notion that LIPI-1 was present in the common ancestor of all these species (Figure 1) rather than obtained through horizontal gene transfer after the speciation process.

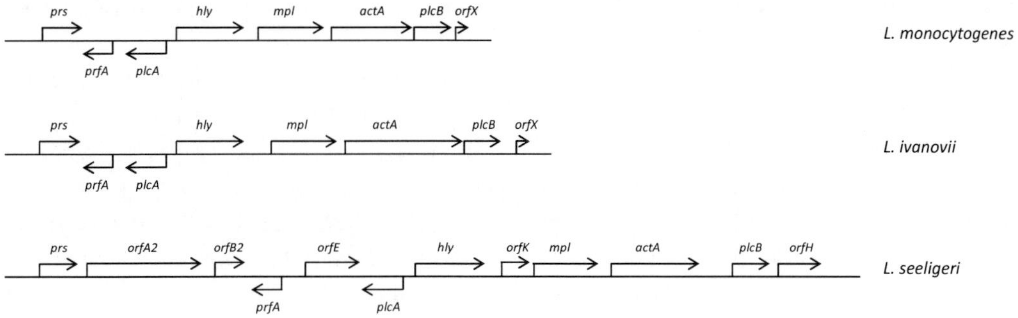

Figure 2. Physical organization of the LIPI-1 in *L. monocytogenes*, *L. ivanovii* and *L. seeligeri*. Direction of transcription is indicated by the respective arrows.

Transcription of all genes present in the LIPI-1 is mainly controlled by the transcriptional activator PrfA, a member of the cyclic AMP (cAMP) receptor protein (CRP)/FNR family of bacterial regulators that is encoded by *prfA*. Additionally, the expression of more than 140 genes is directly or indirectly influenced by PrfA, highlighting the importance of this molecule to *L. monocytogenes* homeostasis.

Two phospholipases C are encoded in LIPI-1: PlcA and PlcB. The former is a phosphatidylinositol phospholipase C (PI-PLC) and encoded by *plcA*. Expression of *plcA* has linked with pathogenic *L. monocytogenes* and *L. ivanovii* and is considered a reliable marker for discrimination between pathogenic and non-pathogenic *Listeria* spp. (Aurora et al., 2008). Therefore, Agar Listeria according to Ottaviani and Agosti as well as RAPID′L.mono are based on detecting PI-PLC activity. PlcB is a phosphatidylcholine phospholipase C (PC-

PLC), a broad spectrum phospholipase C, which is encoded by *plcB*. Both phospholipases cooperate with listeriolysin O (LLO) for effective escape from the phagocytic vacuole allowing thus the escape into the cytoplasm. PlcB can also promote lysis of the primary vacuole in the absence of LLO (Grundling et al., 2003); however both phospholipases are required for virulence in mice (Schluter et al., 1998).

mpl encodes a thermolysin-related zinc-metalloprotease involved in the maturation of pro-PlcB (Domann et al., 1991; Raveneau et al., 1992). However, it has been exhibited that two independent pathways are involved in pro-PlcB activation in vacuoles: an Mpl-dependent and an Mpl-independent with the latter being sensitive to inhibitors of cysteine proteases. Interestingly enough, proPC-PLC activation by either pathway has been reported as sensitive to bafilomycin A_1, a specific inhibitor of vacuolar ATPase, suggesting that activation was dependent on acidification of the vacuolar compartment and therefore is compartment specific and controlled by a combination of bacterial and host factors (Marquis et al., 1997).

hly encodes a cholesterol-binding, pore-forming toxin (listeriolysin O, LLO) essential for the bacterial escape from phagosomes. However, recent studies are revisiting the role of LLO during infection and are revealing new insights into the action of LLO, in particular before bacterial entry (Hamon et al., 2012). It was the first virulence gene identified in *L. monocytogenes* and the concomitant characterization of the respective genomic locus led to the identification of LIPI-1 (Dussurget 2008). Absence of LLO essentially leads to avirulence (Kathariou et al., 1987; Cossart et al., 1989). It has been reported that vacuolar perforation by LLO is pH dependent and takes place preferentially at acidic vacuolar pH (4.9-6.7). On the contrary, a frequency reduction or even complete prevention has been reported upon treatment with the lysosomotropic agents ammonium chloride or bafilomycin A_1 (Beauregard et al., 1997).

actA encodes the ActA protein, a surface protein that appears to be a multifunctional virulence factor. It is necessary for actin-based intra-and inter-cellular motility. Moreover, it plays a major role in epithelial cell invasion: a decrease in invasiveness has been reported upon ActA inactivation that was fully restored upon recovery by trans-complementation of the Δ*actA* mutant. On the other hand, heterologous expression of actA in *L. innocua* resulted in the transformation of the otherwise non-invasive species to invade epithelial cells (Suarez et al., 2001). Finally, *actA* expression has been linked to autophagy evasion during colonization of macrophages. However, ActA was recognized as only one of multiple mechanisms, such as PI-PLC and PC-PLC, that *L. monocytogenes* utilizes to avoid destruction by the autophagy system (Birmingham et al., 2007).

Sequence analysis of LIPI-1 as a whole as well as of the individual genes and the intergenic regions has been performed to study the diversity and the evolution of the *L. monocytogenes* lineages. An extended variability has been exhibited in the studies conducted by Orsi et al., (2008) and Tsai et al., (2011). In the later study, strong evidence has been presented regarding the divergence between lineages III and IV and it has been stated that those two lineages are highly divergent and monophyletic, supporting thus the division of the former linage III to III and IV (den Bakken et al., 2010; Orsi et al., 2011; Ward et al., 2008). In both studies, lineages I, II, III and IV were represented by 18, 21, 8 and 4 strains, respectively; the number of alleles of the whole *prfA* cluster was 12, 12, 8 and 4, respectively. *hly* and *prfA* have been characterized as highly conserved, at least regarding lineages I and II, whereas *mpl*, *actA* and *plcA* as diverse that have been evolved by positive selection, the first in lineages I and II and the rest in all lineages. Moreover, the promoter region of *hly* and *plcA* also seem to have evolved by positive selection, at least in lineages I and II. Finally, strains belonging to lineages II and III exhibited higher degree of recombination that is consistent with observations referring to other chromosomal regions (Meinersmann et al., 2004; Orsi et al., 2007; Levin and Cornejo 2009) and may explain their genetic diversity.

Table 1. Example of PCR assays used for the detection of virulence associated genes of LIPI-1

Target gene	Primers	Product size (bp)	Reference
prfA	F: 5′-GGTATCACAAAGCTCACGAG-3′	571	Cooray et al., (1994)
	R: 5′-CCCAAGTAGCAGGACATGCTAA-3′		
hlyA	F: 5′-CACTCAGCATTGATTTGCCA-3′	276	
	R: 5′-ATTTTCCCTTCACTGATTGC-3′		
plcB	F: 5′-GCAAGTGTTCTAGTCTTTCCGG-3′	795	
	R: 5′-ACCTGCCAAAGTTTGCTGTGA-3′		
plcA	F: 5′-CTGCTTGAGCGTTCATGTCTCATCCCCC-3′	1484	Rawool et al., (2007)
	R: 5′-CATGGGTTTCACTCTCCTTCTAC-3′		
prfA	F: 5′-CTGTTGGAGCTCTTCTTGGTGAAGCAATCG-3′	1060	
	R: 5′-AGCAACCTCGGTACCATATACTAACTC-3′		
hlyA	F: 5′-GCAGTTGCAAGCGCTTGGAGTGAA-3′	456	
	R: 5′-GCAACGTATCCTCCAGAGTGATCG-3′		
iap	F: 5′-ACAAGCTGCACCTGTTGCAG-3′	131	
	R: 5′-TGACAGCGTGTGTAGTAGCA-3′		
actA	F: 5′-CGCCGCGGAAATTAAAAAAAGA-3′	839	
	R: 5′-ACGAAGGAACCGGGCTGCTAG- 3′		
plcA	F: 5′-CTCGGACCATTGTAGTCATCTT-3′	326	Lomonaco et al., (2012)
	R: 5′-CACTTTCAGGCGTATTAGAAACGA-3′		
actA	F: 5′-CCAAGCGAGGTAAATACGGGA-3′	650	
	R: 5′-GTCCGAAGCATTTACCTCTTC-3′		

It has been recognized very early that *L. monocytogenes* strains exhibit significant variability, especially regarding pathogenicity. Therefore, a variety of laboratory tests have been developed in order to assess virulence potential. Detection of virulence associated genes seemed as an adequate alternative to overcome the disadvantages of *in vivo* bioassays and *in vitro* cell assays, namely, high cost and use of animals regarding the former and lack of reproducibility regarding the latter. Virulence genes belonging to LIPI-1 have been the target of both simplex and multiplex PCR assays (Table 1). However, it should be mentioned that detection of a gene does not necessarily implies its expression and, on the other hand, inability to detect a gene may reveal allelic polymorphisms rather than absence, especially in the case of LIPI-1. In any case, when evolutionary events are suggested they should be supported by further evidence.

A relationship between the virulence level of *L. monocytogenes* strains and their detectability was established in a study combining virulent, hypovirulent and avirulent strains and PALCAM, Oxford, RAPID′L. mono and ALOA media (Gracieux et al., 2003). ALOA medium succeeded in recovering all virulent and 92% of hypo- and avirulent strains. On the contrary, up to 60% of hypo- and avirulent strains were not or poorly recovered in 2 days on PALCAM, Oxford and RAPID′L. mono media. However, with a 3 day incubation period the proportion of strains not detected with PALCAM medium fell to 8% suggesting that a longer incubation period of this medium would allow improving detection of hypo- and avirulent strains. It was suggested that higher susceptibility to specific components in the selective media, and more accurately the antimicrobial components, was responsible for poor detection of low-virulence strains and was unrelated to PrfA (Roche et al., 2009).

Significant amount of research has taken place in order to correlate point mutations in key virulence genes, mainly *prfA*, and virulence level. A spontaneous mutation of a *L. monocytogenes* serovar 4b strain, more accurately a point mutation in codon 145 of *prfA* leading to a serine substitution of glycine, resulted in a constitutive overexpression of *prfA* and PrfA-dependent virulence genes and remained high under culture conditions in which virulence factor expression is downregulated in wild-type *L. monocytogenes* (Ripio et al., 1997). This overexpression has been attributed to a conformational rearrangement of the HTH motif that resulted in the adoption of a more favorable conformation for DNA binding; comparable to cAMP-induced Cap (Eiting et al., 2005). The effect of PrfA E77K and G155S spontaneous mutations on the pathogenic potential of the carrying mutant strains has been described by Shetron-Rama (2003). Strains containing either mutation were found to have levels of *actA* expression that resembled the levels normally only observed for bacteria within the cell cytosol but reduced motility. Moreover, PrfA G155S strains were approximately fivefold more virulent in mouse models of infection than the PrfA E77K mutant. In a study of 26 field strains characterized as hypovirulent by a plaque-forming assay as well as by a subcutaneous inoculation test in mice it has been concluded that low virulence was due to a) inactive PrfA proteins due to either a single amino acid substitution, i.e., PrfAK220T, or the production of truncated PrfA, b) low PI-PLC and/or PC-PLC activities due to multiple substitutions in the respective gene, c) poor colonization of mice spleens of unspecified reason (Roche et al., 2005).

The strains exhibiting no PI-PLC activity were further studied and a Thr262Ala substitution has been held responsible. Moreover, these strains exhibited low invasion rate due to a nonsense codon leading to a lack of InlA protein synthesis, and to an Ala117Thr substitution in the leucine-rich repeat of InlB, which altered the interaction with the Met receptor (Temoin et al., 2008). In this context, low virulence has been attributed to point mutations in a number of virulence genes. A mutation that promoted *L. monocytogenes* entry into the cytosol but inhibited actin-based bacterial motility, thereby attenuating bacterial virulence, has also been described (Miner et al., 2008). Although the exact mechanism by which the PrfA Y154C mutation prevented efficient actin polymerization have not been elucidated, it has been suggested that either decreased PrfA protein stability in cytosolic bacteria or failure to undergo necessary conformation shift following bacterial entry into the cytosol may account for the low levels of *actA* expression. Finally, another mutation that resulted in a constitutively activated form of the PrfA protein and high-level virulence gene expression in bacteria grown outside of host cells, conferring some degree of toxicity upon bacterial cells while making them defective for swimming motility has been identified by Wong and Freitag (2004). Strains with PrfA L140F mutation exhibited no significant increase in either *actA* or *hly* expression suggesting that the expression of PrfA-dependent genes is most sensitive to variations in PrfA protein levels when PrfA is present in its nonactivated state.

LIPI-1 TRANSCRIPTIONAL ORGANIZATION

LIPI-1 is organized in three major transcriptional units (Figure 3). The central unit consists of the *hly* gene, downstream of which lies the *mpl-actA-plcB* operon that may be expressed either as a single transcript or in shorter ones: one for *mpl* and a bicistron for *actA* and *plcB*. Upstream of *hly*, the *plcA-prfA* operon is located that may be expressed as a bicistron or separately.

Transcription of the genes in LIPI-1 is largely regulated by PrfA, which requires the presence of a 14-bp sequence of dyad symmetry, with the consensus sequence TTAACANNTGTTAA, mostly referred to as 'PrfA-box'. The PrfA-box is centered at position -41 relative to the transcriptional start site and partially overlaps the -35 promoter regions. In Figure 3 occurrence of the PrfA-box in a promoter sequence for each gene in LIPI-1 is designated by a closed box. The PrfA-box in front of the genes *hly* and *plc*A exhibits a perfectly symmetric sequence while the respective of the other genes contains one or two mismatches and therefore PrfA binds less efficiently (Dickneite et al., 1998; Boeckmann et al., 2000; Lalic-Muelthaler et al., 2001). Moreover, the distance between PrfA-box and -10 box as well as interspace sequence and the sequence downstream of the -10 box are also of great importance. Indeed, Luo et al., (2005) reported that the RNA polymerase binding site as and the sequence immediately downstream of the putative -10 site were responsible for the inhibition of PrfA-dependent transcription by P_{aroA2}.

prfA expression and concomitant regulation of the LIPI-1 genes has been the subject of intensive research due to the apparent importance on *L. monocytogenes* virulence and the complexicity of PrfA regulation has been highlighted.

Synthesis of monocistronic *prfA* mRNA is initiated by two promoter sites designated $P1_{prfA}$ and $P2_{prfA}$. The former is recognized by the housekeeping σ^A. The latter consists of two overlapping promoters, one recognized by σ^A and one by σ^B, and contains a PrfA box (Mengaud et al., 1991; Nadon et al., 2002; Rauch et al., 2005; Scwab et al., 2005; Ollinger et al., 2009). Moreover, bicistronic *plcA-prfA* mRNA is initiated by the *plcA* promoter that also contains a PrfA-box (Figure 3). However, $P2_{prfA}$ has been reported to negatively regulate PrfA expression, forming thus a negative feedback loop. However this system seems to be dispensable for virulence (Greene and Freitag, 2003).

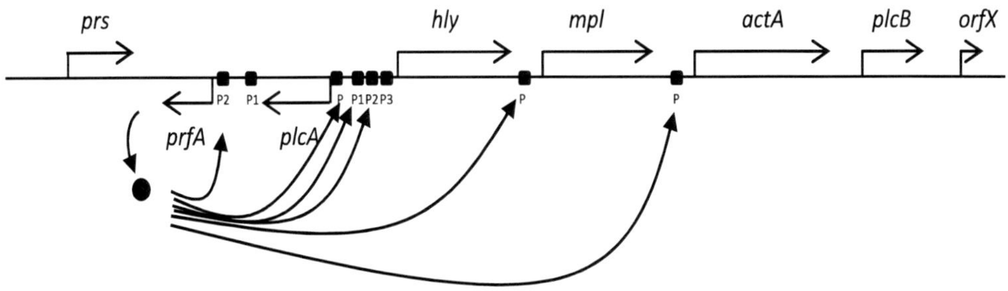

Figure 3. Transcriptional organization of the LIPI-1 in *L. monocytogenes*. *P* indicates known promoters. Closed boxes indicate presence of 'PrfA-box' and arrows indicate PrfA-dependent regulation at transcriptional level. (modified by Kleft and Vazquez-Boland, 2001).

prfA expression is also thermoregulated. It has been shown that at temperatures below 37°C the total amount of PrfA is reduced due to the absence of the *plcA-prfA* bicistronic message (Leimeister-Waechter et al., 1992) while transcripition of *prfA* carries on from $P2_{prfA}$ that do not exhibit temperature dependence. Taken into consideration that $P2_{prfA}$ is, at least partially, controlled by σ^B, a mechanism for σ^B-dependent production of PrfA is evident, even at temperatures encountered outside a mammalian host. Inactivation of $P2_{prfA}$ has no effect on virulence in intra-gastrically inoculated guinea pigs or cell-to-cell spread in, at least some, mammalian cell lines (Chaturongakul et al., 2008).

Cellobiose, a plant disaccharide relatively abundant in soil, has also been reported to affect expression of PrfA-dependent genes, most probably post-translationally as the total amount of PrfA was not affected. Indeed, in a study by Renzoni et al., (1997), presence of cellobiose had no effect on the expression of the *prfA* gene and thus the total amount of PrfA present was not reduced. However, due to the concomitant reduction to the amounts of listeriolysin O and ActA, the hypothesis that cellobiose causes a reduction to the levels of active PrfA synthesis seemed to be supported. This effect has also been verified by Milenbachs et al., (1997) and extended to several other mono- and di- saccharides (such as glucose, fructose, maltose, trehalose), although in a strain dependent manner, suggesting that this effect may be correlated with the metabolic utilization of the carbohydrates rather than an environmental signaling mechanism that was initially proposed. These studies have been integrated by Milenbachs Lukowiak et al., (2004), in which it was proposed that complete abrogation of virulence regulation by cellobiose requires mutations in two unlinked loci, namely *gcr* and *csr*. Strains carrying mutations only in the former locus retain cellobiose-mediated repression but lose repression mediated by other sugars while strains carrying mutations only in the latter locus exhibit partial relief of cellobiose repression. More recently, Joseph et al., (2008) used glycerol as a carbon source and reported that the highest transcriptional upregulation was observed for all PrfA-dependent genes during early and late logarithmic growth phase, suggesting that PrfA activity may be correlated with the phosphorylation state of PTS permeases.

Another regulatory loop has been revealed by the study conducted by Loh et al., (2009). More accurately, the S-adenosyl-methionine riboswitches SreA and SreB have been shown to down-regulate *prfA* expression post-transcriptionally by binding to the distal side of the *prfA*-untranslated protein. Since SreA expression is PrfA dependent, a regulatory loop is formed.

Survival of *L. monocytogenes* within the host requires, among others, two pleiotropic regulatory elements, namely PrfA and the potential chaperone ClpC that is involved in tolerance to environmental stress. It has been reported by Ripio et al., (1998) that PrfA negatively affected *clpC* expression. More accurately, in a mutant that constitutively overexpresses PrfA (PrfA*) and PrfA-dependent genes, *clpC* transcription was reduced to basal levels during exponential growth at 42°C but not during stationary phase. Furthermore, culture in charcoal-treated medium, which triggers in wild-type strains the transcriptional activation of the PrfA regulon (Ermolaeva et al., 2004), also caused a strong downregulation of *clpC*. In addition, *clpC* transcription during exponential growth at 37°C was clearly enhanced only when a *prfA* deletion mutant was used.

It has been suggested that the DNA sequence of PrfA binding sites determines the levels of expression of certain virulence genes, such as *hly* and *mpl*. Other virulence genes, such as *actA* and *plcB*, may depend upon additional factors for full regulation of gene expression (Williams et al., 2000). Moreover, the activity of PrfA in *L. monocytogenes* (PrfALm), *L. ivanovii* (PrfALi) and *L. seeligeri* (PrfALs), i.e., the listeriae in which it is chromosomally present, has been studied by Mauder et al., (2006). It has been reported that amino acid differences in the C-terminal domain of PrfALs, compared to the respective of PrfALm and PrfALi, might contribute to the reduced efficiency to form ternary complexes and initiate transcription at P_{hly} and P_{actA}. On the contrary, the efficiencies of PrfALm, PrfALi and PrfA*Lm, were similar.

It has been demonstrated that increased PrfA synthesis occured during adherence to mammalian cells and in cell extracts due to enhanced transcription of *prfA* gene through *plcA* promoter as well as enhanced activity of itself (Renzoni et al., 1999). However, in a study by Kazmierczak et al., (2006) involving exponential growth phase *L. monocytogenes* exposed to environmental stresses or present in the vacuole or cytosol of human intestinal epithelial cells, although total PrfA activity, as reflected by *plcA* transcript levels, was up to 95-fold higher in intracellular *L. monocytogenes* as compared to non-stressed bacteria, read-through transcription from the upstream *plcA* promoter was very low under all conditions. On the contrary, transcription was initiated by the σ^A-dependent $P1_{prfA}$ and the σ^A and σ^B -dependent $P2_{prfA}$ with relative contribution ranging from 17% to 30% and 70% to 82%, respectively. However, σ^B was primarily activated during environmental stress and seemed to have no contribution to PrfA activation in intracellular *L. monocytogenes*.

Transcription of *hly* is initiated by three promoter sites: $P1_{hly}$, $P2_{hly}$ and $P3_{hly}$. The two former contain PrfA box and therefore are PrfA-dependent whereas the latter is PrfA-independent (Domann et al., 1993).

Transcription of *hly*, along with *actA*, has been demonstrated to take place preferentially during intracellular growth. More acurately, Moors et al., (1999) reported that LLO was produced in both broth cultures and cytosol while ActA production was much lower in the former and significantly larger in the latter, revealing that many factors may play significant in regulating the levels of ActA compared to LLO.

hly expression is also affected by carbohydrates and extracellular pH value. It has been reported that several sugars down-regulate *hly* expression (Milenbachs et al., 1997) despite the unchanged levels of PrfA, leading to the hypothesis that regulation of *hly* expression by sugars is an aspect of global catabolite regulation rather than signature molecule mediated signal transduction unique to cellobiose. Moreover, carbohydrate regulation was found to act in a strain-specific manner and therefore it was suggested that at least two sugar sensing pathways occur simultaneously in *L. monocytogenes*. The observation made by Behari and Youngman (1998) that *hly* expression was also regulated by low extracellular pH suggested that the down-regulation of *hly* expression in the presence of sugars may be due to a combination of effects produced by changes in pH or growth rate or by specific regulatory proteins. It seemed that not only several environmental factors regulate virulence gene expression but that any one factor may affect this regulation by several different mechanisms (Behari and Youngman 1998).

Regarding *mpl* promoter (P*mpl*), an overlapping PrfA - dependent and – independent activity has been demonstrated by Luo et al. (2004). Both promoters share the same -10 box, explaining, thus the dual mode of expression, i.e., PrfA – independent and – dependent that has been reported (Lalic-Muelthaler et al., 2001; Freitag et al., 1993; Williams et al., 2000) *in vitro* and *in vivo*, respectively. In the same study, the effect of GTP on the transcription mode of *mpl* has been reported and a model has been proposed: at low GTP PrfA is either absent or in inactive form and transcription is initiated by a complex consisting of RNA polymerase loaded with SigA and the PrfA-independent promoter. On the other hand, at high GTP concentration, PrfA is complexed with RNA polymerase and the PrfA-dependent promoter and transcription is initiated; at the same time PrfA-independent activity is suppressed (Luo et al., 2004).

The transcriptional response of *inlA*, *prfA*, *plcA* and *hly* during *L. monocytogenes* inactivation by nisin was studied by Shi et al., (2013). Overall, expression of these virulence genes responded uniformly to different nisin concentrations at different temperatures. More accurately, expression decreased rapidly and remained constant at both 37 and 4°C. With cell count increase, expression of virulence genes also increased.

The effect of heat treatment of a multicomponent matrix, such as BHI broth, on virulence gene expression has been studied by Sheikh-Zeinoddin et al., (2000). It was demonstrated that although growth of *L. monocytogenes* was not affected by the heat treatment, virulence gene expression was decreased. It was thus concluded that specific inhibitors of the virulence gene operon were present and correlated to the increase of physicochemical factors that are considered to enhance the Maillard reaction that were also found to increase the levels of suppression.

In a genome wide study by Stasiewicz et al., (2011) it was reported that virulence gene transcription was induced as a response to lactate and diacetate; i.e., organic acids often used to inhibit *L. monocytogenes* growth in ready-to-eat foods. However, elucidation of the exact mechanisms requires further study.

The effect of several food constituents on *L. monocytogenes* virulence gene expression has been studied. However, only limited amount of studies address virulence gene expression in actual food matrix. Generally, the effect of the strain used seemed to be at least similar to the respective of the growth conditions studied.

Comparative expression of key virulence genes (*hly*, *actA* & *inlA*) of two strains (one of high and one of low virulence) while grown on salmon has been studied by Duodu et al., (2010). It has been exhibited that no significant expression of those genes occurred for the highly virulent strain while significant overexpression of *hly* at 20°C compared to 4°C at both 24 and 48 hours, of *actA* and *prfA* at 20°C compared to 4°C at 24 hours have been recorded.

In a study conducted by Schrama et al., (2013), the expression of *hly*, *actA* and *inlA* of four *L. monocytogenes* strains as a response to adaptation to low pH and salt (pH 5.5; 3.5% NaCl), represented by a simulated cheese medium, was assessed. It seemed that gene expression was affected in a strain dependent manner. No correlation was possible between gene expression and virulence potential as assessed using the *Galleria mellonella* infection model.

Significant effect on the expression of *plcA*, *iap* and *hly* by the food matrix during storage at 4 and 12°C in a strain specific manner has been reported by Rantsiou et al., (2012a). More accurately, fermented sausage, soft cheese, UHT milk and minced meat were inoculated and gene expression was assessed after 48 h at 4 or 12°C. *plcA* and *iap* expression showed significant variability but no general trend could describe the response of all three strains. On the other hand, *hly* expression was upregulated in almost all cases with significant differences between the matrices.

In a microarray-based study by Rantsiou et al., (2012b) the comparative expression of 54 genes according to different strains inoculated on minced meat and fermented sausage juices and incubation at 4°C after 30 min and 48 h was assessed. It was reported that *mpl* and *plcB* were mostly downregulated while *plcA*, *actA*, *hly* and *prfA* were differentially affected after incubation in meat juices comparing to growth in BHI broth at 37°C, overnight. When comparing to growth in BHI broth at 4°C for 30 min, *plcA* and *hly* were downregulated while *mpl*, *prfA*, *actA* and *plcB* were differentially expressed. However, comparison with growth in BHI broth at 4°C for 48 h revealed that *hly*, *plcA* and *mpl* were upregulated while *plcB actA* and *prfA* were differentially expressed.

Standard liver pâté as well as three variants with reduced NaCl content, one of which has also been supplied with organic acids (Ca-acetate and Ca-lactate) were used to study *prfA*, *inlA*, *sigB* and *clpC* expression of three *L. monocytogenes* strains. Strain-dependent effects have been recorded; more accurately a reduction in *prfA* for one strain and a reduction in *inlA* and *sigB* expression for two strains were reported when grown in standard liver pâté compared to growth in BHI broth. Reduction of NaCl content of the standard liver pâté did not change relative transcription levels of *prfA*, *inlA*, *sigB* or *clpC* (except for *prfA* in one strain and *sigB* in another) (Olesen et al., 2010).

Response of *L. monocytogenes* to mechanical stress injury has been studied by Bowman et al., (2008) with the application of high hydrostatic pressure processing, a process currently used for certain foods to control the presence of food-borne pathogens. It has been concluded that the whole SigB and PrfA regulons have been downregulated.

Kastbjerg et al., (2010) studied eleven disinfectants representing four different groups of active components for their effect on virulence gene expression, namely *prfA*, *plcA*, *inlA* and *hly*. It has been concluded that sublethal concentrations of disinfectants routinely used in the food industry affect virulence gene expression in *L. monocytogenes*, and the effect depends on the active components of the disinfectant expression. More accurately, gene expression was reduced by peroxy and chlorine compounds and induced by triclosan and quaternary ammonium compounds.

Eleven strains of *L. monocytogenes* were used by Alessandria et al., (2013) in a comparative study on the expression of virulence and stress response genes (*plcA*, *hly*, *iap* and *sigB*) in synthetic laboratory medium and in a soft cheese at two temperatures (4 and 12°C) and for different times (24 and 48 h). Significant expression differences emerged at 12°C after 48 h. The results obtained highlighted the differential expression in different strains, underlining the different pathogenic potentials among different isolates. The differences observed could not be associated neither with the origin of the strains nor their serotype.

Acknowledgments

The authors gratefully acknowledge funding from the European Union Seventh Framework Programme (FP7/2007-2013) under grant agreement n. 289719 (Project QUAFETY).

References

Alessandria, V; Rantsiou, K; Dolci, P; Zeppa G; Cocolin, L. A comparison of gene expression of *Listeria monocytogenes in vitro* and in the soft cheese Crescenza. *Int. J. Dairy Technol.*, 2013, 66, 83-89.

Aurora, R; Prakash, A; Prakash, S; Rawool, DB; Barbuddhe SB. Comparison of PI-PLC based assays and PCR along with in vivo pathogenicity tests for rapid detection of pathogenic *Listeria monocytogenes*. *Food Control*, 2008, 19, 641–647.

Beauregard, KE; Lee, KD; Collier, RJ; Swanson, JA. pH-dependent perforation of macrophage phagosomes by listeriolysin O from *Listeria monocytogenes*. *J. Exp. Med.*, 1997, 186, 1159-1163.

Behari, J; Youngman, P. Regulation of *hly* expression in *Listeria monocytogenes* by carbon sources and pH occurs through separate mechanisms mediated by PrfA. *Infect. Immun.*, 1998, 66, 3635–3642.

Birmingham, CL; Canadien, V; Gouin, E; Troy, EB; Yoshimori, T; Cossart, P; Higgins, DE; Brumell, JH. *Listeria monocytogenes* evades killing by autophagy during colonization of host cells. *Autophagy*, 2007, 3, 442–451.

Boeckmann, R; Dickneite, C; Goebel, W; Bohne, J. PrfA mediates specific binding of RNA polymerase of *Listeria monocytogenes* to PrfA-dependent virulence gene promoters resulting in a transcriptionally active complex. *Mol. Microbiol.*, 2000, 36, 487–497.

Bowman, JP; Bittencourt, CR; Ross, T. Differential gene expression of *Listeria monocytogenes* during high hydrostatic pressure processing. *Microbiol.*, 2008, 154, 462–475.

Chaturongakul, S; Raengpradub, S; Wiedmann, M; Boor, KJ. Modulation of stress and virulence in *Listeria monocytogenes*. Trends Microbiol., 2008, 16, 388-396.

Cooray, KJ; Nishibori, T; Xiong, H; Matsuyama, T; Fujita, M; Mitsuyama, M. Detection of multiple virulence-associated genes of *Listeria monocytogenes* by PCR in artificially contaminated milk sample. Appl. Environ. Microbiol., 1994, 60, 3023-3026.

Cossart, P; Vicente, MF; Mengaud, J; Baquero, F; Perez-Diaz, JC; Berche, P. Listeriolysin O is essential for virulence of *Listeria monocytogenes*: Direct evidence obtained by gene complementation. *Infect. Immun.*, 1989, 57, 3629–3636.

den Bakker, HC; Bundrant, BN; Fortes, ED; Orsi, RH; Wiedmann, M. A population genetics-based and phylogenetic approach to understanding the evolution of virulence in the genus *Listeria. Appl. Environ. Microbiol.*, 2010, 76, 6085–6100.

den Bakker, HC; Manuel, CS; Fortes, ED; Wiedmann, M; Nightingale, KK. Genome sequencing identifies *Listeria fleischmannii* subsp. *coloradonensis* subsp. nov., isolated from a ranch. *Int. J. Syst. Evol. Microbiol.*, 2013, 63, 3257-3268.

Dickneite, C; Boeckmann, R; Spory, A; Goebel, W; Sokolovic, Z. Differential interaction of the transcription factor PrfA and the PrfA-activating factor (Paf) of *Listeria monocytogenes* with target sequences. *Mol. Microbiol.,* 1998, 27, 915–928.

Domann, E; Leimeister-Wachter, M; Goebel, W; Chakraborty, T. Molecular cloning, sequencing, and identification of a metalloprotease gene from *Listeria monocytogenes* that is species specific and physically linked to the listeriolysin gene. *Infect. Immun.*, 1991, 59, 65-72.

Domann, E; Wehland, J; Niebuhr, K; Haffner, C; Leimeister-Wachter, M; Chakraborty, T. Detection of PrfA-independent promoter responsible for listeriolysin gene expression in mutant *Listeria monocytogenes* strains lacking the PrfA regulator. *Infect. Immun.*, 1993, 61, 3073-3075.

Duodu, S; Holst-Jensen, A; Skjerdal, T; Cappelier, JM; Pilet, MF; Loncarevic S. Influence of storage temperature on gene expression and virulence potential of *Listeria monocytogenes* strains grown in a salmon matrix. *Food Microbiol.*, 2010, 27, 795-801.

Dussurget O. New insights into determinants of *Listeria monocytogenes* virulence. *Int. Rev. Cell Mol. Biol.,* 2008, 270, 1–38.

Eiting, M; Hagelueken, G; Schubert, WD; Heinz, DW. The mutation G145S in PrfA, a key virulence regulator of *Listeria monocytogenes*, increases DNA-binding affinity by stabilizing the HTH motif. *Mol. Microbiol.*, 2005, 56, 433–446.

Ermolaeva, S; Novella, S; Vega, Y; Ripio, MT; Scortti, M; Vazquez-Boland, JA. Negative control of *Listeria monocytogenes* virulence genes by a diffusible autorepressor. *Mol. Microbiol.*, 2004, 52, 601–611.

Freitag, NE; Rong, L; Portnoy, DA. Regulation of the prfA transcriptional activator of *Listeria monocytogenes*: multiple promoter elements contribute to intracellular growth and cell-to-cell spread. *Infect. Immun.*, 1993, 61, 2537–2544.

Gracieux, P; Roche, SM; Pardon, P; Velge, P. Hypovirulent *Listeria monocytogenes* strains are less frequently recovered than virulent strains on PALCAM and Rapid' L. mono media. *Int. J. Food Microbiol.*, 2003, 83, 133–145.

Graves, LM; Helsel, LO; Steigerwalt, AG; Morey, RE; Daneshvar, MI; Roof, SE; Orsi, RH; Fortes, ED; Milillo, SR; den Bakker, HC; Wiedmann, M; Swaminathan, B; Sauders BD. *Listeria marthii* sp. nov., isolated from the natural environment, Finger Lakes National Forest. *Int. J. Syst. Evol. Microbiol.*, 2010, 60, 1280-1288.

Greene, SL; Freitag, NE. Negative regulation of PrfA, the key activator of *Listeria monocytogenes* virulence gene expression, is dispensable for bacterial pathogenesis. Microbiol., 2003, 149, 111–120.

Grundling, A; Gonzalez, MD; Higgins, DE. Requirement of the *Listeria monocytogenes* broad-range phospholipase PC-PLC during infection of human epithelial cells. *J. Bacteriol.,* 2003, 185, 6295–6307.

Hamon, MA; Ribet, D; Stavru F; Cossart P. Listeriolysin O: the Swiss army knife of *Listeria. Trends Microbiol.*, 2012, 20, 360-368.

Joseph, B; Mertins, S; Stoll, R; Schaer, J; Umesha, KR; Luo, Q; Mueller-Altrock, S; Goebel W. Glycerol metabolism and PrfA activity in *Listeria monocytogenes. J. Bacteriol.*, 2008, 190, 5412–5430.

Kastbjerg, VG; Larsen, MH; Gram, L; Ingmer H. Influence of sublethal concentrations of common disinfectants on expression of virulence genes in *Listeria monocytogenes. Appl. Environ. Microbiol.*, 2010, 76, 303–309.

Kathariou, S; Metz, P; Hof, H; Goebel, W. Tn916-induced mutations in the hemolysin determinant affecting virulence of *Listeria monocytogenes. J. Bacteriol.*, 1987, 169, 1291–1297.

Kazmierczak, MJ; Wiedmann, M; Boor, KJ. Contributions of *Listeria monocytogenes* σ^B and PrfA to expression of virulence and stress response genes during extra- and intracellular growth. *Microbiol.*, 2006, 152, 1827–1838.

Lalic-Muelthaler, M; Bohne, J; Goebel, W. *In vitro* transcription of PrfA-dependent and -independent genes of *Listeria monocytogenes. Mol. Microbiol.,* 2001, 42, 111–120.

Lang Halter, E; Neuhaus, K; Scherer, S. *Listeria weihenstephanensis* sp. nov., isolated from the water plant *Lemna trisulca* taken from a freshwater pond. *Int. J. Syst. Evol. Microbiol.*, 2013, 63, 641-647.

Leclercq, A; Clermont, D; Bizet, C; Grimont, PAD; Le Fleche-Mateos, A; Roche, SM; Buchrieser, C; Cadet-Daniel, V; Le Monnier, A; Lecuit, M; Allerberger F. *Listeria rocourtiae* sp. nov. *Int. J. Syst. Evol. Microbiol.*, 2010, 60, 2210-2214.

Leimeister-Waechter, M; Domann, E; Chakraborty, T. The expression of virulence genes in *Listeria monocytogenes* is thermoregulated. *J. Bacteriol.*, 1992, 947-952.

Levin, BR; Cornejo, OE. The population and evolutionary dynamics of homologous gene recombination in bacterial populations. *PLoS Genet.* 2009, 5, e1000601.

Loh, E; Dussurget, O; Gripenland, J; Vaitkevicius, K; Tiensuu, T; Mandin, P; Repoila, F; Buchrieser, C; Cossart, P; Johansson J. A trans-acting riboswitch controls expression of the virulence regulator PrfA in *Listeria monocytogenes. Cell*, 2009, 139, 770–779.

Lomonaco, S; Patti, R; Knabel, SJ; Civera, T. Detection of virulence-associated genes and epidemic clone markers in *Listeria monocytogenes* isolates from PDO Gorgonzola cheese. Int. J. Food Microbiol., 2012, 160, 76–79.

Luo, Q; Herler, M; Mueller-Altrock, S; Goebel, W. Supportive and inhibitory elements of a putative PrfA-dependent promoter in *Listeria monocytogenes. Mol. Microbiol.*, 2005, 55, 986–997.

Luo, Q; Rauch, M; Marr, AK; Mueller-Altrock, S; Goebel, W. *In vitro* transcription of the *Listeria monocytogenes* virulence genes *inlC* and *mpl* reveals overlapping PrfA-dependent and -independent promoters that are differentially activated by GTP. *Mol. Microbiol.*, 2004, 52, 39–52.

Marquis, H; Goldfine, H; Portnoy. DA. Proteolytic pathways of activation and degradation of a bacterial phospholipase C during intracellular infection by *Listeria monocytogenes. J. Cell Biol.*, 1997, 137, 1381–1392.

Mauder, N; Ecke, R; Mertins, S; Loeffler, DIM; Seidel, G; Sprehe, M; Hillen, W; Goebel, W; Mueller-Altrock, S. Species-specific differences in the activity of PrfA, the key regulator of Listerial virulence genes. *J. Bacteriol.*, 2006, 188, 7941–7956.

Meinersmann, RJ; Phillips, RW; Wiedmann, M; Berrang, ME. Multilocus sequence typing of *Listeria monocytogenes* by use of hypervariable genes reveals clonal and recombination histories of three lineages. *Appl. Environ. Microbiol.*, 2004, 70, 2193–2203.

Mengaud, J; Dramsi, S; Gouin, E; Vazquez-Boland, JA; Milon, G; Cossart. P. Pleiotropic control of *Listeria monocytogenes* virulence factors by a gene that is autoregulated. *Mol. Microbiol.*, 1991, 5, 2273–2283.

Milenbachs Lukowiak, A; Mueller, KJ; Freitag, NE; Youngman, P. Deregulation of *Listeria monocytogenes* virulence gene expression by two distinct and semi-independent pathways. *Microbiol.*, 2004, 150, 321–333.

Milenbachs, AA; Brown, DP; Moors, M; Youngman, P. Carbon-source regulation of virulence gene expression in *Listeria monocytogenes. Mol. Microbiol.*, 1997, 23, 1075-1085.

Miner, MD; Port, GC; Bouwer, HGA; Chang, JC; Freitag NE. A novel prfA mutation that promotes *Listeria monocytogenes* cytosol entry but reduces bacterial spread and cytotoxicity. *Microb Pathogenesis*, 2008, 45, 273–281.

Moors, MA; Levitt, B; Youngman, P; Portnoy, DA. Expression of listeriolysin O and ActA by intracellular and extracellular *Listeria monocytogenes Infect. Immun.*, 1999, 67, 131–139.

Nadon, CA; Bowen, BM; Wiedmann, M; Boor, KJ. σ^B contributes to PrfA-mediated virulence in *Listeria monocytogenes. Infect. Immun.*, 2002, 70, 3948–3952.

Olesen, I; Thorsen, L; Jespersen, L. Relative transcription of *Listeria monocytogenes* virulence genes in liver pâtés with varying NaCl content. *Int. J. Food Microbiol.*, 2010, 141, S60–S68.

Ollinger, J; Bowen, B; Wiedmann, M; Boor, KJ; Bergholz, TM. *Listeria monocytogenes* σ^B modulates PrfA-mediated virulence factor expression. *Infect. Immun.*, 2009, 77, 2113–2124.

Orsi, RH; den Bakker, HC; Wiedmann, M. *Listeria monocytogenes* lineages: Genomics, evolution, ecology, and phenotypic characteristics. *Int. J. Med. Microbiol.*, 2011, 301, 79–96.

Orsi, RH; Maron, SB; Nightingale, KK; Jerome, M; Tabor, H; Wiedmann M. Lineage specific recombination and positive selection in coding and intragenic regions contributed to evolution of the main *Listeria monocytogenes* virulence gene cluster. *Infect. Genet. Evol.*, 2008, 8, 566–576.

Orsi, RH; Ripoll, DR; Yeung, M; Nightingale, KK; Wiedmann, M. Recombination and positive selection contribute to evolution of *Listeria monocytogenes inlA*. *Microbiol.*, 2007, 153, 2666–2678.

Rantsiou, K; Greppi, A; Garosi, M; Acquadro, A; Mataragas, M; Cocolin, L. Strain dependent expression of stress response and virulence genes of *Listeria monocytogenes* in meat juices as determined by microarray. *Int. J. Food Microbiol.*, 2012b, 152, 116–122.

Rantsiou, K; Mataragas, M; Alessandria, V; Cocolin, L. Expression of virulence genes of *Listeria monocytogenes* in food. *J. Food Safety*, 2012a, 32, 161–168.

Rauch, M; Luo, Q; Muller-Altrock, S; Goebel, W. SigB-dependent *in vitro* transcription of *prfA* and some newly identified genes of *Listeria monocytogenes* whose expression is affected by PrfA *in vivo. J. Bacteriol.*, 2005, 187, 800–804.

Raveneau, J; Geoffroy, C; Beretti, JL; Gaillard, JL; Alouf, JE; Berche, P. Reduced virulence of a *Listeria monocytogenes* phospholipase-deficient mutant obtained by transposon insertion into the zinc metalloprotease gene. *Infect. Immun.*, 1992, 60, 916-921.

Rawool, DB; Malik, SVS; Shakuntala, I; Sahare, AM; Barbuddhe, SB. Detection of multiple virulence-associated genes in *Listeria monocytogenes* isolated from bovine mastitis cases. *Int. J. Food Microbiol.*, 2007, 113, 201–207.

Renzoni, A; Cossart, P; Dramsi, S. PrfA, the transcriptional activator of virulence genes, is upregulated during interaction of *Listeria monocytogenes* with mammalian cells and in eukaryotic cell extracts. *Mol. Microbiol.*, 1999, 34, 552-561.

Renzoni, A; Klarsfeld, A; Dramsi, S; Cossart, P. Evidence that PrfA, the pleiotropic activator of virulence genes in *Listeria monocytogenes*, can be present but inactive. *Infect. Immun.*, 1997, 65, 1515–1518.

Ripio, MT; Dominguez-Bernal, G; Lara, M; Suarez, M; Vazquez-Boland, JA. A Gly145Ser substitution in the transcriptional activator PrfA causes constitutive overexpression of virulence factors in *Listeria monocytogenes. J. Bacteriol.,* 1997, 179, 1533-1540.

Ripio, MT; Vazquez-Boland, JA; Vega, Y; Nair, S; Berche, P. Evidence for expressional crosstalk between the central virulence regulator PrfA and the stress response mediator ClpC in *Listeria monocytogenes. FEMS Microbiol. Lett.*, 1998, 158, 45-50.

Roche, SM; Gracieux, P; Milohanic, E; Albert, I; Virlogeux-Payant, I; Temoin, S; Grepinet, O; Kerouanton, A; Jacquet, C; Cossart, P; Velge P. Investigation of specific substitutions in virulence genes characterizing phenotypic groups of low-virulence field strains of *Listeria monocytogenes. Appl. Environ. Microbiol*, 2005, 71, 6039–6048.

Roche, SM; Gracieux, P; Velge, P. Poor detection of low-virulence field strains of *L. monocytogenes* is related to selective agents in selective media and is unrelated to PrfA. *Food Microbiol.*, 2009, 26, 21–26.

Schluter, D; Domann, E; Buck, C; Hain, T; Hof, H; Chakraborty, T; Deckert-Schluter, M. Phosphatidylcholine-specific phospholipase C from *Listeria monocytogenes* is an important virulence factor in murine cerebral listeriosis. *Infect. Immun.*, 1998, 66, 5930–5938.

Schrama, D; Helliwell, N; Neto, L; Faleiro, ML. Adaptation of *Listeria monocytogenes* in a simulated cheese medium: effects on virulence using the *Galleria mellonella* infection model. *Lett. Appl. Microbiol.*, 2013, 56, 421-427.

Schwab, U; Bowen, B; Nadon, C; Wiedmann, M; Boor, KJ. 2005. The *Listeria monocytogenes prfA*P2 promoter is regulated by σ^B in a growth phase dependent manner. *FEMS Microbiol. Lett.*, 2005, 245, 329–336.

Sheikh-Zeinoddin, M; Perehinec, TM; Hill, SE; Rees, CED. Maillard reaction causes suppression of virulence gene expression in *Listeria monocytogenes. Int. J. Food Microbiol.*, 2000, 61, 41–49.

Shetron-Rama, LM; Mueller, K; Bravo, JM; Bouwer, HGA; Way SS; Freitag NE. Isolation of *Listeria monocytogenes* mutants with high-level in vitro expression of host cytosol-induced gene products. *Mol. Microbiol.*, 2003, 48, 1537–1551.

Shi, H; Trinh, Q; Xu, W; Luo, Y; Tian, W; Huang, K. The transcriptional response of virulence genes in *Listeria monocytogenes* during inactivation by nisin. *Food Control,* 2013, 31, 519-524.

Stasiewicz, MJ; Wiedmann, M; Bergholz, TM. The transcriptional response of *Listeria monocytogenes* during adaptation to growth on lactate and diacetate includes synergistic changes that increase fermentative acetoin production. *Appl. Environ. Microbiol.*, 2011, 77, 5294–5306.

Suarez, M; Gonzalez-Zorn, B; Vega, Y; Chico-Calero, I; Vazquez-Boland, JA. A role for ActA in epithelial cell invasion by *Listeria monocytogenes*. *Cell. Microbiol.*, 2001, 3, 853–864.

Temoin, S; Roche, SM; Grepinet, O; Fardini Y; Velge P. Multiple point mutations in virulence genes explain the low virulence of *Listeria monocytogenes* field strains. *Microbiol.*, 2008, 154, 939–948.

Tsai, YHL; Maron, SB; McGann, P; Nightingale, KK; Wiedmann, M; Orsi RH. Recombination and positive selection contributed to the evolution of *Listeria monocytogenes* lineages III and IV, two distinct and well supported uncommon *L. monocytogenes* lineages. *Infect. Genet. Evol.*, 2011, 11, 1881–1890.

Vazquez-Boland, JA; Dominguez-Bernal, G; Gonzalez-Zorn, B; Kreft, J; Goebel, W. Pathogenicity islands and virulence evolution in *Listeria*. *Microb. Infect.*, 2001, 3, 571-584.

Ward, TJ; Ducey, TF; Usgaard, T; Dunn, KA; Bielawski, JP. Multilocus genotyping assays for single nucleotide polymorphism-based subtyping of *Listeria monocytogenes* isolates. *Appl. Environ. Microbiol.*, 2008, 74, 7629–7642.

Williams, JR; Thayyullathil, C; Freitag, NE. Sequence variations within PrfA DNA binding sites and effects on *Listeria monocytogenes* virulence gene expression. *J. Bacteriol.*, 2000, 182, 837–841.

Wong KKY; Freitag NE. A novel mutation within the central *Listeria monocytogenes* regulator PrfA that results in constitutive expression of virulence gene products. *J. Bacteriol.*, 2004, 186, 6265–6276.

In: *Listeria monocytogenes*
Editor: Edmund C. Hambrick

ISBN: 978-1-63117-054-6

Chapter 14

PREVALENCE OF *LISTERIA MONOCYTOGENES* AND OCCURRENCE OF LISTERIOSIS FROM READY-TO-EAT FRESH FRUITS AND VEGETABLES

Agni Hadjilouka, Spiros Paramithiotis *
and Eleftherios H. Drosinos
Laboratory of Food Quality Control and Hygiene, Department of Food Science and Human Nutrition, Agricultural University of Athens, Athens, Greece

ABSTRACT

Human foodborne listeriosis has been mainly linked to the consumption of poultry, beef and dairy products. However, the number of studies reporting incidence of *L. monocytogenes* in ready-to-eat fruits and vegetables has been dramatically increased over the last decade; prevalence has been reported to range from 0.04% to 36.8% to a variety of products, such as carrots, cabbage, cauliflower, celery, cucumber, various herbs, lettuce, rocket, strawberries etc. At the same time listeriosis outbreaks epidemiologically associated with the consumption of such products have also increased, in both number and severity, with the 2011 US multistate outbreak linked to whole cantaloupes being the most characteristic. In this chapter prevalence of *L. monocytogenes* in unprocessed and minimally processed fruits and vegetables as well as listeriosis outbreaks linked to the consumption of such products are addressed.

INTRODUCTION

Presence of pathogenic microorganisms in raw fruits and vegetables is an issue that has generated great interest in scientific community over the last thirty years. Numerous surveys have been carried out in many countries in order to determine the prevalence of pathogenic microorganisms in these products. In some cases bacterial pathogens have not been detected whereas in other very high prevalence of microbial pathogens has been estimated.

* Corresponding author. E-mail: sdp@aua.gr; Tel.: +30 210 5294705 Fax: +30 210 5294683.

Listeria monocytogenes has been implicated as the causative agent in several outbreaks of food-borne listeriosis. Specifically, in North America and Europe listeriosis outbreaks are often reported, contrary to the fact that in Africa, Asia and South America listeriosis cases are rarely reported (Low and Donachie, 1997). An apparent increase in the incidence of listeriosis has been recorded in the past 20 years, even if awareness of listeriosis and improved techniques have increased diagnosis. Most human listeriosis cases seem to be sporadic; however, there is a great possibility that a portion of these cases might have the same unrecognized common-source (Ciesielski et al., 1988). Although listeriosis is rare, illness is frequently sever and causing high hospitalization and mortality rates. In European Union (EU) in 2009, 1645 listeriosis cases were reported -the cases increased 19.1% compared to 2008- with 270 human deaths. In 2010 the number of listeriosis cases remained almost at the same level; 1601 cases were reported. In 2011, 1470 human cases were reported, with a mortality rate of 12.7% (EFSA, 2011; 2012; 2013). According to Wesley (2009) United States account about 2500 listeriosis cases, 2289 hospitalizations and 449 deaths each year with the mortality rate ranging at about 28% -the highest of all food-borne pathogens.

The Canadian coleslaw outbreak in 1981, that led to recognition of listeriosis as food-borne human disease prompted an intense research activity. After that, investigations are being immediately organized in order to find the original sources of *L. monocytogenes* when a food-borne listeriosis illness appears, especially when hospital-acquired food-borne illness outbreaks are involved. Despite the improved techniques and the developed technology, investigation of hospital-acquired food-borne illness outbreak has many difficulties. One of the most important is that foods consumed by patients are rarely reported in medical reports. Moreover, there aren't any standard guidelines concerning consumption of ready-to-eat produce by immune-compromised hospital patients. Cokes et al., (2011) conducted a survey of New York hospitals to characterize their policies regarding food-borne illness risk management guidelines. According to the survey, 57% of hospitals do not keep records of the foods consumed by patients. Fifty-four of 61 hospitals (89%) responded to the survey and all of them (100%) reported serving prepared ready-to-eat salads to patients. Based on hospital reports, ready-to-eat salads were served at 94%, 89% and 73% to pregnant women, patients receiving immunosuppressive drugs, and patients undergoing chemotherapy, respectively. The above survey confirmed that usually patients are not placed on a diet that restricts foods commonly associated with listeriosis, a fact that increases the possibilities for disease outbreaks.

1. Listeriosis Outbreaks Associated with Vegetables

Although *L. monocytogenes* was described as human pathogen in the 1920s (mistakenly thought to be the cause of infectious mononucleosis), the first documented outbreak associated with this microorganism was in 1979 while the first confirmed outbreak linking listeriosis cases to food consumption was in 1981 (Public Health Agency of Canada). As early as 1967, Blenden and Szatalowicz reported that 731 cases of human listeriosis had been documented between 1933 and 1966, stating that plant products such as lettuce or other fresh vegetables contaminated with *L. monocytogenes* might be responsible for some of these cases.

Human food-borne listeriosis has been mainly linked to the consumption of poultry, beef and dairy products. However, the number of studies reporting incidence of *L. monocytogenes* in ready-to-eat fruits and vegetables has been dramatically increased over the last decade. One of the most important reasons is that advances in agronomic practices, processing, preservation, distribution and marketing leading to globalization of food supply disseminates pathogens over wide geographical areas introducing important hazards. Moreover, thanks to their increasing popularity - due to consumers' tendency to eat healthier- fresh fruits and vegetables have become important vehicles in food-borne illness statistics (Beuchat and WHO, 2002). On the other hand, *L. monocytogenes* is widely distributed in soil surface, in decaying vegetation soil sewage, in animal feces, in river and canal waters, in fertilizers, in plants and in animals and it is known to survive in plant material for 10-12 years (Beuchat, 1996) and in soil up for 295 days (Welshimer, 1960). Therefore fruits and vegetables can become easily contaminated with the pathogen while growing in fields or greenhouses or during harvesting or post-harvest handling process (Nguyen-the and Carlin, 1994). *L. monocytogenes* is also known to grow on a variety of vegetables at refrigeration temperatures (Berrang and Brackett, 1989; Brackett, 1999; Beuchat and Brackett, 1990). The above microorganism's characteristics in combination with the change in consumers' dietary habits and globalization of food supply, expound the reason the frequency of outbreaks epidemiologically associated with raw fruits and vegetables has increased.

1.1. Listeriosis Outbreak Caused by Celery, Tomatoes and Lettuce –Boston, 1979

In 1986 Ho et al., reported an outbreak of *L. monocytogenes* infection that involved 23 patients from eight Boston hospitals in 1979. Twenty (87%) of the isolates were serotyped during the 26 months preceding the outbreak and only nine (33%) had been 4b. Five of the case patients finally died. Case patients (patients with 4b *Listeria* infection) differed from control patients (patients with sporadic *Listeria* infection). More of the case patients had hospital-acquired infection [15/20 (75%) vs. 4/18 (22%)], had received antacids or cimetidine before the onset of listeriosis [12/20 (60%) vs. 3/18 (22%)], and had gastro-intestinal tract symptoms that began at the same time as fever [17/20 (85%) vs. 4/18 (22%)] (Ho et al., 1986). Case patients preferred more often tuna fish, chicken salad and cheese than control patients did. However, the only common foods served with these foods were celery, tomatoes and lettuce. It was concluded that the consumption of these vegetables had probably caused the listeriosis outbreak. At the time of the outbreak no attempt was made to *isolate L. monocytogenes* strains from the suspect vegetables in order to confirm the presence of the microorganism,

1.2. Listeriosis Outbreak Caused by Coleslaw –Canada, 1981

The first outbreak of *L. monocytogenes* that could be definitely linked to food consumption was caused by commercially prepared coleslaw in 1981 in the Maritime Provinces (Prince Edward Island, Nova Scotia and New Brunswick) in Canada (Schlech et al., 1983). Forty-one cases (7 adult and 34 perinatal) occurred between 1st March 1981 and 1st

September 1981 with the attack rate for perinatal listeriosis at a maternity hospital being approximately 1.3% during May, June and July (Beuchat, 1996). Of the 34 perinatal cases, there were 9 stillbirths, 23 live births of an ill infant, with a subsequent 27% mortality rate, and 2 live births of a well infant. The adult mortality rate was 28.6% (Farber and Peterkin, 1991). A survey between case and control patients revealed that case patients were more likely to have consumed coleslaw during the three months preceding the onset of the outbreak. Ingestion of radishes was associated with coleslaw consumption but not with illness. Coleslaw obtained from the refrigerator of a patient was positive for *L. monocytogenes* serotype 4b, which was the serotype of the epidemic strain and the strain isolated from the patient's blood. The coleslaw was obtained from wholesalers and local farmers and was commercially prepared with cabbage and carrots. Two unopened packages of coleslaw purchased from two different supermarkets of Halifax in Nova Scotia were positive for the presence of *L. monocytogenes* serotype 4b and both of them were produced by the same processor.

An investigation of the sources of cabbage identified a farmer who, in addition to raising cabbage also maintained a flock of sheep. According to a review on farmer's practices two of his sheep had died from listeriosis in 1979 and 1981, and the farmer used to apply both composted and fresh sheep manure in fields in which cabbages were grown. From the latest harvest in October through the winter and early spring, cabbage was kept in a cold-storage shed. During the period of the outbreak, a shipment of cabbage from that shed was traced to the implicated coleslaw processor, leading to the suggestion that coleslaw was the source of the 1981 Canadian outbreak (U.S. FDA, 2013). A statistical association of listeriosis with the coleslaw was supported also by identification of the 4b strain in the implicated food (Beuchat, 1996).

1.3. Listeriosis Outbreak Caused by Diced Celery -Texas, 2010

In 2012 Gaul et al., investigated a listeriosis outbreak that lasted seven months and involved five hospitals in Texas. Ten outbreak-related patients (5 female) were identified. Ages ranged from 56 to 93 years old, with a median age of 80 years old. Nine of the patients lived in central Texas and one in south Texas. Nine of them were hospitalized in one hospital for 5 to 23 days while 1 patient had visited 3 hospitals during the typical incubation period for listeriosis. All of them either remained in a hospital (5 patients) or had a new hospital admission for listeriosis (5 patients). Five patients died within 3 months of the first collection date of samples that confirmed the presence of *L. monocytogenes*. Listeriosis seemed to have contributed to 2 of the deceased patients' death. For the rest 3 deceased patients, listeriosis was identified on death certificates as an instant or contributing cause. The last patient was identified in August 2010.

Reviews of patients' medical records were studied and patients and surrogates were interviewed. Medical reviews didn't identify any common medication, procedure, or other possible exposure. According to most patients' medical records – from which only a static menu based on type of diet (eg. diabetic, renal) was available- patients were not restricted to consume foods commonly associated with listeriosis. In addition, according to patients' medical history, all of them had visited or been admitted to an acute care hospital within 70 days (the maximum incubation period) before listeriosis sample collection. These clues led to the hypothesis that all patients had acquired listeriosis from a food item supplied in multiple hospitals. This hypothesis was confirmed when a survey determined that the 5 hospitals with outbreak–related patients used the same distributor for the majority of food items (meat, poultry, prepared and frozen foods). In the end, nine of the outbreak cases were linked to the produce processing facility. The patients either visited or been admitted to these hospitals had consumed the facility's diced celery during the listeriosis incubation period. Only one case was not linked to celery, since the hospital where the patient had been admitted, received items other than diced celery from the produce processing facility.

The company closed permanently in February 2011, and no additional cases have been detected so far. However the original source of *L. monocytogenes* wasn't determined. The celery used in the facility was grown in another state, came from multiple companies and farms, and was distributed nationally.

1.4. Listeriosis Outbreak Caused by Cantaloupes – Colorado, 2011

The worst food borne listeriosis outbreak in the United States' history, since the CDC began tracking outbreaks in the 1970s, was linked to whole cantaloupes from a single farm in Colorado. The multistate outbreak of listeriosis occurred from July through October, 2011. According to the final report on August 27th, 2012, the outbreak resulted in 143 hospitalizations, 33 deaths and 1 miscarriage. Totally 147 infected persons from 28 states, with any of the five outbreak-associated subtypes of *L. monocytogenes,* were reported to CDC. The number of infected persons identified in each state was as follows: Alabama (1), Arkansas (1), California (4), Colorado (40), Idaho (2), Illinois (4), Indiana (3), Iowa (1), Kansas (11), Louisiana (2), Maryland (1), Missouri (7), Montana (2), Nebraska (6), Nevada (1), New Mexico (15), New York (2), North Dakota (2), Oklahoma (12), Oregon (1), Pennsylvania (1), South Dakota (1), Texas (18), Utah (1), Virginia (1), West Virginia (1), Wisconsin (2), and Wyoming (4). According to people from whom information was available, the first case was recorded on July 31st, 2011, and the last one on October 27th, 2011. Among the 145 ill persons with available information on whether they were hospitalized, 143 (99%) were hospitalized. Fifty-eight percent of ill persons were female. Ages ranged from <1 to 96 years with a median age of 78 years, while most of the infected people were over 60 years old. Thirty-three deaths were reported totally: Colorado (9), Indiana (1), Kansas (3), Louisiana (2), Maryland (1), Missouri (3), Montana (1), Nebraska (1), New Mexico (5), New York (2), Oklahoma (1), Texas (2), and Wyoming (2). Age of the people who died ranged from 48 to 96 years old, with a median age of 81 years. Seven of the illnesses were related to a pregnancy. Specifically three were diagnosed in newborns and four were diagnosed in pregnant women with one reported miscarriage owing to the illness (CDC, 2012).

2. Prevalence of *Listeria monocytogenes* in Fresh Fruits and Vegetables

Prevalence of *L. monocytogenes* has been reported to vary from 0.04% (2/50) in various juices to 36.8% (25/68) in radish (Sado et al., 1998; Francis et al., 1999). Generally it appears to be very variable and it can be as low as 0.3% (1/297), 0.7% (2/300) and 1.6% (1/63) in lettuce–iceberg samples, in fresh, minimally processed fruit, vegetables and sprout samples and in mixed salads, respectively, or even as high as 25.4% (88/347), in frozen vegetable salads and 33.7% in raw mixed salads (U.S. FDA, 2013; Abadias et al., 2008; Lin et al., 1996; Cordano and Jacquet, 2009; Kramarenko et al., 2013). In some cases a small number of samples was obtained and the reported prevalence was most of the times as high as 50% (10/20), 80% (4/5) and 85% (6/7) in lettuce, cucumber and bean sprouts samples, respectively (Gunasena et al., 1995; Arumugaswamy et al., 1994).

Surveys of fresh products have revealed the presence of *L. monocytogenes* in various vegetables and fruits, such as beets, carrots, cabbage, cauliflower, celery, cucumber, roots, bulbs, tomato, spinach, broccoli, leek, various herbs, lettuce, rocket, lamb's lettuce, endive, potatoes, radishes, bean sprouts, coleslaw, mushrooms, radicchio, bean sprouts, leeks, pepper, rocket and strawberries.

2.1. Presence of *L .monocytogenes* in categories of fruits and vegetables

2.1.1. Prevalence of L. monocytogenes in Juices and Raw Fruits

Juices

Fifty various juices collected from retail markets in US, 1996, were analyzed for the presence of pathogenic bacteria including *L. monocytogenes*. Two of them (4%), an apple juice and an apple raspberry blend, were positive for *L. monocytogenes*' presence (Sado et al., 1998).

Papaya - Watermelon

Vahidy et al., (1992) tested 150 samples of fresh fruits and vegetables collected from Karachi city, in Pakistan. Out of 30 samples of papaya and watermelon, *L. monocytogenes* was isolated from 2 samples of papaya (6.7%) and 1 sample of watermelon (3.3%).

2.1.2. Prevalence of L. monocytogenes in Raw Vegetables

Bean Sprouts

Bean sprouts were positive for presence of *L. monocytogenes* in France and in Malaysia. According to U.S. FDA (2013), *L. monocytogenes* was detected in 1 of the 31 (3.1%) tested samples collected at production and in 19 of 102 (19%) tested samples collected at retail markets. Six of 7 samples (86%) taken from refrigerated supermarkets and open wet markets in Malaysia were also positive for microorganisms' presence (Arumugaswamy et al., 1994).

Lettuce

Lettuce samples were tested in Canada (2 surveys), France, Malaysia (2 surveys), Sri Lanka, UK, US (3 surveys) and Australia. Positive samples were detected in Canada, Sri Lanka, US, Australia and Malaysia's samples. Specifically, *L. monocytogenes* was detected in 5 of 39 (13%) chopped lettuce samples collected from hospitals in Canada. Samples were stored at 4°C or 10°C for up to eleven days (Odumeru et al., 1997). Two surveys in Malaysia and one in Sri Lanka, reported prevalence 3.6% (1/28) & 22.7% (5/22) and 50% (10/20) respectively (Francis et al., 1999; Arumugaswamy et al., 1994; Gunasena et al., 1995). In Australia lettuce samples were cut and packaged and obtained over an eight month period. Prevalence of these samples was 2.5% (3/120) (Szabo et al., 2000). On the other hand, none of the samples was positive in Canada (0/50), France (0/35) and UK (0/151) surveys (Farber et al., 1989; Gras et al., 1994; Little et al., 1999). According to unpublished but reviewed data from industries in US, 297 samples of lettuce-iceberg and 320 samples of lettuce romaine were collected from December 1992 to February 1993. Prevalence of *L. monocytogenes* was 0.3% (1/297) in lettuce–iceberg and 0% in lettuce romaine (U.S. FDA, 2013).

Broccoli

Broccoli samples were also tested in Canada and US for *L. monocytogenes* presence. In Canada 35 samples were collected from hospitals and 2 of them were positive (5.7%) (Odumeru et al., 1997). Unlikely to Canada, in US 92 samples were collected from retail markets but no one was positive for microorganism's presence (Heisick et al., 1989).

Cabbage

Heisick et al., (1989) examined 92 samples of cabbage collected from two supermarkets in US. One of them was positive for *L. monocytogenes* presence and prevalence was 1.1%. According to unpublished industrial data (U.S. FDA, 2013) 1016 samples of green and 399 samples of red cabbage were collected between December 1992 and February 1993 from a processing facility. Prevalence was 7.2% (73/1016) and 2.5% (10/399) in green and red cabbage respectively. In Canada *L. monocytogenes'* prevalence was 2.2% (2/92) and 6.7% (1/15) in two different surveys (Schlech et al., 1983; Odumeru et al., 1997). Finally, a survey in Sri Lanka reported prevalence 33% (6/18) (Gunasena et al., 1995).

Carrots

Ninety-two carrot samples were obtained from two different supermarkets in US. None of them was positive for *L. monocytogenes'* presence (prevalence 0%) (Heisick et al., 1989). Similarly, all of the 35 samples collected from hospitals in Canada and all of the 15 samples collected in Karachi city were negative for the microorganism (Odumeru et al., 1997; Vahidy et al., 1992).

Cauliflower- Celery- Radicchio

None of the 131 cauliflower, 69 celery and 180 radicchio samples were positive for the presence of *L. monocytogenes.* The cauliflower samples, 35 and 92, were obtained from US retail markets and Canada hospital, respectively (Odumeru et al., 1997; Heisick et al., 1989). Celery samples were collected in Canada. Thirty celery samples were obtained from retail markets and 39 were collected from hospital (Farber et al., 1989; Odumeru et al., 1997).

Radicchio samples were gathered from processing facility between December 1992 and February 1993 (U.S. FDA, 2013).

Coleslaw

Four prevalence studies took place in Canada (2 studies), Singapore and UK. According to Schlech et al., (1983), prevalence of *L. monocytogenes* in coleslaw samples collected in Canada was 2.2% (2/92). Among the 35 samples collected from Canada hospitals only 1 was positive and prevalence was finally 2.9% (Odumeru et al., 1997). In a survey that lasted from January 1992 to June 1993, 50 samples of coleslaw were investigated for the presence of *L. monocytogenes* in Singapore. Two of them were positive and prevalence was estimated 4% (Doris et al., 1995). In UK 39 samples were collected from four supermarkets and prevalence was estimated 7.7% (MacGowan et al., 1994).

Cucumber

Cucumber samples tested were collected from retail markets in Pakistan (15 samples) and US (92 samples) and from street vendors (sliced cucumbers) in Malaysia (5 samples). Prevalence was 6.7%, 2.2% and 80% respectively (Vahidy et al., 1992; Heisick et al., 1989; Arumugaswamy et al., 1994).

Green Pepper

A survey held in Canada with samples collected from hospital, reported prevalence 2.9% (1/35) (Odumeru et al., 1997).

Leafy Vegetables

Arumugaswamy et al. (1994) examined for the presence of *L. monocytogenes* in 22 samples of leafy vegetables collected from refrigerated supermarkets and open wet markets in Malaysia. Five of them were positive for *L. monocytogenes'* presence (prevalence 23%).

Leeks

A study of *Listeria* spp. on raw foodstuff collected from markets and other establishments in the city of Barcelona was conducted by de Simon et al. (1992) from September 1989 to March 1990. *L. monocytogenes* was isolated from 2 samples of leeks and prevalence was 20%.

Potatoes

Potato samples were studied for presence of *L. monocytogenes* in US and Spain. In US 19/70 and 28/132 potato samples were positive for *L. monocytogenes'* presence and prevalence was 27.1% and 21.2% respectively (Francis et al., 1999; Heisick et al., 1989). De Simon et al. (1992) reported 16.7% (2/12) prevalence of *L. monocytogenes* in potato samples collected from markets and other establishments in Barcelona.

Radish

Two studies conducted in US on radish samples reported the prevalence rates 36.8% (25/68) and 14.4% (19/132) (Francis et al., 1999; Heisick et al., 1989). On the other hand, according to Farber et al., (1989), all of the 10 radish samples studied for the presence of *L.*

monocytogenes were negative -samples were grown either in the US or Canada and obtained from retail markets. Similarly, no one of the 15 samples collected in Karachi city was positive (Vahidy et al., 1992).

Tomatoes

A study held in Pakistan reported prevalence 13.3% (2/15) for *L. monocytogenes* in tomato samples. Samples were collected from various localities of Karachi city (Vahidy et al., 1992). Contrary to this result, 92 tomato samples, grown either in Canada or in the US, and 20 tomato samples, obtained from two supermarkets in Canada, were studied in two surveys held in US and Canada respectively. The reported prevalence for all tested samples was 0% (Farber et al., 1989; Heisick et al., 1989).

Unspecified Vegetables

According to the study of de Simon et al., (1992) the number of positive vegetable samples, as far as presence of *L. monocytogenes* was concerned, was 8. A total of 103 vegetables was collected including lettuces, potatoes, tomatoes, cabbages, leeks, mushrooms, chards, celery, carrots, fresh garlic, asparagus and radishes. Prevalence was estimated to be 7.8% totally. Ninety samples of vegetables were obtained from four supermarkets in UK. Five (5.6%) of them were positive for *L. monocytogenes'* presence (MacGowan et al., 1994). A survey in Italy reported that prevalence of *L. monocytogenes* was 6.9% (7/102) in vegetable samples tested (Beuchat, 2002), while a review in incidence of *L. monocytogenes* in food samples in Italy between 1990 and 1999 reported prevalence 33% in vegetables (Gianfranceschi et al., 2003).

2.2. Presence of *L. monocytogenes* in Salads

2.2.1. Prevalence of L. monocytogenes in Various Categories of Salads

Pre-packed Mixed Salads

L. monocytogenes was isolated from pre-packed salads in two surveys. Twenty-one samples were collected in Northern Ireland and 3 of them were positive for *L. monocytogenes'* presence (14.3%) (Harvey and Gilmour, 1993). In UK's survey prevalence was reported 6.7% (4/60) (Francis et al., 1999).

Mixed Salads

Various *L. monocytogenes* prevalence surveys have been conducted in mixed salads in Germany, Netherlands, Northern Ireland, UK (2 surveys), Canada, Spain, US and Estonia. According to Breer and Baumgartner (1992) 263 samples were tested in Germany and 6 (2.3%) of them were positive for presence of *L. monocytogenes*. In Spain, Canada, Northern Ireland, UK and US the reported prevalence ratios were 30% (21/70), 40% (6/15), 25% (4/16), 1.8% (2/108) and 1.6% (1/63), respectively (Garcia-Gimeno et al., 1996; Odumeru et al., 1997; Harvey and Gilmour et al., 1993; Velani and Roberts, 1991; U.S. FDA, 2013). In Netherlands, study in fresh cut salad vegetables reported 44% (11/25) prevalence (Beckers et al., 1989) and in UK MacGowan et al., (1994) reported 3.8% (1/26) prevalence in prepared vegetables. Finally, in Estonia 550 samples of mixed salads were collected from 2008 to 2010. Mixed salads contained different fruits and vegetables. One hundred eighty seven samples were collected in 2008, 227 samples were collected in 2009, 136 samples were collected in 2010 and prevalence was 33.7%, 9.3% and 13.2%, respectively. In total, 102 samples were positive for *L. monocytogenes'* presence and prevalence was 18.5% (Kramarenko et al., 2013).

Frozen Salads

In Navarra, northern Spain, 1750 frozen vegetables samples were collected from retail outlets within 4 years and examined for *L. monocytogenes'* presence. The products obtained were potatoes, carrot, spinach, broccoli, string-beans, peas, artichoke and cauliflower and the reported prevalence was 1.8% (31/1750) (Vitas et al., 2004).

2.3. Prevalence of *L. monocytogenes* in Various Vegetable Products

Three types of vegetable salads were purchased from supermarkets in Santiago, Chile, and tested between 2000 and 2005 for *L. monocytogenes'* presence. A total of 717 samples were collected: (i) 347 samples of frozen salads, intended for cooking, (ii) 216 samples of either cooked or raw ready-to-eat vegetable salads and (iii) 154 industrial minimally processed, ready-to-eat raw vegetable salads. All three types of salads were made from single or mixed vegetables. Frozen salads consisted of 63 samples with mixed vegetables and 284 samples from single products. Namely, single product samples were: sweet corn (39), green beans (37), broad beans (33), green peas (33), broccoli (21), fresh kidney beans (20), cauliflower (17), carrot (17), mushrooms (14), spinach (13), onion (12), asparagus (12), artichoke (8), and Brussels sprouts (8). Ready-to-eat vegetable salads were: mixed vegetables (87), broad beans (20), sweetcorn (15), raw celery (13), beetroot (10), raw tomato (10), raw carrot (6), mushrooms (6), seaweed (6), green peas (6), fresh kidney beans (6), cauliflower (6), broccoli (5), bean sprouts (5), raw cabbage (5), green beans (5) and raw lettuce (5). Finally, minimally processed ready-to-eat raw vegetable salads were either single vegetables such as carrot (21), lettuce (21), celery (18), green cabbage (17), red cabbage (16) and spinach (16) or mixed vegetables (45). *L. monocytogenes* was isolated from 110 out of 717 (15.3%) samples. A total of 88 (25.4%) frozen vegetable salad samples and 22 (10.2%) fresh ready-to-eat salads were contaminated with the microorganism whereas 154 samples from industrially prepared minimally processed salads that were negative *for L. monocytogenes'* presence. Analytically, *L. monocytogenes'* prevalence in frozen salads was: 48.6% in green beans, 42.9% in mixed salads, 36.4% in green peas, 30% in fresh kidney beans, 25.6% in sweet corn, 23.5% in cauliflower, 21.2% in broad beans, 9.5% in broccoli, 8.3% in asparagus, 7.7% in spinach and 7.1% in mushrooms. *L. monocytogenes* was not isolated from carrots, onion, artichokes or Brussels sprouts. Moreover, *L. monocytogenes'* prevalence in fresh ready-to-eat salads was: 50% in cooked seaweed, 33.3% in cooked green beans, 20% in raw tomato, 20% in cooked broccoli, 16.7% in cooked kidney beans, 15.4% in raw celery, 10.3% in mixed salads, 10% in cooked beetroot, 6.6% in cooked sweet and 5% in cooked broad beans (Cordano and Jacquet, 2009).

Crepet et al., (2007) reviewed 165 studies extracted from 51 papers and reports regarding the contamination of fresh and minimally processed vegetables and reported 3% (753/25,078) total prevalence of *L. monocytogenes*. The studies were published between 1988 and 2005 and included 25,078 samples. Analytically, samples were: 8,511 minimally processed vegetables and 3,934 unprocessed vegetables (beets, carrots, cabbage, cauliflower, celery, cucumber, roots, bulbs, tomato, spinach, broccoli, leek, various herbs, and mixed vegetables), 7,511 samples minimally processed leafy salad vegetables and 4,337 unprocessed leafy salad vegetables (lettuce, rocket, lamb's lettuce, and endive) and 785 sprouts and germinated seed, mung bean, alfalfa, and bean sprouts.

Table 1. Reported outbreaks of food-borne disease (listeriosis) associated with vegetables

Location	Year	Source	Source Location	No. of cases	No. of deaths	Reference
Boston	1979	tomatoes, lettuce, celery	NR	20	5	Ho et al., 1986; Schlech et al., 1983
Canada	1981	coleslaw	Canada	41	17	U.S. FDA, 2013; Schlech et al.,1983
Texas	2010	celery	NR	10	5	Gaul et al., 2012
USA-Multistate	2012	cantaloupes	Colorado	143	33	CDC, 2012

*NR: Not Reported.

Acknowledgments

The authors gratefully acknowledge funding from the European Union Seventh Framework Programme (FP7/2007-2013) under grant agreement n. 289719 (Project QUAFETY).

References

Abadias, M; Usall, J; Anguera, M; Solsona, C; Vinas I. Microbiological quality of fresh, minimally-processed fruits and vegetables, and sprouts from retail establishments. *Int. J. Food Microbiol.*, 2008, 123, 121-129.

Arumugaswamy, RK; Ali, GRR; Abdul-Hamid, SNB. Prevalence of *Listeria monocytogenes* in foods in Malaysia. *Int. J Food Microbiol.*, 1994, 23, 117-121.

Beckers, HJ; In't Veld, PH; Soentoro, PSS; Delfgou-van Ash, EHM. The occurrence of *Listeria* in foods. 1989. In: Foodborne Listeriosis. Proceeding of a Symposium on September 7, 1988, 83-97, Wiesbaden, Hamburg, FRG.

Berrang, ME; Brackett, RE. Growth of *Listeria monocytogenes* on fresh vegetables stored under controlled atmosphere. *J. Food Prot.*, 1989, 52, 702–705.

Beuchat, LR. *Listeria monocytogenes*: incidence on vegetables. *Food Control*, 1996, 7, 223-228.

Beuchat, LR. Ecological factors influencing survival and growth of human pathogens on raw fruits and vegetables. *Microbes Infect.*, 2002, 4, 413–423.

Beuchat, LR; Brackett, RE. Inhibitory effects of raw carrots on *Listeria monocytogenes*. *Appl. Environ. Microbiol.*, 1990, 56, 1734-1742.

Beuchat LR. World Health Organization (WHO), Surface decontamination of fruits and vegetables eaten raw: a review. Geneva: World Health Organization, 2002.

Blenden, DC; Szatalowicz, FT. Ecological aspects of listeriosis. *J. Am. Vet. Med. Assoc.*, 1967, 151, 1761-1766.

Brackett, RE. Incidence, contributing factors, and control of bacterial pathogens on produce. *Postharvest Biol. Tec.*, 1999, 15, 305–331.

Breer, C; Baumgartner, A. Occurrence and behavior of *Listeria monocytogenes* in salads, vegetables and fresh vegetables juices. *Arch Lebensmittelhyg.*, 1992, 43, 108–110.

Centers of Disease Control and Prevention (CDC). Multistate outbreak of listeriosis linked to whole cantaloupes from Jensen farms, Colorado. 2012,http://www.cdc.gov/listeria/outbreaks/cantaloupes-jensen-farms/082712/index.html Last updated: 04/09/2012

Ciesielski, CA; Hightower, AW; Parsons, SK; Broome, CV. Listeriosis in the United States: 1980-1982. *Arch. Intern. Med.*, 1988, 148, 1416-1419.

Cokes, C; France, AM; Reddy, V; Hanson, H; Lee, L; Kornstein, L; Stavinsky, F; Balter, S. Serving high-risk foods in a high-risk setting: survey of hospital food service practices after an outbreak of listeriosis in a hospital. *Infect. Cont. Hosp. Ep.*, 2011, 32, 380–386.

Cordano, AM; Jacquet, C. *Listeria monocytogenes* isolated from vegetable salads sold at supermarkets in Santiago, Chile: Prevalence and strain characterization. *Int. J. Food Microbiol.*, 2009, 132, 176–179.

Crepet, A; Albert, I; Dervin, C; Carlin, F. Estimation of microbial contamination of food from prevalence and concentration data: application to *Listeria monocytogenes* in fresh vegetables. *Appl. Environ. Microbiol.*, 2007, 73, 250–258.

de Simon, M; Tarrago, C; Ferrer, MD. Incidence of *Listeria monocytogenes* in fresh foods in Barcelona (Spain). *Int. J. Food Microbiol.*, 1992, 16, 153-156.

Doris, LKNg; Seah, HL. Isolation and identification of *Listeria monocytogenes* from a range of foods in Singapore. *Food Control*, 1995, 6, 171-173.

EFSA - European Food Safety Authority. The European Union summary report on trends and sources of zoonoses, zoonotic agents and food-borne outbreaks in 2009. *EFSA Journal* 2011, 9, 2090.

EFSA - European Food Safety Authority. The European Union summary report on trends and sources of zoonoses, zoonotic agents and food-borne outbreaks in 2010. *EFSA Journal* 2012, 10, 2597.

EFSA - European Food Safety Authority. *Listeria*. 2013. http://www.efsa.europa.eu/en/topics/topic/listeria.htm. Last updated: 27 June 2013.

Farber, JM; Peterkin, PI. *Listeria monocytogenes*, a food-borne pathogen. *Microbiol. Reviews*, 1991, 55, 476-511.

Farber, JM; Sanders, GW; Johnston, MA. A survey of various foods for the presence of *Listeria* species. *J. Food Prot.*, 1989, 52, 456-458.

Francis, GA; Thomas, C; O'Beirne, D. The microbiological safety of minimally processed vegetables. *Int. J. Food Sci. Tech.*, 1999, 34, 1-22.

Garcia-Gimeno, RM; Zurena-Cosano, G; Amaro-Lopez, M. Incidence, survival and growth of *Listeria monocytogenes* in ready-to-use mixed salad vegetables in Spain. *J. Food Safety*, 1996, 16, 75-86.

Gaul, LK; Farag, NH; Shim, T; 3 Kingsley, MA; Silk, BJ; Trees, EH. Hospital-acquired listeriosis outbreak caused by contaminated diced celery- Texas, 2010. *Clin. Infect. Dis.*, 2012, 56, 20–26.

Gianfranceschi, M; Gattuso, A; Tartaro, S; Aureli, P. Incidence of *Listeria monocytogenes* in food and environmental samples in Italy between 1990 and 1999: Serotype distribution in food, environmental and clinical samples. *Eur. J. Epidemiol.*, 2003, 18, 1001–1006.

Gras, MH; Druet-Michaud, C; Cerf, O. La flore bacterienne des feuilles de salade fraiche (Bacterial flora of salad leaves). *Sci. Aliment.*, 1994, 14, 173-188.

Gunasena, DK; Kodikara, CP; Ganepola, K; Widanaparthirana, S. Occurrence of *Listeria monocytogenes* in food in Sri Lanka. *J. Natl. Sci. Counc. Sri Lanka*, 1995, 23, 107–114.

Harvey, J; Gilmour. A. Occurrence and characteristics of *Listeria* in foods produced in Northern Ireland. *Int. J. Food Microbiol.*, 1993, 19, 193-205.

Heisick, JE; Wagner, DE; Nierman, ML; Peeler, JT. *Listeria* spp. found on fresh market produce. *Appl. Environ. Microbiol.*, 1989, 55, 1925-1927.

Ho, JL; Shands, KN; Friedland, G; Eckind, P; Fraser, DW. An outbreak of type 4b *Listeria monocytogenes* infection involving patients from eight Boston hospitals. *Arch. Intern. Med.*, 1986, 146, 520–524.

Kramarenko, T; Roasto, M; Meremäe, K; Kuningas, M; Põltsama, P; Elias, T. *Listeria monocytogenes* prevalence and serotype diversity in various foods. *Food Control*, 2013, 30, 24-29.

Lin, CM; Fernando, SY; Wei, CI. Occurrence of *Listeria monocytogenes*, *Salmonella* spp., *Escherichia coli* and *E. coli* O157:H7 in vegetable salads. *Food Control*, 1996, 7, 135-140.

Little, C; Roberts, D; Youngs, E; de Louvois, J. Microbiological quality of retail imported unprepared whole lettuces: a PHLS Food Working Group Study, *J. Food. Prot.*, 1999, 62, 325-328.

Low, JC; Donachie, W. Review of *Listeria monocytogenes* and Listeriosis. *Vet. J.*, 1997, 153, 9-29.

MacGowan, AP; Bowker, K; McLauchlin, J; Bennett, PM; Reeves, DS. The occurrence and seasonal changes in the isolation of *Listeria* spp. in shop bought food stuffs, human faeces, sewage and soil from urban sources. *Int. J. Food Microbiol.*, 1994, 21, 325-334.

Nguyen-the, C; Carlin, F. The microbiology of minimally processed fresh fruits and vegetables. *Crit. Rev. Food Sci. Nutr.*, 1994, 34, 371–401.

Odumeru, JA; Mitchell, SJ; Alves, DM; Lynch, JA; Yee, AJ; Wang, SL; Styliadis, S; Farber, JM. Assessment of the microbiological quality of ready-to-use vegetables for health-care food services. *J. Food Prot.*, 1997, 60, 954-960.

Sado, PN; Jinneman, KC; Husby, GJ; Sorg, SM; Omiecinski, CJ. Identification of *Listeria monocytogenes* from unpasteurized apple juice using rapid test kits. *J. Food Prot.*, 1998, 61, 1199-1202.

Schlech, WE; Lavigne, PM; Bortolussi, RA; Allen, AC; Haldane, WV; Wort, AJ; Hightower, AW; Johnson, SE; King, SH; Nichols, ES; Broome, CV. Epidemic listeriosis - Evidence for transmission by food. *New Engl. J. Med.*, 1983, 308, 203-206.

Szabo, EA; Scurrah, KJ; Burrows, JM. Survey for Psychrotrophic bacterial pathogens in minimally processed lettuce. *Lett. Appl. Microbiol.*, 2000, 30, 456-460.

U.S. Food and Drug Administration (FDA). Outbreaks associated with Fresh Produce: Incidence, growth, and survival of pathogens in fresh and fresh-cut produce. 2013. http://www.fda.gov/Food/FoodScienceResearch/SafePracticesforFoodProcesses/ucm091268.htm Last updated: 24/04/2013.

Vahidy, R; Jahan, F; Nasim, R. Isolation of *Listeria monocytogenes* from fresh fruits and vegetables. *Hortscience*, 1992, 27, 628.

Velani, S; Roberts, D. *Listeria monocytogenes* and other Listeria spp. in pre-packed mixed salads and individual salad ingredients. *PHLS Microbiology Digest*, 1991, 8, 21-22.

Vitas, AL; Aguado, V; Garcia-Jalon, I. Occurrence of *Listeria monocytogenes* in fresh and processed foods in Navarra (Spain). *Int. J. Food Microbiol.*, 2004 90, 349–356.

Welshimer, H.J. Survival of *Listeria monocytogenes* in soil. *J. Bacteriol.*, 1960, 80, 316-320.

Wesley, I. Public health impact of food-borne illness: impetus for the international food safety effort. 2009. In N. Heredia, I. Wesley, & S. Garcia (Eds.), Microbiologically safe foods (pp. 3-13). United States of America: Wiley, A John Wiley & Sons, Inc.

INDEX

A

B

C

D

E

F

G

H

I

J

K

L

M

N

O

P

T